Time and Space in Ancient Myth, Religion and Culture

MythosEikonPoiesis

Herausgegeben von
Anton Bierl

Band 10

Time and Space in Ancient Myth, Religion and Culture

Edited by Anton Bierl, Menelaos Christopoulos and Athina Papachrysostomou

DE GRUYTER

ISBN 978-3-11-053419-1
e-ISBN (PDF) 978-3-11-053515-0
e-ISBN (EPUB) 978-3-11-053422-1
ISSN 1868 – 5080

Library of Congress Cataloging-in-Publication Data
A CIP catalog record for this book has been applied for at the Library of Congress.

Bibliographic information published by the Deutsche Nationalbibliothek
The Deutsche Nationalbibliothek lists this publication in the Deutsche Nationalbibliografie; detailed bibliographic data are available in the Internet at http://dnb.dnb.de.

Printing and binding: CPI books GmbH, Leck
♾ Printed on acid-free paper
Printed in Germany

MIX
Papier aus verantwortungsvollen Quellen
FSC® C083411

www.degruyter.com

Table of Contents

Preface

The present volume on *Time and Space in Greek and Roman Myth, Religion and Culture* is the elaborate outcome of a project recently undertaken by the "Center for the Study of Myth and Religion in Greek and Roman Antiquity", which is affiliated with the Department of Philology, University of Patras. This volume constitutes the logical sequence of the Center's previous publication, *Light and Darkness in Greek Mythology and Religion*, edited by M. Christopoulos, E. Karakantza and O. Levaniouk (Lanham 2000: Lexington Books).

Since the Center's establishment in 2007, many and diverse events and activities have taken place (talks, colloquiums, workshops, and conferences), in the attempt to explore as fully and deeply as possible various aspects of Greek and Roman myth and religion. What kept emerging from our discussions and research was the necessity to approach, examine, and try to interpret myth and religion from the point of view of two distinct parameters that essentially permeate the literary tradition, that is *time* and *space*. Hence, in July 2015 an International Conference was organised at Patras University, on "Time and Space in Myth and Religion". This four-day Conference brought together scholars (classicists, archaeologists, and historians) from Europe, the US and Canada, who dealt with a wide range of notions (past, present, future; here, there, elsewhere) and themes (genealogies, time structure, topography, succession and flux), in multiple literary genres (epic and lyric poetry, tragedy, comedy, satyr drama and philosophy), and even in archaeology. The Conference adopted a cross-generic approach; its goal was to tackle key cases and discuss exemplary passages originating from the entire spectrum of the surviving material.

The present volume features the select and peer-reviewed proceedings of this International Conference. The chapters included herein were selected not only for their originality and academic merit, but also because they constitute crucial paradigms of how manifestations of myth and religion in Greek and Latin texts can dynamically engage in a variety of literary relations with the notions of time and space, thus leading to multifarious results (regarding theme, language, style, and narrative), such as rhetorical emphasis and pathos, narrative restructuring and dramaturgical function, intensification of literary patterns and motifs, time- and space-travelling, anachronisms and déjà vus. In this regard, Mikhail Bakhtin's (1895–1975) "chronotope", the term of how time and space interfere with each other in specific temporal and spatial configurations, has perhaps proven the unifying concept of the entire volume.

The volume is divided in four parts: (i) epos, (ii) drama, (iii) empirical and imaginary chronotopes, and (iv) shifting chronotopes. The first chapter of Part

I features a number of "Strange Instances of Time and Space in Odysseus' Return" by M. Christopoulos, focusing on some unexpected aspects of time and space in Homeric *Odyssey*. The author studies the theme of Odysseus' return to the same point from which he had departed, a theme which occurs four times in the *Odyssey* and seems to undermine the concept of the *nostos* itself. He also examines an interesting aspect of complementarity, in terms of time and space, between a sailing Odysseus and his weaving wife, Penelope, through the notion of sailing and weaving in which they mutually excel. Finally, the chapter explores some inconsistencies in Telemachus' expected coming of age and access to manhood, issues which are removed towards an ulterior and undefined instance of time by the sole fact of Odysseus' return.

The second chapter of Part I, by C. Antypas, deals with "Calculating the Mythical Dimension: Time and Distance in Homeric Navigation". The author discusses how Homeric fiction needs to respect the actual experience of the audience, in order to construct a solid frame of realism for myth. Since the audience of rhapsodes were seashore people, the Homeric information about seafaring reflects the actual conditions of navigation of the era. Already by 800 BC, Greek sailors had restarted the exploration of the western Mediterranean. Probably, Polis Bay on Ithaca was the last port before the Greek sailors cross the Ionian Sea heading to Italy or Sicily and the first port of their return (*nostos*), after the wanderings in the western Mediterranean basin. In addition, Homeric information about night-faring, distances, and time of journey are revealed to reflect the reality of navigation by 8th century BC, although under extremely favorable weather and sea conditions.

The third chapter, by J.S. Burgess, explores aspects of "Land and Sea in the *Odyssey* and the *Telegony*", from the perspective of spatial theory. The polarity of land and sea is of great significance in Odyssean narratives. The *Odyssey* features its hero's account of his wanderings at sea, striving to return to Ithaca, an island in the sea. The Homeric poem indicates that Ithaca has various politico-economic connections to the mainland, and some of Odysseus' lying tales feature Thesprotia on the mainland. Tiresias imposes a geographically obscure "inland" journey on Odysseus, and his prediction of death "from the sea" for Odysseus centers on the spatial binary of land/sea. In the *Telegony* the hero journeys to both Elis and Thesprotia, before returning to Ithaca, where death arrives "from the sea". In ancient stories of Odysseus the polarity of land and sea is prominent, and the spatial status of Ithaca as both land and sea, with all the ambiguity and paradox that this involves, is essential to understanding myth and literature about Odysseus.

The following, fourth, chapter of Part I, by D. Bouvier, studies "The Correlation of Fountains and Altars in Archaic Greek Poetry". In ancient hexametric poetry several descriptions may be found of altars built close to a source (*pege*) or a fountain (*krene*). This chapter examines the possible relation between the pure water emerging from earth and the blood poured on the altar. Far from proposing a single answer, the author observes how different poetical genres suggest different ways of approaching this correlation; for example, at the beginning of Hesiod's *Theogony*, the fountain and the altar around which Muses dance appear to be an essential indication of good order.

The fifth chapter of Part I, by E. Peraki-Kyriakidou, discusses the figure of "Iris as Messenger and Her Journey: Speech in Space and Time", as presented in Homer, Apollonius' *Argonautica*, Virgil's *Aeneid*, and Ovid's *Metamorphoses*. In the *Iliad* what matters is Iris' swiftness; neither her visual characterization nor the space she travels through is of importance. In Apollonius' *Argonautica*, the travelling space is again of secondary importance. In the *Aeneid* Iris' visual characterization is more important, while there is a shift of interest to the travelling space itself. The space now is blown up with colours left by Iris herself as she passes; in a way, Iris becomes the space itself. Ovid, in his *Metamorphoses*, goes one step further: the poet presents Iris as travelling 'invested' in colours, which are not her own. Since Iris is recognized in scholarship as an 'epic voice' and an embodiment of tradition, in a metapoetic reading the visualization of the original word is affected by countless different influences in its course.

In the final chapter of Part I, S. Kyriakidis discusses "The Patronymics *Pelides* and *Aenides:* Past, Present and Future in Homeric and Virgilian Genealogical Catalogues". The author shows how every addition of a patronymic in a genealogical catalogue plays with time and space. On the one hand a patronymic adds the temporal space of a generation to a catalogue and on the other it has a minimal presence on the page and in the reading time of the whole catalogue. In this chapter, *Pelides,* the most acclaimed Greek patronymic, is used as a case study for showing the multi-faceted function of a patronymic. This function mainly depends on the context of the catalogue and on whether the patronymic is applied as a personal attribute by the hero or his opponent or even by the poet himself. In principle, it constitutes an attribute of praise to its bearer but on occasion it may become a device for neutralizing personal features of the hero (in our case, Achilles), thus allocating to him the common fate to all humans. Whereas Homer did not create for *Pelides* Achilles any *Achilleides* successor, Virgil did so in his *Aeneid*; the descendant of Aeneas, Ascanius, is divinely called *Aenides*. Thus, he vouchsafes the glorious continuation of the Trojan race. Furthermore, the patronymic *Pelides* acquires clear Roman 'undertones' in the *Aeneid*, since its use as an attribute to Achilles' son, Neoptolemus, offers the poet the means to play down Achilles and thus disregard the hero's contribution to the Trojan war.

Part II, "Drama", opens with A. Bierl's study on "The Bacchic-Chor(a)ic Chronotope: Dionysus, *Chora* and Chorality in the Fifth Stasimon of Sophocles' *Antigone*". The Bakhtinian chronotope, Lacanian theory, and Kristeva's *chora*, along with concepts from Derrida and others, combined with research into *choreia*, Dionysus, rituality, metatheatre, performativity and, last not least, philological and literary close-reading, all aided in determining a specific Bacchic chronotope in archaic and classical choral song culture, a new structural interplay of time and space in Greek literature. Dionysus as the total Other, the unconscious, the god of the middle-ground, of mediation and transformation, serves, in a way, as the emblem of this chor(a)ic constellation. From *chora* emerges *choros,* his special medium of vitality in performativity. The Bacchic chronotope is permeable, hybrid, fluid and shifting. Moreover, Dionysus' *chora* is a space of arrival in procession where the god transgresses boundaries in sudden epiphany. After identifying briefly the major features of the Bacchic chronotope in the parodos of *Bacchae*, the main focus of the paper is on the fifth stasimon of *Antigone*, an exemplary tragic song in this regard. This Bacchic chronotope is a whirl of concentric choruses extending even to the cosmic level. It is highly metatragic and a powerful *mise en abyme*, since it references back to the choral performance executed in honour of Dionysus and displayed in the orchestra of the Athenian theatre of Dionysus. The chapter concludes in giving an account of the web of Dionysiac references and Bacchic patterns that constitute Sophocles' *Antigone*, a most Dionysiac play. All things considered, the author argues that the Bacchic chronotope, the special configuration of space and time, is a vital element to understand texts performed in a Dionysian context and occasion.

The following chapter, by E. Iakovou, analyses "The Re-enactment of the Past in the Present and the Transformation of Space in Sophocles' *Oedipus Tyrannus*". The author shows how the past and the events that happened before the play are re-evoked within the play's action, and also how the significant places of Oedipus' life gain special weight as they are retrospectively surveyed during the play's action. Cithaeron, Corinth, Delphi, the crossroads, the palace of Laius, and the polis of Thebes each provide crucial information about significant events in Oedipus' past. When Oedipus brings those distant events together, the past overtakes the present and Oedipus discovers his true identity. Finally, the author discusses the spatio-temporal poetics of the play, which suggest an association between Oedipus' human condition and the Sphinx and the solution of its riddle.

Part II concludes with a chapter on Comedy by A. Papachrysostomou, who explores how "Time- and Space-Travelling in Greek Middle Comedy" works. It is the current scholarly belief that myth and contemporary reality were inextricably intertwined in Greek Middle Comedy. This chapter seeks on the one hand to corroborate further this hermeneutic approach (via analysis of exemplary cases) and on the other hand to demonstrate that this intricate interweaving of myth and reality is the natural result of an idiosyncratic dramaturgical application of the notions of "time" and "space" within the comic genre, during especially the fourth century BC. Through close reading of five comic fragments, the present chapter exemplifies how in Middle Comedy various mythical figures (for reasons of comic effect) are – irrationally and abruptly – pulled out from their mythical context and "travel" in time and in space, so that they land in a different era, the fourth century BC, and in a different place, Athens. The chapter studies the representative cases of four such mythical "travellers": Athamas, Odysseus, Peirithous and Oinopion – as exemplified in five fragments of Middle Comedy.

Part III, "Empirical and Imaginary Chronotopes", comprises seven chapters. The first chapter, by M. Meyer, studies, from an archaeological point of view, "The *martyria* of the Strife for Attica – *martyria* of Changes in Cult and Myth. Space and Time in the West Pediment of the Parthenon". When Herodotus visited the Acropolis, he saw an olive tree and a *thalassa* (salt water), which Athena and Poseidon had set as *martyria*, when they quarrelled about the land. Thus, Herodotus links a mythical tale to visual marks in the sanctuary; his remark is the earliest literary evidence for the competing interests of these two gods, and the west pediment of the Parthenon is the first visual representation of this strife. The author argues that the marks served as visual "proofs" for a narrative which was not much older than its first manifestation in word and image – a tale which was a combination of a common motif (Poseidon's claim to a country) and a local Athenian myth (the successful defence against invaders, with Erechtheus and Eumolpus, protégés of Athena and Poseidon respectively, as protagonists). Interestingly, all attempts to present a coherent narrative of the gods' competition for Attica fail to do so. The physical marks on the Acropolis might have a long tradition; the myth they allegedly testify, however, does not.

The second chapter, by C. Nobili, focuses on "Cattle-raid Myths in Western Peloponnese". Several episodes in Greek myth connect Messenia with cattle-raid. Especially Pylos (although its location shifts between Messenia and Elis) seems to be extremely relevant for cattle-raid myths. Scholars have investigated the importance of cattle-raid in Indo-European societies, in which it has an economic function and often plays an important role in the initiation of young warriors. This phenomenon is very common in the archaic Greek world, too, and the fertile plains of Western Peloponnese seem to offer the best location for this sort of myths. This chapter collects myths of cattle-raid located in Western Peloponnese and analyzes the economic and cultural reasons which may explain such an accumulation in that locus.

In the third chapter E. Pellizer explores "Time and Space in Argolic Traditions: From Ocean to Europe". Argolic mythic traditions include a genealogy going back to Ocean and to the river Inachus, as well as the vicissitudes of Phoroneus, Io, and Epaphus. The author studies how the chronology and geography of these traditions, being fundamental to the myths that define them, offer the possibility of discovering the hypothetical relations between Greek and Eastern populations in the second millennium BC. In fact, Herodotus masterfully inserted these fantastic traditions, dominated by a bovine symbolism, into a de-mythicizing context, in order to explain many centuries of proto-history (and history) of the eastern Mediterranean.

In the fourth chapter of Part III, P.D. Scirpo talks "About the Boeotian Origin of the Emmenidai's Genos: An Indication from Gela". From the recent analysis of the literary sources carried out by G. Adornato, it is clear that we need to distinguish two strands of tradition about Akragas, one concerning the origin of the polis (sub-colony of Gela, founded around 580 BC) and one concerning the origin of the Emmenidai's genos. Based on Pindar's Second Olympic Ode about Theron, his victory in 476 BC and his lineage (the genos of "Emmenidai"), the author shows how a marble base of *louterion* bearing an inscription, found in Gela in the early 20th century by P. Orsi, could provide the missing link for the correct reconstruction of the eventful past of the Emmenidai's genos.

In the fifth chapter, "Fighting on the River: the Alpheus and the *Pylian Epic*", G. Zanetto discusses Nestor's narrations in the *Iliad* about his past glorious accomplishments (cattle raids, battles, athletic competitions). Scholars think that these passages are what remains of the "Pylian Epic", i.e. of the epic songs which in Mycenaean age celebrated the glory of the lords of Pylos. The author focuses on the presence of a river, which is an important element of the scenery: the river is the line of contact between the armies or the boundary beyond which the stolen herd must be pushed, the plain of the river is the setting of the horse-race. Hence, he argues that in the songs performed in the Palace of Pylos the river was often the background of the action. When the Pylians moved to Kakovatos of Triphylia, they had to adapt their traditional stories to the new geographical context. The Alpheus, the river which marks the border between Triphylia and Elis, became the new setting of the Pylian exploits. This is the reason why the Alpheus is so often referred to in Nestor's speeches in the *Iliad* and in the archaic texts which reflect the poetic traditions of the "Pylian Epic".

In the penultimate chapter of Part III, N. Villagra discusses aspects of "Time and Space in the Myth of Byblis and Caunus". The myth on the incestuous passion of the siblings Byblis and Caunus is well known since Hellenistic times. The author analyses the symbolic value of different elements related to space and time in the version of the myth found in Parthenius of Nicaea's *Erotica Pathemata*, which combines mythographical prose with poetic fragments.

In the final chapter of Part III, A. Michalopoulos studies the nature and function of "Mythological Time and Space in Ovid's Exile Poetry". From his place of exile, Tomi, on the Black Sea coast, Ovid sent back to Rome nine books of elegies (*Tristia* and *Epistulae ex Ponto*). In these poems Ovid speaks about his miserable life in exile and vainly tries to achieve his recall to Rome. The chapter discusses Ovid's reception of myth and his use of myth as a means of enriching his arguments and of constructing his own exilic *persona*. It focuses on elegy *Tr.* 1.3, in which Ovid elaborates on mythological examples that cover a vast span, both spatial and temporal (Greece, Troy, Rome, Theseus, Trojan war, Roman history, Ovid's present). Different periods and places provide proper mythological background for Ovid's life and exile, and various mythological strings run parallel or converge (Aeneas, Priam, Theseus and Mettus, to name but a few). Furthermore, the author attempts to defend the authenticity of lines *Tr.* 1.3.75–76 on the grounds of the suitability of Mettus' mythological *exemplum* for Ovid's case.

Part IV, entitled "Shifting Chronotopes", begins with J. Andrews chapter on "*Kairos*: The Appropriate Time, Place and Degree in Protagoras's Myth of Origins". The myth by which the Platonic Protagoras explains for Socrates Athenian beliefs about political virtue is also an account of how excellence in deliberation and action arises from an understanding of *kairos:* "the time or place at which, or degree in which, something is appropriate". Taking liberties with the standing Prometheus myth, Protagoras has the gods order the titan and his brother to make an appropriate distribution, assigning to each species a power of survival duly compensating for that species' natural weakness and making all species equally viable. Viewed in terms of the timely address of the necessary and appropriate, the first attempt (Epimetheus') ends in error. The second, Prometheus', is a signal success. Prometheus is thus presented as the mythical and prototypical master of *kairos*. Given that Protagoras has moments earlier indicated that the subject of his own teaching is excellence in decision-making, we may infer that this myth of timely and strategic decision-making also functions as a protreptic to the study of Protagorean *euboulia*. Intended for Hippocrates and other such prospective students of the sophist, the myth shows that *euboulia* is mastery over *kairos*, and establishes Prometheus as the mythical predecessor of the great fifth-century sophist.

"From Here to Eternity: *Mythologein* in Plato's *Phaedo*" is the title of the second chapter of Part IV, by C.A. Zafiropoulos. Marked by the presence of *mythoi* at its two ends – the fables of Aesop at the introduction and the concluding great eschatological myth on the afterlife of the soul – *Phaedo* is also notable for Socrates' blending of *logoi* with strongly illustrative fictitious accounts, similar to myths, as discursive tools in his inquiry for true knowledge with regard to the afterlife of the soul. In the end, philosophical *mythologein* proper is employed and its transcendent account helps the philosopher overcome the limitations of human reasoning that are caused by the emotional frailty of materiality and by man's pessimistic perception of time and space as a continuum to be nullified by death.

In the third chapter, M. Garani deals with "Ovid's Temple(s) of Vesta (*Fasti* 6.249–460)". In the sixth book of his *Fasti*, in which Ovid delves into the origins of the Roman festivals that take place in the month of June, the poet accounts for the shape of the temple of Vesta in the Forum Romanum, the etymology of the Goddess' name, as well as the origins of the Vestalia, i.e. the people's festival that was held in her honour in the 9th of June (*Fast.* 6.249–460). In this chapter the author argues that by his elegiac treatment, Ovid questions and destabilizes the recent Augustan integration of Vesta, the ancient guarantor of Roman safety, into the Roman space and the calendar. Ovid employs a number of proleptic and retrospect references that result in her temporal fragmentation over the Ovidian months as well as her temporal projection into the Roman past. Whereas Ovid underscores the immobility of Vesta's temple in the Forum as well as its lack of a cult image, the Goddess herself constantly oscillates between her ancient dwelling and the one recently founded on the Palatine hill. The author concludes that (on the basis that old and new, i.e. the popular and the Augustan Vesta, were also so separate topographically) Ovid keeps throughout his account the interplay between the two facets of the Goddess, ultimately undermining her imperial credentials.

In the fourth chapter of Part IV, S. Papaioannou focuses on "Carmenta in the *Fasti:* A Tale of two Feasts". Starting with Julius Caesar and continuing with Augustus the Roman calendar became fixed, but at the same time, several new feasts were added in celebration of the imperator's accomplishments or in honour of the members of his family, and new meanings were offered to traditional festivals. Augustus' control of the new calendar epitomised his progressive control of the State. Ovid's *Fasti* represents an informed reply to Augustus' reconstruction of the Roman calendar. Ovid seemingly sides with the Augustan effort to bring back to life long-forgotten cults whose precise ritual and origin had to be devised a new. In reality, he exposes the imperial practice of devising festal aetiologies convenient to the version of Roman time engineered by the regime. The two feast days of the *Carmentalia* put together a case-study on Augustus' policy of reconstructing time and civic life, in order to control it. Especially the second feast, set at *Fasti* 1.617–636, which associates the Roman goddess Carmentis and the *carpenta*, the covered two-wheeled carriages, by means of a fictitious etymology, alludes to Augustus' policy of restructuring Roman time, and, along with it, Roman social and religious life.

In the final chapter of this volume F. Létoublon discusses "The Decisive Moment in Mythology: The Instant of Metamorphosis". The author analyzes the process of metamorphosis in Greek "mythographers" as the "decisive moment" when a person is transformed. The chapter begins with examples drawn from Antoninus Liberalis for showing the role of verbal aspect in the narrative. The study of the "instant before" shows the importance of pursuit and impossible flight. Incestuous loves appear in Antoninus Liberalis, but with more frequency in Parthenius of Nicaea, which allows to imagine that Freud could have found benefit studying these texts for his theory, especially with the narrative of Periander's mother and the expression of pleasure felt by the son in the relation with his mother (he does not know then who she is). The author also analyzes the kinship between metamorphosis and metaphor, important for poetry and visual arts, and concludes with the link between metamorphosis and the notion of *rites of passage*. Pursuit and impossible flight eventually appear as a means of escaping and yielding for the pursued girl, whereas for the pursuer as a means of giving up sexual possession and keeping forever a substitute as the syrinx or the laurel.

The editors would like to express their thanks to Serena Pirrotta, Michiel Klein-Swormink, Katharina Legutke and the entire De Gruyter team for their constant support to make the volume appear in the series "MythosEikonPoiesis". MEP represents an almost ideal publishing framework for these papers, bringing felicitously together the interests of the series with Patras' "Center for the Study of Myth and Religion in Greek and Roman Antiquity".

The editors would also like to express their gratitude and sincere thanks to Sotiris Karampelas (currently a PhD candidate at the Department of Philology, University of Patras) for his invaluable assistance in editing and typesetting the material of this volume.

It is the editors' hope that throughout this volume the readers will find rich and interesting analyses, useful methodological approaches and up to date bibliographical references on a variety of topics, aspects and figures related to Greek and Roman myth, religion and ritual concerning Time and Space, that is the two fundamental categories and parameters of human existence.

Anton Bierl
Menelaos Christopoulos
Athina Papachrysostomou

Part I **Epos**

Menelaos Christopoulos

Strange Instances of Time and Space in Odysseus' Return

Although the *Odyssey* is usually read and studied by well advised readers, who are perfectly aware of the poet's major device to start the narrative *in medias res*, the structure of time and space throughout the poem often conveys to the audience subtle and unexpected aspects which speak in favour of this article's title. The purpose of this article is, therefore, to focus on some of these aspects which appear to be particularly significant.[1]

An important one seems to be what I will further call a 'corrupted nostos', an idea underlying or parallel to the general narrative of Odysseus' return. One remembers that four times in the *Odyssey* Odysseus returns to the same point from which he had departed. The first time occurs at the very beginning of Odysseus' return and is mentioned by Nestor when Telemachus visits Pylos (3.13–166). The Pylian king tells Telemachus that after the sack of Troy, Odysseus initially sailed with Nestor from Troy up to Tenedos but then he returned back to Troy to join Agamemnon who had stayed there to offer sacrifices to Athena. The second time is Odysseus' return to the island of Aeolus after Odysseus' companions thoughtlessly opened the skin bag that Aeolus had offered to Odysseus (10.46–76). The third time is Odysseus' return to the island of Circe after having visited the Underworld on Circe's advice (12.1–36). The fourth time is Odysseus' return to the strait of Skylla and Charybdis after the shipwreck of his ship due to the impious slaughter of Helios' cattle by Odysseus' companions (12.426–446).[2] Of these four returns, the first (Tenedos-Troy) and the third (Underworld-Circe) are due to Odysseus' own intentions and have no negative consequences for him nor is there any differentiation perceived in the circumstances he had already experienced in his first contact with these places. But in the second and the fourth occasion (return to Aeolus' island and to the strait of Skylla and Charybdis respectively) the situation is totally different. The first time that Odysseus crosses the strait of Skylla and Charybdis he is on his boat, with his companions and the danger he encounters comes from Skylla. The second time Odysseus is

1 One should mention here the concept of belatedness, i.e. the fact that in several occasions in Odysseus' wanderings and tales Odysseus is portrayed as a belated traveller; on this see Burgess 2012, 269–290.

2 The same theme, in miniature, can also be detected in the return of Odysseus' ship to the Cyclops' shore due to the rock thrown by Polyphemus in Book 9.

DOI 10.1515/9783110535150-001

without his boat and without his companions; plus, he enters the strait from the opposite edge and the danger he encounters comes from Charybdis. In contrast to Jason, the hero of the pre-homeric Argonautic epic, who crossed the strait of the Clashing Rocks one way only, Odysseus crosses twice and both ways a strait with two different types of danger coming from each one of the strait's two rocks. Further on, this second crossing is for Odysseus the first danger he experiences alone, after the death of his companions. The circumstances of the second crossing are definitely worse than those of the first. The return to Aeolus' island leads also to a worse departure, since, instead of being welcomed, hosted and gifted, Odysseus is this time cursed, rejected and expelled.

Where does this theme of reverse *nostos* or second departure lead us? It seems to me that here we have something more than the occurrence of one of our usual, frequent and well identified tools often registered in Homeric poetry: this is neither a simple narrative doublet nor a narrative retardation of the plot. Actually, these reverse *nostoi* are highly interesting as to their own content and their own context. First of all, we observe in them the persistence of the number 2 (two departures from the same place, four occurrences of the theme, two crossings from two entries of the strait, two dangers from two rocks), whereas the usual number pertaining to Homeric epics is traditionally stated to be either 3 or some multiple of 3.[3] Secondly, one tends to see in these reverse *nostoi* and repeated departures an ironic undermining of the concept of the *nostos* itself, by forwarding first a return to the departure instead of an arrival to the end – something that I would tend to call a 'corrupted *nostos*' – and then a second departure from the same place under much worse circumstances than those prevailing the first time. Further on, since these corrupted *nostoi* and their unhappy outcome are always the result of the companions' unwise initiative, one wonders whether the purpose of these narratives on Odysseus' second departure from the same place is not only to echo but actually to enhance and further develop the idea announced in line 5 of the prooimion, where we are told that Odysseus did not finally save his companions, although he very much wished to do so. In other words, these corrupted *nostoi* could merely illustrate the idea that the companions were not saved, because they simply did not 'deserve' to be saved; whether this distinction is to be considered against the background of aristocratic values prevailing in the *Odyssey* is a question to be answered in another paper or in another book, but it certainly has to do with the way the idea of 'collective' and the idea of 'individual' are perceived in the epic. What these narratives say (which the prooimon does not say) is not only that the companions are ruined by

3 See Blom 1932; Germain 1954; see also schol. in *Il.* 6.174.

their own ἀτασθαλίαι, but mainly that their existence is ruinous to Odysseus' own salvation; by the middle of the trip to Ithaca this major incompatibility has already become clear: it is the companions' salvation against Odysseus' salvation, it is them or him, themselves versus himself.

In Odysseus' return there are not only significant departures and arrivals, there are also significant stays. The longest of these stays is, of course, his sojourn on the island of Calypso, from where the narrative of his return begins. Strangely enough, this idleness of Odysseus on the island of Ogygia allows a remarkable spread of his Trojan *kleos*. The wreck of Odysseus' ship after Thrinakia and the loss of his companions is by many aspects an important shift in Odysseus' return, since the *nostos* is no longer collective but individual and, consequently, heroic *kleos* is individually ascribed to him.[4] But from that very moment and further on, the nature of the deeds related to his person changes and light is now increasingly shed on his Trojan exploits. Until the arrival to the island of Calypso, Odysseus is active at a present time and this action concerns a collective *nostos*, his own return, as well as his companions'. But with the stay in Ogygia, the prevalence of present time stops. As long as Odysseus remains inactive in Calypso's island, witnesses of his Trojan *kleos*, such as Nestor, Menelaus and Helen, have the time to go home and propagate it, so that the *kleos* itself precedes Odysseus when he regains his activity and listens to Demodocus' song on the Wooden Horse. In terms of time and space then, as Odysseus' *nostos* brings him from Troy back to Ithaca, the renown of his *kleos* brings him from Ithaca back to Troy, an achievement of an itinerary which he initially had refused to undertake.

Much has been said about the character complementarity between Odysseus and Penelope, mainly in matters of mental capacity;[5] formulaic persistence speaks in favour of this complementarity: περίφρων Penelope is the suitable spouse of πολύμητις Odysseus. She even manages to fool him when she tests his identity by pretending that his bed, nailed on the root of an olive tree, can be easily removed to another room of the palace. Further formulaic evidence can perhaps strengthen this concept of complementarity also in terms of time. Each time that Odysseus' trip continues, we read / listen that he and his companions (or he alone when he leaves Calypso's island), "raised the mast and fixed the sail on it" and as long as the sail is raised on the mast, on the *histos*, it brings him closer to his destination and, then, the course of time is positive for his time-

4 That is why in a previous publication I argued that not only *kleos* but also time is individually perceived; see Christopoulos 2001, 93–105.

5 On Penelope in the *Odyssey*, see indicatively Katz 1991; Felson 1994; Papadopoulou-Behlmedi 1994; Felson and Slatkin 2004, 91–115.

ly arrival. What happens with Penelope? She also has her own *histos*, the loom. But as long as the cloth, the shroud for Laertes, is woven on this *histos*, the course of time is negative for Odysseus' timely arrival. As he sails for three years and spends other seven waiting in Ogygia, she weaves for three years and spends other seven waiting in Ithaca.[6] Finally, when both *histoi* stop functioning for both characters, the husband and the wife, the course of time seems to lead to an impasse. Penelope's trick is revealed and she is now forced to move towards an unwished marital life, while Odysseus is trapped within an unwished substitute of marital life on Calypso's island, where he had been brought to grasping to the remains of his broken *histos*.

The idea presented above is perhaps another aspect of complementarity, in terms of time and space, between a sailing Odysseus and his weaving wife who both know how to "weave malice" (δόλους).[7] What kind of son can be the offspring of such an alliance? Another strange occurrence related to time is Odysseus' arrival at Ithaca, just a little earlier than his son's own return. Several points of view have been developed, as far back as antiquity, as to the utility and the contribution of Telemachus' trip related to the economy and structure of the whole poem.[8] To understand this trip better in relation to Odysseus' return, one should perhaps consider the whole issue of time in the case of Telemachus and his coming of age. As early as the *Iliad* we find two references to Telemachus, both made by Odysseus, both in a strange context and both completely out of time. The first one concerns Odysseus' threat to punish Thersites in Book 2.260: the hero claims that he will punish Thersites "or if I don't", he says, "let me not be called the father of Telemachus any longer". The second reference to Telemachus concerns Odysseus' answer to Agamemnon in Book 4.353–355; Agamemnon accuses Odysseus of opportunism, greediness and shunning battle (4.339–346) and Odysseus refutes these accusations by presenting himself as "Telemachus' beloved father" (4.354).[9] Aristarchus' view was that in this passage the poet has in mind the Odysseus of the *Odyssey* rather than the Iliadic hero. In any case, however, Odysseus' reference to Telemachus in these two passages of the *Iliad* (2.260 and 4.354) shows clearly that the story of Telemachus was known in the epic tradition some time before its being related in the *Odyssey*. But what

6 *Odyssey* 2.87–110.

7 On "weaving malice" in the *Odyssey*, see indicatively 5.356, 9.422, 19.137.

8 See schol. in *Od*. 1.93: ἄτοπος δοκεῖ Τηλεμάχου ἡ ἀποδημία ... οὐκ ὠφελοῦσα τὴν ζήτησιν τοῦ πατρός. See also Apthorp 1980, 1–22. On the role of the sea in Telemachus' trip see Troncoso 2016, 523–535.

9 This is contrary to the usual mode of heroic self-presentation, through the hero's father's name (in the case of Odysseus, cf. the frequent formula 'Οδυσσεὺς Λαερτιάδης).

story exactly? At the point that Odysseus mentions his son in the *Iliad* Telemachus is about nine years old and whatever is related about him can only refer to the events of his childhood. In the epic tradition the role of Telemachus at this stage is only to provide an argument against Odysseus' participation to the Trojan expedition and, indeed, the hero's reluctance to join the army was obviously a well-established epic motif[10] possibly explaining the reason of Agamemnon's accusation in *Iliad* 4 and clearly evoked in *Odyssey* 24.115–119 again by Agamemnon when he mentions the difficulties he had convincing Odysseus to join the Trojan expedition. If this is so, then the story of Odysseus feigning madness and sowing salt to avoid this task is probably underlying these passages and, consequently, the device used by Palamedes, who unmasked Odysseus by putting the infant Telemachus in front of the plough, could also be alluded to. In such a case, many scholars' strong conviction that Palamedes is totally absent from both Homeric epics would be considerably weakened.[11]

Telemachus' age is also the issue of a famous 'error' in the Odyssean narrative, namely the information given by Antikleia to Odysseus when he encounters his mother in the Underworld. In *Odyssey* 11.185–187, Antikleia informs Odysseus that Telemachus is already a man who joins men's gatherings and banquets. Although the listener / reader has been watching Telemachus' almost 'adult' activity during the first four books of the *Odyssey*, by the time this information is given, Telemachus is barely twelve years old and manhood is yet far away. When is finally Telemachus about to reach manhood – or is he at all?

It is characteristic of Telemachus' portrayal in the *Odyssey* that all his initiatives, either spontaneous or directed by the gods, remain incomplete. This is true for the Assembly of the Ithacians, for the trip to Pylos and Sparta, and also for the effort to defend his household from the suitors;[12] it is perhaps a significant reversal of time order in the 'Telemachy' the fact that in Sparta Menelaus is naturally giving his daughter to marriage whereas in Ithaca Telemachus is strangely proclaiming the marriage of his own mother – another unachieved task within the frame of Telemachus' potential expedience in the *Odyssey*. Finally, the process of achieving manhood through a series of appropriate and emblematic actions is definitely and surprisingly cancelled for Telemachus by the return of his own father. Odysseus, transformed until then by Athena into an old beggar, suddenly regains a form of juvenile manhood

---

10 See, for instance, Proclus' summary of the *Cypria*.

11 I have elsewhere argued on possible traces of Palamedes' story in Homeric epics; see Christopoulos 2014, 153–166; see also Scodel 2009, 15–16, 108–109.

12 On Telemachus asserting his authority in the *Odyssey*, see Gottesman 2014, 31–60.

and a promised wife from long ago and takes *ipso jure* things in his hands; by so doing, he removes Telemachus' expected and imminent access to manhood towards an ulterior and undefined instance of time. If we knew more about the *Telegony*, the lost epic of the Epic Cycle, we could more easily guess what later happened to Telemachus. I have my own doubts as to the archaic character of this double marriage, Telemachus to Circe and Telegonus to Penelope, announced at the end of the relevant summary of Proclus. But at the present state of our knowledge, I feel that the end of the *Odyssey* leaves us with two – at least – suspended issues: political stability in Ithaca in view of Odysseus' announced new departure and, mainly, Telemachus' coming of age and manhood which, whenever it occurred, was not necessarily rooted in the mythographic tradition related to Odysseus' kingship in Ithaca.

Bibliography

Apthorp, M.J. 1980. The Obstacles in Telemachus' Return. *CQ* 30: 1–22.
Blom, J.W.S. 1932. *De typische getallen bij Homeros en Herodotos.* Nijmegen.
Burgess, J.S. 2012. Belatedness in the Travels of Odysseus. In *Homeric Contexts: Neoanalysis and The Interpretation of Oral Poetry*, eds. F. Montanari, A. Rengakos and C. Tsagalis (Trends in Classics, Suppl. 12), 269–290. Berlin.
Christopoulos, M. 2001. Nostos by Sea and Poetic Structure in the *Odyssey.* In *Eranos. Proceedings of the 9th International Symposium on the Odyssey*, ed. M. Païzi-Apostopoloulou, 93–105. Ithaca.
Christopoulos, M. 2014. Odysseus, Diomedes, Dolon and Palamedes. Crimes of Mystery and Imagination. In *Crime and Punishment in Homeric and Archaic Poetry*, eds. M. Christopoulos and M. Païzi-Apostolopoulou, 153–166. Ithaca.
Felson, N. 1994. *Regarding Penelope.* Princeton.
Felson, N. and L. Slatkin. 2004. Gender and Homeric Epic. In *The Cambridge Companion to Homer*, ed. R. Fowler, 91–115. Cambridge.
Germain, G. 1954. *Homère et la mystique des nombres.* Paris.
Gottesman, A. 2014. The Authority of Telemachus. *CA* 33: 31–60.
Katz, M. 1991. *Penelope's Renown. Meaning and Indeterminacy in the Odyssey.* Princeton.
Papadopoulou-Behlmedi, I. 1994. *Le chant de Pénélope.* Paris.
Scodel, R. 2009. *Listening to Homer. Tradition, Narrative and Audience.* Ann Arbor.
Troncoso, V.A. 2016. La Télémachie et la mer: rites et épreuves d'un apprenti héros. *Historika* 5: 523–535.
West, S. 1998. *A Commentary on Homer's Odyssey*, vol. 1. Oxford.

Constantin Antypas

Calculating the Mythical Dimension: Time and Distance in Homeric Navigation

By the end of the 23rd century BCE, an Egyptian scribe committed to papyrus the first known fiction narrative in Mediterranean: the *Tale of Shipwrecked Sailor.*[1] This text of Early Bronze Age is a legend about magic creatures and supernatural events; on the other hand, it is a valuable source for nautical history: the description of the huge merchantman hired by the narrator, the ship's dimensions, the number of the crew, the requisite skills of the sailors, even the political correctness of the expression 'king's mines' – a propagandistic term for the Cypriot copper mines – reflect the reality, or, better, the experience of a cultivate pharaoh's subject around 2200 BCE.

In this early text, a constant feature of successful fiction narrative is already distinctive: the mythical nucleus of a tale is acceptable, if the narrative elements concerning the daily life and the common experience of the audience are both realistic and accurate.

Homeric poems – at least, their oral version – probably took an almost final form on the east Aegean coast, in Ionia,[2] some time between the 8th and the 6th centuries BCE.[3]

The audience of the rhapsodes, in that era and that region, were seashore people, habitants of port towns, who were familiar with navigation and its language. If Homeric fiction aimed to legalize its fables about giants, witches and monsters in the Mediterranean wild far west, it needed to respect the living experience of its listeners; ships, distances, winds, sea currents, seamen behavior and technical data should construct a solid frame of realism supporting the fiction; if so, the Homeric information about seafaring reflects – more or less – the nautical experience and the actual conditions of navigation in the era of the epic's final formation.

---

1 Matthews 2002, 6. Published translations of the *Tale:* Golénischeff 1912; Lichtheim 1975, 211–215; Parkinson 1997, 89–101; Simpson 2003³, 45–53.

2 Frame 2009, 551–620.

3 Nagy 1980, 391; Nagy 1990, 53; Powell 1991, 231–237; Nagy 1996, 29–63; Nagy 1999, Introduction.

DOI 10.1515/9783110535150-002

Exit Points

The Homeric ships sailed in the Mediterranean, but the Mediterranean was not conceived as *a* sea: it was *the Sea*, the one and sole saltwater space permitting human navigation. The Sea was divided into two separate districts: the eastern basin was the καθ' ἡμᾶς θάλασσα, the inner sea of the familiar and known world; the western basin was the home of the *Others*, a world almost forgotten after the collapse of the Mycenaean palatial societies[4] and the subsidence of sea trade.[5] By 800 BCE, after a lapse of two hundred years, Greek sailors had already restarted exploring, discovering and exploiting these western seas.[6]

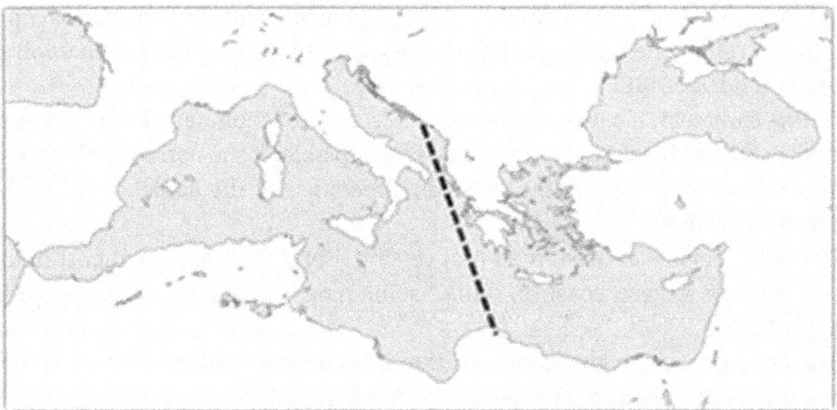

Map 1: The two Mediterranean basins:
The eastern καθ' ἡμᾶς θάλασσα and the western *Sea of Others*

The *Odyssey* gives some clear indications of the exit points from καθ' ἡμᾶς θάλασσα to the world of the *Others:*

When the current and the gale deterred Odysseus from weather cape Maleas, he had only one choice: let his flotilla sail south, before the northerly wind (Βορέης – *Od.* 9.80–81), waiting for a southerly air current and then set the course northwest, to cape Tainaron. Actually, Odysseus' sailing strategy is a typ-

4 For the first phase of the collapse, see Iakovides 2001, 145; Castleden 2005, 219; Kelder 2010, 34; Cline 2014, 129–130. For the final collapse and its causes, see Mylonas 1966, 227–232; Drews 1993; Dickinson 2006, 43–56; Demand 2011, 193–209; Hall 2014², 44–55.

5 Drews 1993, 85–90; Deger-Jalkotzy 2008; Cazzella and Recchia 2010, 36–37.

6 Tsetskhladze 2006; Tsetskhladze 2008; Antypas 2014, 242–266.

ical routine, even for modern sailboats; if a skipper decides that the straight course from cape Maleas to cape Tainaron is not effectible due to adverse winds, he tries to reach Tainaron moving in 'triangle', from point A (Maleas) to point C (Sea of Cythera) and then back to point B (Tainaron) (see map 2).

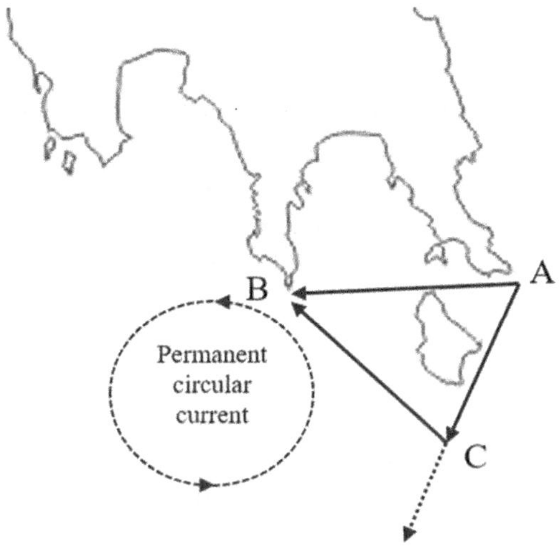

Map 2: From A (Maleas) to B (Tainaron): the straight course from Maleas to Tainaron under (rare) favourable weather and sea conditions.
From A (Maleas) to C (Sea of Cythera) and then to B (Tainaron): the typical sailing strategy to weather Maleas and reach cape Tainaron.
After point C, Odysseus and his companions lost visual contact with Cythera.

So, the first exit point, the Sea of Cythera, is the gate to nightmare; but which is the regular gate to the West?

Until the first decades of the 8th century BCE and before the foundation of Corcyra by Corinthians,[7] a ship sailing from the Aegean Sea to Italy or Sicily should, necessarily, cross the Ionian Sea, following a route depending on the prevailing winds and the surface sea currents.

The crossing from the eastern harbours of the Ionian Sea to its western coasts of Apulia, Gulf of Taranto and Sicily was a 250-mile adventure in open

7 Corinthians founded the colony of Corcyra on 733 or 706/705 BCE (Gehrke and Wirbelauer 2004, 360; for archaeological evidence see Coldstream 1968, 251; Kalligas 1984–1985; de Fidio 1995, 90–94; Lang 1996, 301).

sea, a three or four-day journey.[8] It is obvious that the last port in the καθ' ἡμᾶς (eastern) Mediterranean basin, before the perilous jump, should meet three requirements:

a) offer the possibility of water supply;
b) offer the possibility of weather forecast;
c) offer a sacred place where sailors could pray for good fortune.[9]

Assume that an Aegean merchantman heading to Sicily has reached the NW end of Peloponnese, possibly Cyllene; thence, the captain should choose the optimal course to the optimal last port before he crosses the Ionian Sea.

The first choice was a course to SW coast of Cephalonia. This route is extremely dangerous: there is no safe natural harbour in this part of Cephalonia; the sandy beaches of Scala, Kateleios and Lourdas are wide open to westerly wind, without any alternative protection; moreover, the ship should cross the sea strait between cape Schinari of Zante and cape Munda of Cephalonia, a region where irregular sea currents and unforeseeable gusts caused a great deal of shipwrecks.[10] A heavy ancient merchantman, with a shallow keel, a quadrilateral sail, and only partially docked, had no reason to venture in this dangerous and harbourless strait.

A second choice would be a harbour on Leucas, probably in Vassiliki Bay. The distance from Cyllene to Leucas is, more or less, 50 miles. An ancient sail vessel could run in summertime a maximum of 40–45 miles from dawn till dusk,[11] under normal conditions; the Aegean ship, in order to find a harbour in Vassiliki Bay, should keep sailing into the Echinades archipelago by night, passing by – almost, scratching – the sharp rocks of inhabit islets like Oxeia or Arkoudi. It is evident that no crew wanted to take risks and prolong their trav-

8 Livy *Ab Urbe Condita* 42.48.9: *C. Lucretius praetor ab Neapoli profectus, superato freto [Siculo], die quinto in Cephallaniam transmisit* ("C. Lucretius sailed from Naples, crossed the strait of Sicily, and reached Cephalonia in five days"). Procopius relates a 16-day journey from Zante to Sicily, but in extremely unfavourable weather conditions (Procop. *De Bellis* 3.13.22–23). An average speed of 3 knots, under normal conditions, could be acceptable for an ancient sailboat (for literary evidence see Casson 1995, 282–291; for experimental archaeological evidence see Katzev 1990).

9 For a further discussion and parallels in nautical history, see Antypas 2014, 254–255.

10 Meteorological data: weather stations of Cyllene (Peloponnese), Spartia (Cephalonia), Vathy (Ithaca), airport of Cephalonia (source: *Archives of Hellenic Meteorological Service*).

11 A daily sea journey in summertime could not be less than 13 hours or more than 16 hours (in Ionian Sea: 16 daylight hours on summer solstice). The distance of 40–45 miles is calculated by assuming average speed of 3 knots (see n. 9).

el without any tangible benefit. From a nautical point of view, the optimal route to Italy and Sicily could not include Leucas as a port of call.

The third possible route crosses the Strait of Ithaca. Sailing in almost straight line and after a 10 – 14 hours' journey, always under normal conditions, the Aegean ship would reach the northern exit of the Strait, having run a distance of 40 miles. Until mid-July, the ship would not wait more than 2–4 days in Cyllene harbour before meeting a favourable fair southern wind. The weather condition is usually predictable in this area. The masses of Zante and Cephalonia protect against the western gales. And, finally, our ship could get easily a mooring somewhere in the Strait, in case of danger.

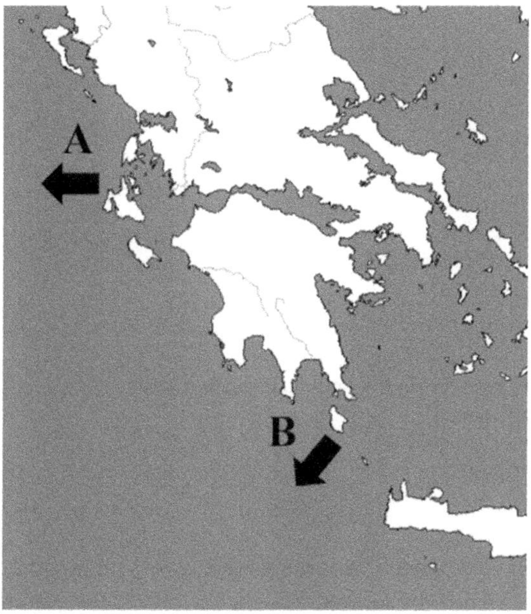

Map 3: The *exit points* to the western Mediterranean basin
A: Eastern Ionian Sea
B: Sea of Cythera

But, which is this Last Port in the Strait of Ithaca?

Here, a digression is necessary. Our concept of orientation issues from the use of map and compass; for a modern sailor, north and west, east and south, are fixed points; consequently, our point of view dictates that Cephalonia lays in the west relatively to Ithaca. But, for a pre-compass sailor, horizon sectors were not fixed points; were large areas defined by the relative sun's position: ζόφος direction was not a clear western direction; it was conceived as a region

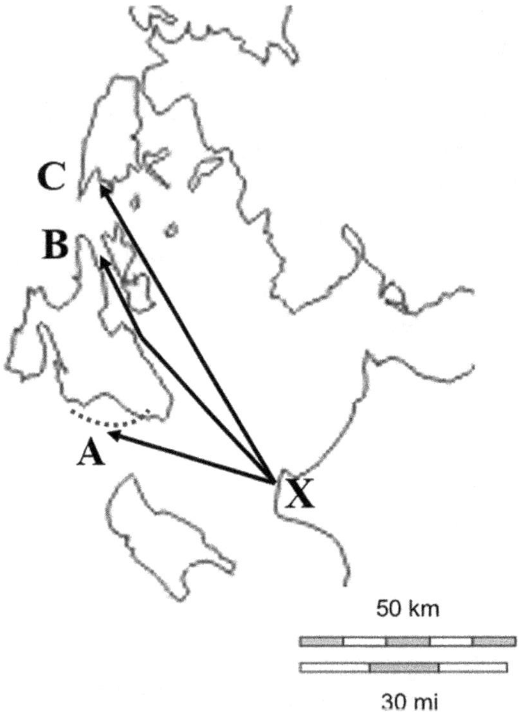

Map 4: The possible courses before the Last Port in the καθ' ἡμᾶς eastern basin:
1.From X (Cyllene) to A (SW coast of Cephalonia)
2. From X (Cyllene) to B (Strait of Ithaca)
3. From X (Cyllene) to C (Vassiliki Bay, Leucas)

extended from northwest to southwest, since the sun sets northwest on summer solstice and southwest on winter solstice. In summertime, a mariner sailing from Same Bay (map 5, area c) to Fiscardo or Polis Bay (map 5, area d) realizes that ζόφος, the sunset, the darkness,[12] lays beyond the exit point of the Strait: this 'Last Port to the West', the port πρὸς ζόφον,[13] should be either Fiscardo or Polis Bay.

Although Fiscardo is a very safe harbour, no source of potable water is located a short distance away and, up to present day, there is no attested archaeological or literary evidence of a marine cult in this area of northern Cephalonia.

---

12 Ζόφος: a) "nether darkness"; b) "darkness"; c) "the wester quadrant, the West" (LSJ s.v.).
13 *Od.* 9.25–26: [Ἰθάκη] εἰν ἁλὶ κεῖται πρὸς ζόφον ("Ithaca lies in the sea, towards the gloom").

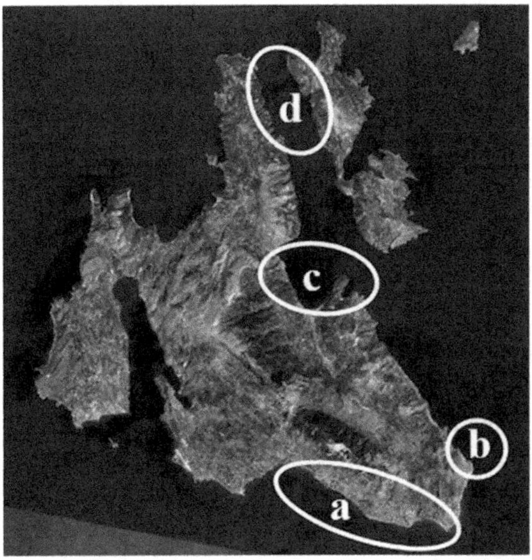

Map 5: Possible Last Port areas:
a. Scala – Kateleios – Lourdas
b. Poros
c. Same Bay – Antisamos
d. Fiscardo – Polis Bay

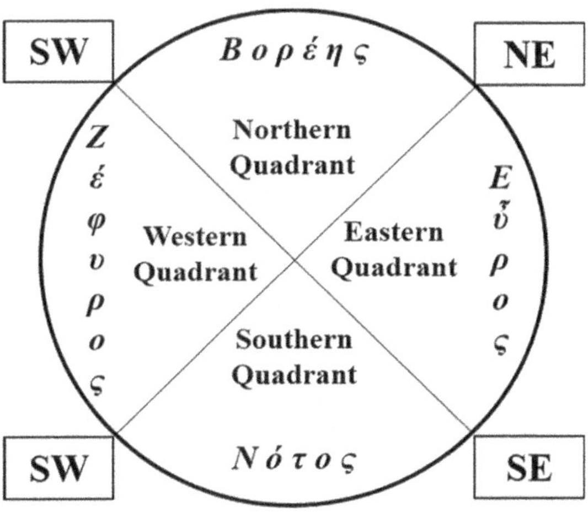

Figure 1: The Homeric concept of four winds system and the main horizon points

Polis Bay, on Ithaca, is probably the fittest Last Port. The small bay offers an excellent harbour which was even safer (and larger) in Homeric era: today, an alluvial plain has reduced considerably its ancient surface area. Polis is the only place in Ithaca with a source of potable water – actually, the only place on the two coasts of the Ithaca Strait with a spring or a well of potable water. Moreover, the archaeological findings in *Loizos' Cave* indicate that Polis Bay was a place of cult, probably a cult of aquatic deities.[14] And finally, the hill of Exogi nearby Polis is a decent weather observatory; on this hill, the captain of the Aegean ship could observe and interpret the τέρατα, the weather signs of the gods, and decide to set sail or not.

The two exit points to the west Mediterranean basin had not equivalent functions: the exit of Ithaca was a gate to a partially unknown, sometimes dangerous, but always desirable world of commercial and colonial enterprises; conversely, the return to Ithaca, after an adventure in western seas, was the first step of νόστος, the home-coming, the re-entrance to the familiar world. Maybe the sight of Ithaca, after the wandering and the misfortunes in the western Mediterranean basin was a yearning not of Odysseus only, but of all Greek sailors venturing in the Adriatic or Tyrrhenian Seas, or even further, in waters adjoining Oceanus.

On the other hand, the exit of the Cythera Sea, was a gate to a simply unknown and dangerous world, to the coasts of Libya:[15] a *topos* of Greek and Latin literature is the representation of Sidra and Gabès Gulfs (the ancient Libyan *Syrteis*) as a place of desolation, where gale and current drive the shipwrecks.[16]

Telemachus Set Sail to Pylos

Telemachus, aboard a hired or borrowed ἐεικόσορος (a twenty oars galley), with his ἑταῖροι (companions) as a crew and Athena on his side, set sail for Pylos. After sunset, Telemachus' companions grappled the oars and drove the ship

---

14 Neolithic archaeological findings in *Loizos' Cave:* Souyoutzoglou-Haywood 1999, 7. Early Bronze Age: Benton 1939, 1–6. Middle Bronze Age: Benton 1935, 73; Souyoutzoglou-Haywood 1999, 101. Late Bronze Age: Benton 1928; Catling 1956, 118; Souyoutzoglou-Haywood 1999, 102–103, 108. 11th – 8th centuries BCE: Souyoutzoglou-Haywood 1999, 143; Antypas 2014, 388–390, 393–395.

15 Cyrene, the first of the five Greek colonies in Libya (Cyrenaica), was founded as late as 630 BCE by settlers from Thera (Hdt. 4.150–161), near present-day Sahhat. On the foundation of Cyrene, see Chamoux 1953; Strucchi 1989, 73–84; Laronde 1990, 169–180; Auston 2008, 192–194.

16 Among others: Hdt. 4.152; Thuc. 6.2; Alciphron *Epistles* 1.10.

off the harbour of Ithaca; Athena sent a favourable wind (ἀκραὴς ἄνεμος). Next morning, the small galley approached the sandy beaches of Pylos (*Od.* 2.414–3.5).

Suppose that the point of departure was somewhere in the northern half of Ithaca Strait; in that case, the ship sailed approximately 10–12 hours[17] and covered 100 miles to reach Pylos, with an average speed of 8–10 knots. Given that ἐεικόσορος was a sailboat 40–50 feet long[18] and the top speed of any wooden sailboat of this length does not exceed 10 knots,[19] the average speed of Telemachus' ship was equal to the top speed of this type of vessel. Such a speed would be possible only if the ship was sailing before breeze, for all the twelve hours of its journey. This is not exactly impossible, but statistically improbable; to maintain an average speed of 10 knots, Telemachus needed both extraordinary luck and, literally, the help of the heavens. Of course, with Athena aboard, Odysseus' son had assured the necessary divine assistance.

In any case, according to the *Odyssey*, the ἐεικόσορος ran a distance of 100 miles sailing in the dark of the night. Is this a poetic exaggeration or, maybe, this journey was less dangerous and more effectible than we thought, even without Athena's help?

Telemachus set sail a fine summer evening, during the navigable period. Athena offered him a Ζέφυρος, a wind blowing from the point of sunset. In these days of year, the sunset is located in the northwestern end of the western quadrant; obviously, the specific Ζέφυρος was a northwesterly breeze. A modern sail ship, departing nearby the skerry Dascalio (the supposed Homeric Ἀστερίς island) and moving always in straight line (heading on 145°) will reach the port of Pylos after several (approximately 20) hours.

Although this is a long night-faring of 100 miles and although the ship must cross two sea straits (the Strait of Ithaca and the Strait of Zante), the journey could be accomplished without any change, or even correction of the SE course. In addition, this *night* faring is not a *blind* faring: in a summer night, the stars of the clear sky and the phosphorescence of the sea help the sailors distinguish the

17 Telemachus landed on Pylos beach in the morning, a few hours after sunrise (Nestor and his people are already gathered – actually, mustered – on shore, ready to sacrifice 81 bulls to Poseidon; *Od.* 3.1–8). During the navigable (estival) period, the night is considerably shorter than day (8–10 hours versus 16–14 hours); accordingly, Telemachus and his companions sailed for 10–12 hours before reaching Pylos.

18 Antypas 2014, 149–151. On boat-building data see Casson 1995, 54–55; Coates, McGrail, Brown, Gifford, Grainge, Greenhill, Marsden, Rankov, Tipping and Wright 1995; Morrison, Coates and Rankov 2000².

19 Antypas 2014, 151.

line separating the sea from the land; in addition, the mountain masses on Cephalonia and on Ithaca, with their distinctive shapes, are visible by night, offering the helmsman information about the course and the relative position of his vessel.

It is clear that Telemachus' journey from Ithaca to Pylos by night was not an extraordinary achievement: it was a demanding effort, but after all, the audience of *Odyssey* knew that the job of an experienced helmsman and a competent crew was the accomplishment of demanding efforts.

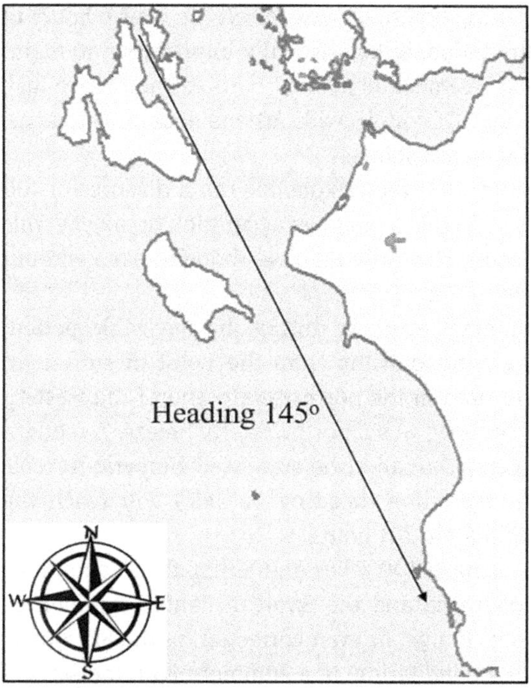

Map 6: From Ithaca to Pylos: a clear, straight course of appr. 100 miles

To Cross or Not to Cross?

After the fall of Troy, a part of Achaeans' fleet moored in Lesbos (*Od.* 3.169), probably in Kalloni Bay. The commanders, Nestor, Menelaus, and Diomedes, wondered about the route of return; a choice was a southern course, to cape Mimas (Karaburun) and into the Strait of Chios (*Od.* 3.170); the *Odyssey* has no need to describe the rest of the route, since every other detail would be a redundancy for an Ionic audience[20] that knew very well the elements of this course. The listeners understood that the fleet, departing from Lesbos and heading south, would cross the Strait of Chios and then, from island to island and from harbour to harbour, would reach the western Aegean coast. In theory, this route is safe because the mariners have always land in sight, sailing by daylight and mooring after dusk. The obvious disadvantage is the great length of this route.

On the other hand, the direct course from Lesbos to Euboea seems like a bold decision: the fleet should run a distance of 110 miles; before an ἀκραής (favourable) wind and moving at high average speed (5 knots), the Greek ships would arrive on Euboea's shores after 20 – 22 hours, sailing a great part of this journey by night; and, eventually, the captains should find their final destination *in* the night. But, is actually a route chosen by Nestor the Wise dangerous?

Besides two reefs in the middle of the distance between Psara and Andros (the *Kalogeroi Reefs*) and the skerry Psarouda about a mile off Cyme, there is no other threat on the direct route from Lesbos to Euboea. The sun and the stars of a clear sky offer the information to plot out a safe course, far away from the rocks. The Greek fleet arrived at Geraistos ἐννύχιος ("in the middle of the night"; *Od.* 3.178), but the skilled navigators had a perfect knowledge of the standard harbours and landed their ships without major problems.

Of course, the commanders of the fleet should have read the τέρατα of the gods, the weather signs (*Od.* 3.173). If those τέρατα predict a favourable wind for several hours and high visibility during the night, the commanders should give the order to set sail. Otherwise, they should postpone the departure or make another choice. Anyway, Nestor, Menelaus and Diomedes read the signs correctly, chose the brief direct route and did not regret it.

The case of Ajax the Lesser is exactly the opposite. Ajax had never been a seaman: his light flax corslet (λινοθώρηξ, *Il.* 2.529), his ability of spear launching (*Il.* 2.530), his swiftness (*Il.* 23.754), all indicate that he was a mountain warrior; on the other hand, he was defiant, foolish (*Il.* 23.783), and the rape of Cassandra blemished his reputation for ever. The ignorance of the sea, the arrogance and the impiety determined his lethal decision.

First of all, he did not join the other three commanders in their decision to cross Aegean by night. The *Odyssey* states ex silentio that Ajax the Lesser followed the route through Chios Strait and then Cyclades: the lengthy daylight route. But in navigation daylight is not tantamount to safety.

---

20 See n. 2.

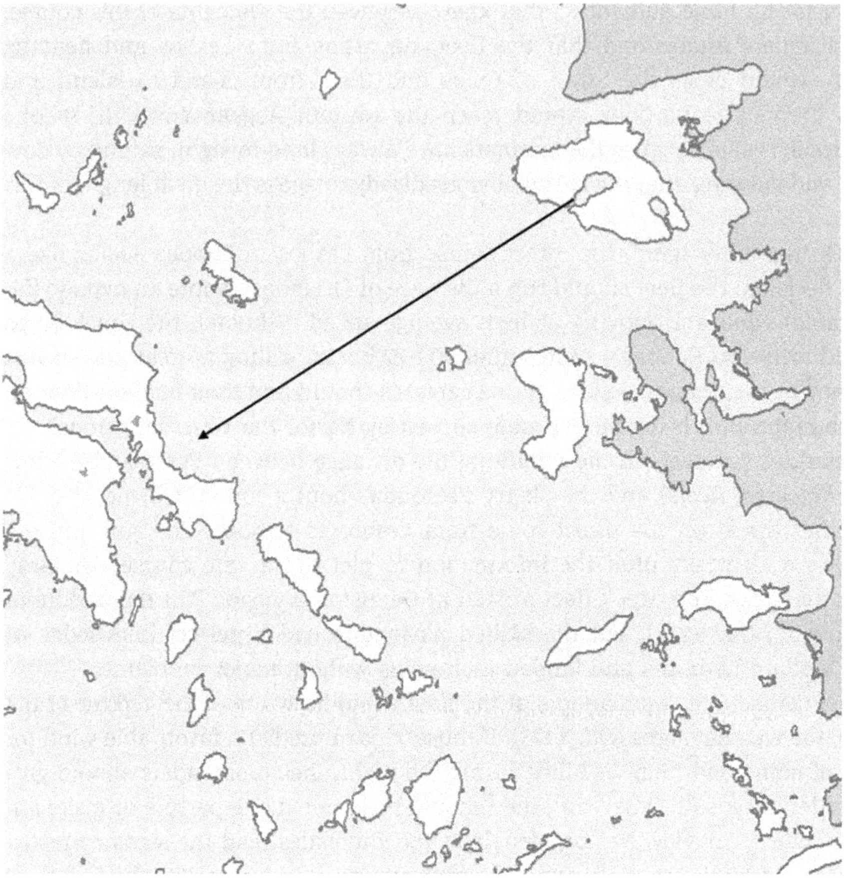

Map 7: From Lesbos to Geraistos: a clear, straight course of 110 miles

Euboea, Andros, Tenos, and Myconos form an insular wall with only three gates of entrance: the Strait of Caphereus, the strait between Andros and Tenos, and the strait between Tenos and Myconos. The Strait of Caphereus is one of the most dangerous place for navigation in eastern Mediterrean. The marine maps indicate as "not navigable" the reefy strait between Andros and Tenos. The strait between Tenos and Myconos is navigable, but also tormented by sudden strong gusts of wind: according to Christopoulos,[21] this is the place of Ajax shipwreck (*Od.* 4.500).

21 Christopoulos 1997, 14 n. 1.

The route of Ajax, king of Locreans, was full of perils: he should sail close by the coast; he should frequently change direction with unfavourable winds;[22] and, above all, he should attempt to enter the Cyclades complex by the gates of the insular wall, a venture much more demanding than the direct night crossing of the Aegean. Eventually, only the ignorance or the necessity would drive a sail-boat skipper to choose the daylight route and move from eastern Aegean to its western coasts passing through the Cyclades archipelago.

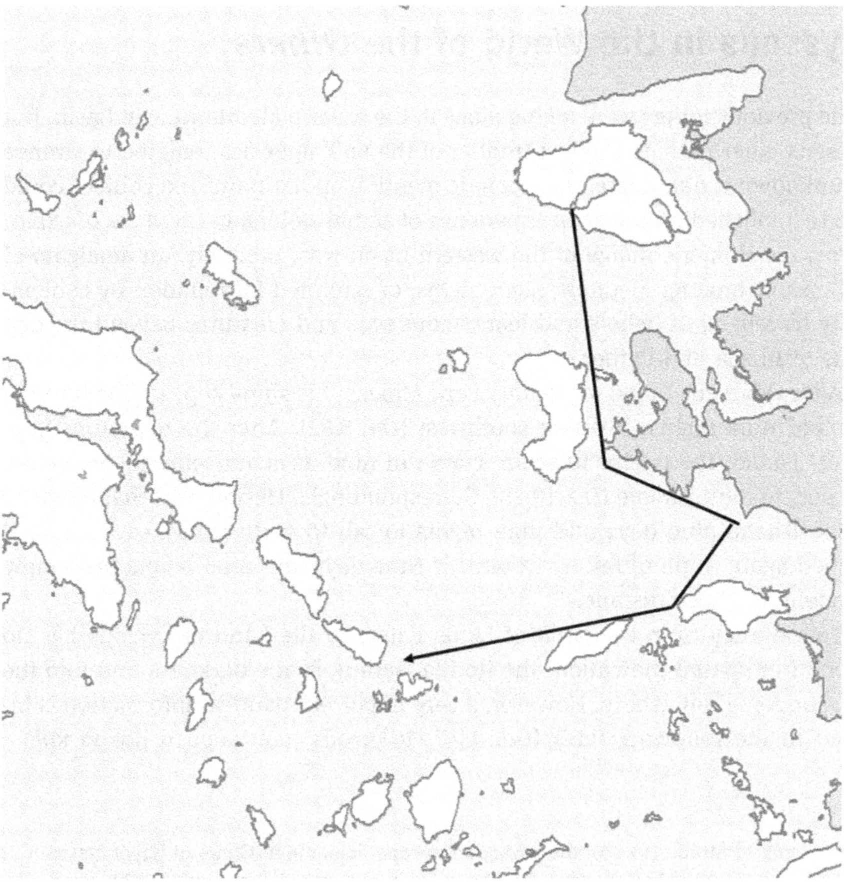

Map 8: The steps of an indirect daylight course: from Lesbos to the strait of Tenos-Myconos

22 In Aegean, the ἐτησίαι ("meltem"), blowing from NE, are the prevailing winds during daylight on summertime, permitting a SW course.

In one of his Cretan lying stories, Odysseus mentioned a five-day journey from Crete to Egypt. This time indication corresponds to the actual distance of 320 miles between eastern Crete and the Nile Delta: an ancient sail ship, departing from Crete and moving under favourable weather needed four days to reach the shore of Egypt.[23] It is obvious that ποντοπορεῖν, "sail in open sea", without a possibility of a night stop or a port of call, was a skill developed by Mediterranean sailors at least fifteen centuries before Homer.[24]

Odysseus in the World of the *Others*

All the previous routes were taking place in the eastern Mediterranean basin. But Odysseus, after passing the last frontier of the καθ' ἡμᾶς sea, entered in strange and unknown waters. Here, the Homeric mention on harbours and courses could not be established on common experience of actual sailors; in these era of explorations, the Homeric image of the western basin was, probably, an amalgam of pre-Greek or broader Mediterranean myths, of scattered information by contemporary travellers, of beliefs and fears about seas and creatures beyond the exit points of Ithaca and Cythera.

After Odysseus failed to weather cape Maleas, his ships kept sailing for nine days and nine nights, heading southwest (*Od.* 9.82). After the departure from Aeolus' island, the twelve Ithacian ships ran nine days and nine nights before Ζέφυρος, to return home (*Od.* 10.28). Correspondingly, Hesiod says that a bronze anvil will take nine days and nine nights to fall to earth from heaven and, if dropped again, nine to fall to Tartarus.[25] Nine days and nine nights are simply an indefinite "huge distance".

The journey from the Land of Lotus Eaters to the Land of Cyclops has no temporal or spatial indication; the flotilla, sailing in the darkness and into the mist, suddenly ran ashore. However, a very important nautical information is included in the following lines (*Od.* 9.147–148): οὔτ' οὖν κύματα μακρὰ κυλιν-

23 According to Strabo (10.4.5), the voyage from cape Samonium (Crete) to Egypt "takes four days and four nights, though some say three". It is not clear if by Egypt Strabo meant the port of Alexandria (approx. 320 miles from cape Samonium) or the port of Marsa Matruh (approx. 220 miles).

24 Early Aegean – Egypt contacts remain unconfirmed, but, despite little textual information, archaeological data confirm Egyptian imports to Crete since Old Kingdom period; in any case, by the 20th century BCE, a direct connection between Crete and Egypt had already been established (Mumford 2001, 359).

25 Hes. *Theog.* 622–625.

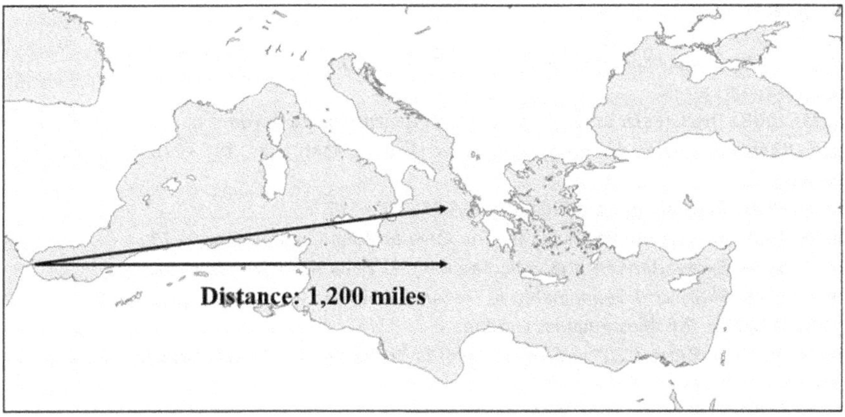

Map 9: From Gibraltar to the frontiers of the eastern Mediterranean basin: a distance of approx. 1,200 miles in straight line

δόμενα προτὶ χέρσον εἰσίδομεν, πρὶν νῆας ... ἐπικέλσαι ("nor did we see long waves rolling on the beach, until we ran our ships ... ashore"). Odysseus wondered why he could not see the foam of the waves – an evidence that a ship approaches a land mass or a shallow bottom; even in a pitch-black night, the effect of bioluminescence (emission of light by plankton) turns visible the foam of the wave and this effect is more intense in the western than in the eastern basin of the Mediterranean.[26] The absence of this well-known sign cancelled all established knowledge and experience of the Greek mariners: they were sailing in a nightmare where nothing was given or predictable.

Odysseus, after Calypso's permission to abandon her and leave Ogygia, built hastily a sea craft, a σχεδίη, to return home (*Od.* 5.238 – 261). Heading east, with a steady favourable wind, the σχεδίη sailed for seventeenth days, before Poseidon's wrath sank it (*Od.* 5.270 – 281). With an average speed of 3 knots, Odysseus' boat covered a distance of about 70 miles per day or a total distance of approximately twelve hundred miles. Twelve hundred miles is the distance from Gibraltar to the frontiers of the eastern Mediterranean basin, the Sea of Cythera or the eastern Ionian Sea. Of course, Homeric universe was not conceivable by numeric data, but an 8th century Ionic audience, hearing about a 17-day sailing journey towards east and before the wind, could understand that Odysseus' craft departed from a place near Oceanus and sank a few miles off the frontiers of the καθ' ἡμᾶς world.

---

26 Haddock, Moline and Case 2010.

Bibliography

Antypas, C. 2014. *Ὑγρὰ Κέλευθα: Πλοία και Ρότες στην Ομηρική Εποχή* (Ph.D. Diss.). University of Patras.

Austin, M. 2008. The Greeks in Libya. In *Greek Colonization. An Account of Greek Colonies and Other Settlements Overseas*, 2 vols., ed. G.R. Tsetskhladze, 187–218. Leiden and London.

Benton, S. 1928. Antiquities from Thiaki. *ABSA* 29: 113–116.

Benton, S. 1935. Excavations in Ithaca III, The Cave at Polis I. *ABSA* 35: 45–73.

Benton, S. 1939. Excavations in Ithaca III, The Cave at Polis II. *ABSA* 39: 1–51.

Casson, L. 1995. *Ships and Seamanship in the Ancient World* (revised edition). London.

Castleden, R. 2005. *The Mycenaeans*. London.

Cattling, H.W. 1956. Bronze Cut-and-Thrust Swords in the Eastern Mediterranean. *PPS* 22: 102–125.

Cazzella, A. and G. Recchia 2010. The 'Mycenaeans' in the Central Mediterranean: a Comparison between the Adriatic and the Tyrrhenian Seaways. *Pasiphae* 3: 27–40.

Chamoux, F. 1953. *Cyrène sous la monarchie des Battiades*. Paris.

Christopoulos, M. 1997. *Θαλασσινά Επεισόδια της Οδύσσειας*. Athens.

Cline, E.H. 2014. *1177 B.C. The Year Civilization Collapsed*. Princeton, NJ.

Coates, J.F., S. McGrail, D. Brown, E. Gifford, G. Grainge, B. Greenhill, P. Marsden, N.B. Rankov, C. Tipping and E. Wright. 1995. Experimental Boat and Ship Archaeology: Principles and Methods. *IJNA* 24: 293–301.

Coldstream, N.J. 1968. *Greek Geometric Pottery: A Survey of Ten Local Styles and their Chronology*. London.

De Fidio, P. 1982. Fiscalità, redistribuzione, equivalenze: per una discussione sull' economia micenea. *SMEA* 23: 83–136.

Deger-Jalkotzy, S. 2008. Decline, Destruction, Aftermath. In *The Cambridge Companion to the Aegean Bronze Age*, ed. C. Shelmerdine, 387–415. Cambridge.

Demand, N.H. 2011. *The Mediterranean Context of Early Greek History*. Chichester.

Dickinson, O. 2006. *The Aegean from Bronze Age to Iron Age: Continuity and Change between the Twelfth and the Eight Centuries BC*. London and New York.

Drews, R.R. 1993. *The End of Bronze Age: Changes in Warfare and the Catastrophe ca.1200 BC*. Princeton, NJ.

Frame, D. 2009. *Hippota Nestor*. Cambridge, MA and London.

Gehrke, H.J. and E. Wirbelauer. 2004. Akarnania and Adjacent Area. In *An Inventory of Archaic and Classical Poleis*, eds. M.H. Hansen and T.H. Nielsen, 351–378. Oxford.

Golénischeff, M.W. 1912. *Le Conte du Naufragé*. Cairo.

Haddock, S.H.B., M.A. Moline and J.F. Case 2010. Bioluminescence in the Sea. *Annual Review of Marine Science* 2: 443–493.

Hall, J.M. 2014². *A History of the Archaic Greek World, ca.1200–479 BCE*. New York.

Iakovides, S.E. 2001. *Gla and the Kopaïs in the 13th Century B.C.* Athens.

Kalligas, P.G. 1984–1985. Ανασκαφές στο Λευκαντί Ευβοίας. *Αρχείον Ευβοϊκών Μελετών* 26: 253–269.

Katzev, M. 1987. An Analysis of Experimental Voyage of Kyrenia II. In *Tropis II. Proceedings of the 2nd International Symposium of Ship Construction in Antiquity, Delphi, 27, 28, 29 August 1987*, ed. H.P. Tzalas, 245–256. Athens.

Kelder, J.M. 2010. *The Kingdom of Mycenae*. Bethesda, MD.

Lang, F. 1996. *Archaische Siedlungen in Griechenland: Struktur und Entwicklung*. Berlin.
Laronde, A. 1990. Greeks and Libyans in Cyrenaica. In *Greek Colonists and Native Populations. Proceedings of the First Australian Congress of Classical Archaeology, Sydney 9–14 July 1985*, ed. J.-P. Descoeudres, 169–180. Oxford.
Lichtheim, M. 1975. *Ancient Egyptian Literature – Volume I: The Old and Middle Kingdoms.* Berkley, Los Angeles and London.
Matthews, R. 2002. *The Liberation of Imagination.* London.
Morrison, J.S., J.F. Coates and N.B. Rankov 2000[2]. *The Athenian Trireme.* Cambridge.
Mumford, G. 2001. Mediterranean Area. In *The Oxford Encyclopedia of Ancient Egypt*, vol. II, ed. D.B. Redford, 358–367. New York.
Mylonas, E.M. 1966. *Mycenae and the Mycenaean World.* Princeton, NJ.
Nagy, G. 1980. An Evolutionary Model for the Text Fixation of Homeric Epos. In *Oral Traditional Literature: A Festschrift for Albert Bates Lord*, ed. J.M. Foley, 390–393. Colombus.
Nagy, G. 1990. *Pindar's Homer.* Baltimore.
Nagy, G. 1996. *Homeric Questions.* Austin.
Nagy, G. 1999. *The Best of the Achaeans. Concepts of Hero in the Ancient Greek Poetry.* Baltimore. (2nd revised and digitized edition, http://www.press.jhu.edu)
Parkinson, R.B. 1997. *The Tale of Sinuhe and Other Ancient Egyptian Poems 1940–1640 BC.* Oxford.
Powell, B. 1991. *Homer and the Origin of the Greek Alphabet.* New York.
Scodel, R. 2000. *Listening to Homer: Tradition, Narrative, and Audience.* Ann Arbor.
Simpson, W.K. (ed.) 2003[3]. *The Literature of Ancient Egypt: An Anthology of Stories, Instructions, and Poetry.* New Haven and London.
Souyoutzoglou-Haywood, C. 1999. *The Ionian Islands in the Bronze Age and Early Iron Age 3000–800 B.C.* Liverpool.
Stucchi, S. 1989. Problems Concerning the Coming of Greeks to Cyrenaica and the Relations with their Neighbors. *MedArch* 2: 73–84.

Jonathan S. Burgess

Land and Sea in the *Odyssey* and the *Telegony*[1]

This paper explores the epics *Odyssey* and *Telegony* from the perspective of spatial theory.[2] The polarity of land and sea is of great significance in Odyssean narratives. The *Odyssey* features its hero's account of his wanderings at sea, striving to return to Ithaca, an island in the sea. The Homeric poem indicates that Ithaca has various politico-economic connections to the mainland, and some of Odysseus' lying tales feature Thesprotia on the mainland. Tiresias imposes a geographically obscure "inland" journey on Odysseus, and his prediction of death "from the sea" for Odysseus centers on the spatial binary of land / sea. In the *Telegony* the hero journeys to both Elis and Thesprotia, before returning to Ithaca, where death arrives "from the sea". Afterwards his corpse is removed to Circe's island, which in the *Odyssey* is far out in the sea, though Aeaea may in the *Telegony* already be localized at the cape Monte Circeo in Italy.

Island in the Sea

The home to which the hero returns is spatially ambiguous.[3] By this I do not refer to modern controversy over the "true" location of Ithaca, since there is no doubt that modern Thiaki was considered to be Homeric Ithaca in antiquity.[4] But we can begin with the passage that has inspired this controversy, Odysseus' description of his homeland to the Phaeacians (*Od.* 9.21–28):

1 A draft of this paper was presented at a conference in Patras in July of 2015; I appreciate the hospitality of the hosts and the interest of the audience. For helpful comments on the death of Odysseus, I thank Jenny Strauss Clay, Robert Bostock and Gregory Nagy. I am grateful to the Social Science and Humanities Research Council for funding, and to the University of Toronto for an Affiliated Fellowship at the American Academy in Rome in May 2016.

2 For an introduction to spatial theory, see Cresswell 2004. For recent spatial approaches to Homer, see Edwards 1993; Purves 2010; Clay 2011; Tsagalis 2012; Burgess 2015a, 115–117; Higgins 2015. For Homer and beyond, see Skempis and Ziogas 2014.

3 See Burgess 2014a; Burgess 2015b.

4 Bittlestone 2005 is in a long line of implausible relocalizations of Ithaca over the past century. See the discussion with bibliography in a section of my website "In the Wake of Odysseus": http://homes.chass.utoronto.ca/~jburgess/rop/pages/ithaca.html·

DOI 10.1515/9783110535150-003

ναιετάω δ' Ἰθάκην εὐδείελον· ἐν δ' ὄρος αὐτῇ
Νήριτον εἰνοσίφυλλον, ἀριπρεπές· ἀμφὶ δὲ νῆσοι
πολλαὶ ναιετάουσι μάλα σχεδὸν ἀλλήλῃσι,
Δουλίχιόν τε Σάμη τε καὶ ὑλήεσσα Ζάκυνθος.
αὐτὴ δὲ χθαμαλὴ πανυπερτάτη εἰν ἁλὶ κεῖται
πρὸς ζόφον, αἱ δέ τ' ἄνευθε πρὸς ἠῶ τ' ἠέλιόν τε,
τρηχεῖ', ἀλλ' ἀγαθὴ κουροτρόφος· οὔ τι ἔγωγε
ἧς γαίης δύναμαι γλυκερώτερον ἄλλο ἰδέσθαι.

I dwell in far-seen Ithaca, and on it is Mt. Neriton of quivering leaves, very conspicuous. Many islands are situated about, very near to one another, Doulichion and Same and wooded Zakynthos. It lies rooted in the sea very far out towards darkness, with others apart towards the sun and dawn, rugged, but a good nourisher of youth. I am not able to imagine anything sweeter than one's land.

One sees where confusion might arise. Though my translation strives to avoid incoherence, one might otherwise suppose that after a conspicuous (ἀριπρεπές) mountain is specified, the island is then described as both "low" and "high" (χθαμαλὴ πανυπερτάτη). The passage also might be thought to imply that other islands encircle it (ἀμφί), though it is westward in contradistinction to other, eastern islands (πρὸς ζόφον, αἱ δέ τ' ἄνευθε πρὸς ἠῶ τ' ἠέλιόν τε), and perhaps "furthest" (πανυπερτάτη) west.

It is well-known that Ithaca in fact is not the furthest west, according to the compass. But as my translation indicates, a close reading with lexical analysis can yield plausible topographical and geographical sense; there is no need for far-fetched speculation about the origins of the island, or the text.[5] That does not mean that the passage provides geographical exactitude in a positivistic sense. The hero's words are more easily understood as signaling the spatial ambiguity of Ithaca. Odysseus' description would seem to play with the shifting qualities of his homeland, with much depending on perspective. Ithaca has a conspicuous mountain, yet is it "rooted" in the sea. The island's superlative location "toward the gloom" marks it as peripheral, yet it is also amidst a network of Ionian Islands.[6] There is also defensive tone about the "rugged" island's nur-

5 See Merry and Riddell 1886, appendix 3; Stubbings 1963. After visiting Kephallenia and Ithaca, I agree with the view that Ithaca can be conceived hodologically as "towards darkness". Luce 1998 provides an autoptic and optimistic exploration of Homeric locations on Ithaca. The argument of Bittlestone 2005 amounts to believing Homer and his audience were content with a vestigial description of a prehistoric part of Kephallenia, hypothesized as an island by geological special pleading.

6 For the purposes of my argument, we can leave aside identification of Same (cf. Samos in the quotation below) and Doulichion; Zakynthos is certainly modern Zante or Zakynthus.

turing abilities; the craggy barrenness of Ithaca is nevertheless fertile enough to nurture those who are raised there.

In the Catalogue of Ships the description of forces from the region of the Ionian Islands (*Il.* 2.625–637) is surprising:

οἳ δ’ ἐκ Δουλιχίοιο Ἐχινάων θ’ ἱεράων
νήσων, αἳ ναίουσι πέρην ἁλὸς Ἤλιδος ἄντα,
τῶν αὖθ’ ἡγεμόνευε Μέγης ἀτάλαντος Ἄρηϊ
Φυλεΐδης, ὃν τίκτε Διῒ φίλος ἱππότα Φυλεύς,
ὅς ποτε Δουλίχιόνδ’ ἀπενάσσατο πατρὶ χολωθείς
τῷ δ’ ἅμα τεσσαράκοντα μέλαιναι νῆες ἕποντο.
αὐτὰρ Ὀδυσσεὺς ἦγε Κεφαλλῆνας μεγαθύμους,
οἵ ῥ’ Ἰθάκην εἶχον καὶ Νήριτον εἰνοσίφυλλον
καὶ Κροκύλει’ ἐνέμοντο καὶ Αἰγίλιπα τρηχεῖαν,
οἵ τε Ζάκυνθον ἔχον ἠδ’ οἳ Σάμον ἀμφενέμοντο,
οἵ τ’ ἤπειρον ἔχον ἠδ’ ἀντιπέραι’ ἐνέμοντο·
τῶν μὲν Ὀδυσσεὺς ἦρχε Διῒ μῆτιν ἀτάλαντος
τῷ δ’ ἅμα νῆες ἕποντο δυώδεκα μιλτοπάρῃοι.

And those from Doulichion and the holy Echinai islands, which are situated across the sea facing Elis, these in turn led Meges alike to Ares, son of Phyleus, whom the horseman Phyleus dear to Zeus bore, who once angered at his father moved to Doulichion. Forty black ships followed him. But Odysseus led the great-hearted Kephallenians, who held Ithaca and quavering leafed Neriton and dwelt in Krokyleia and rugged Aigilips, and who held Zakynthos and dwelt in Samos, and who held the mainland and dwelt in the places opposite. These Odysseus led, equal to Zeus in wile; twelve red-cheeked ships followed him.

Odysseus is a leader, but only of twelve ships. The minor figure of Meges, also from the region, commands forty ships. His troops are from Doulichion and the Echinades, while those led by Odysseus are from Ithaca, Zakynthos, Samos, and an area on the mainland. Without worrying about which of the ancient names is which of the modern islands, I take Meges' area as relatively closer to the mainland, and Odysseus' group as relatively westward and southward.[7] Those led by Odysseus are called Kephallenians, which in Homer seems to be a tribal, not geographic, designation.[8] Odysseus does lead a seemingly broad geographical complex of islands and coastal land, but this area musters relatively fewer forces than the area controlled by Meges, and Ithaca is only a small part of our hero's command.

Neither this Iliadic passage nor the *Odyssey* suggests that Ithaca holds political or military sway over the area, though Ithacan alliances or skirmishes with

7 See the discussion at Petrakis 2006.

8 Kephallenia first designates an island at Herodotus 9.28, Thucydides 2.30.

the mainland are occasionally mentioned, or thought possible, in the *Odyssey.*[9] Our hero's command may result more from his mythological reputation than the importance of Ithaca, historically or mythologically. If Ithacan political or economic hegemony ever existed, it dissolves during the absence of Odysseus when the island falls prey to unwanted suitors from the region. Of the one hundred and eight suitors only twelve are from Ithaca; half are from Doulichion.[10] Do these suitors desire some sort of political power over the area that Odysseus once possessed? This is not unambiguously clear.[11] When a suitor openly wishes that Telemachus not become a ruler on Ithaca (1.386–404), Telemachus concedes that there are other "kings" inhabiting the island, though he affirms his desire to keep at least his father's home and slaves.[12] Various scenarios for the remarriage of Penelope are suggested in Books 1–2; late in the poem Penelope announces she will join a new husband in his home.[13] Odysseus is the most notable mythological figure in the region, but neither the *Iliad* nor the *Odyssey* portrays his island as a political power, since this is implausible.

Hints at economic and political interconnectivity between Ithaca, other islands, and the mainland occasionally arise. The swineherd Eumaeus boasts that Odysseus possessed dozens of herds on the mainland, tended by foreigners as well as his own slaves (14.96–108). The Ithacan Noemon keeps horses at Elis on the mainland (4.634–637); the cowherd Philoetius brings livestock to Ithaca by boat and recalls that Odysseus had placed him in charge of cattle among "Kephallenians" (20.185–225). These passages, rather than suggesting Ithacan control over resources outside the island, seem to be *ad hoc* indications of regional interconnectivity. And the recurring theme is the apparent limitations of the "rugged" island's nurturing abilities, not the power of Ithaca. Revealing is Itha-

9 See 1.174–212, 1.257–264, 10.38–39 (with Danek 1998, 195), 14.314–344, 16.424–430, 19.285–302, 24.353–378, 24.430–431.

10 Cf. 1.245–248 = 16.122–125 = 19.130–133, 16.247–251, 21.346–347.

11 Mülder 1931 argued that "Ithaca" refers to the Ionian islands as a political unit. This must be wrong, but points to the imprecise ambiguity about the issue in Homer.

12 An anonymous suitor supposes (2.232–235) that should Telemachus die the suitors would divide the possessions and the house would go to Penelope or her new husband.

13 19.577–581 = 21.75–79. On the complexity of Penelope's intentions, see Mueller 2007. The ambiguity over arrangement and consequence of a new marriage for Penelope perhaps reflects variant temporal or cultural practices (see Finley 1954; Snodgrass 1974; Westbrook 2005). But the "problems" can also be attributed to the *Odyssey*'s self-conscious unwillingness to be clear about the nature of Ithaca.

ca's lack of space for horses and larger herds.[14] As a small island at sea, it has a degree of resources, but is largely dependent on the larger economy of the other islands and the mainland. We lack real-world evidence about the region for either prehistory (analogous to the heroic age) or for the time of the composition of the *Odyssey* in the early Archaic Age, but historically Ithaca has never been a very important part of economic or political systems in northwest Greece.[15]

Inland / Mainland Journeys

Odysseus is associated with the mainland as well as Ithaca in legend and myth. Readers of the *Odyssey* find this most prominently with the inland journey mandated by the seer Tiresias (*Od.* 11.119 – 137). The hero is to travel inland with an oar on his shoulder until he meets someone who does not know what an oar is:

> αὐτὰρ ἐπὴν μνηστῆρας ἐνὶ μεγάροισι τεοῖσι
> κτείνῃς ἠὲ δόλῳ ἢ ἀμφαδὸν ὀξέϊ χαλκῷ,
> ἔρχεσθαι δὴ ἔπειτα λαβὼν εὐῆρες ἐρετμόν,
> εἰς ὅ κε τοὺς ἀφίκηαι οἳ οὐκ ἴσασι θάλασσαν
> ἀνέρες, οὐδέ θ' ἅλεσσι μεμιγμένον εἶδαρ ἔδουσιν
> οὐδ' ἄρα τοί γ' ἴσασι νέας φοινικοπαρήους
> οὐδ' εὐήρε' ἐρετμά, τά τε πτερὰ νηυσὶ πέλονται.
> σῆμα δέ τοι ἐρέω μάλ' ἀριφραδές, οὐδέ σε λήσει
> ὁππότε κεν δή τοι ξυμβλήμενος ἄλλος ὁδίτης
> φήῃ ἀθηρηλοιγὸν ἔχειν ἀνὰ φαιδίμῳ ὤμῳ,
> καὶ τότε δὴ γαίῃ πήξας εὐῆρες ἐρετμόν,
> ῥέξας ἱερὰ καλὰ Ποσειδάωνι ἄνακτι,
> ἀρνειὸν ταῦρόν τε συῶν τ' ἐπιβήτορα κάπρον,
> οἴκαδ' ἀποστείχειν ἔρδειν θ' ἱερᾶς ἑκατόμβας
> ἀθανάτοισι θεοῖσι, τοὶ οὐρανὸν εὐρὺν ἔχουσι,
> πᾶσι μάλ' ἑξείης.

> But when you kill the suitors in your palace either by deceit or openly with sharp bronze, then go, taking a well-poised oar, until you reach those men who do not know the sea, or eat food flavored with salt, or know even of red cheeked ships or well-poised oars, which are wings for ships. I will tell you of a sign, very noticeable, and it will not escape you: when some wayfarer happening upon you says that you hold a chaff-destroyer on your

14 At 13.242 – 247 the disguised Athena admits Ithaca is not suited for horses, despite other attractive qualities; elsewhere (4.605 – 608) Telemachus states that Ithaca is too rough for horses, but good for goats.

15 For early archaeological evidence, see Souyoudzoglou-Haywood 1999; Waterhouse 1996. Malkin (1998, 120 – 155) would portray Ithaca as an important port and cult site in the early first millennium. For the political powerlessness of Ithaca in post-antiquity, see Burgess 2015b.

> bright shoulder, indeed then fixing the well-poised oar in the ground, and sacrificing fine animals to lord Poseidon, a sheep, a bull, and a boar, mounter of swine, proceed home and accomplish holy hecatombs to the immortal gods who occupy wide heaven, all of them in order.

Notable for our purposes is the dichotomy of land and sea. The journey's motivation is apparently appeasement of Poseidon.[16] Paradoxically, Odysseus will sacrifice to the sea god where the sea is unknown. With the notional transformation of an oar into a winnowing shovel, a cultural and spatial boundary is crossed. In some ways the centrifugal inland to which Odysseus travels post-*nostos* is a doublet of centripetal Ithaca to which Odysseus travels during his *nostos*.[17] Both, to varying degrees, serve as an antithesis to the sea, and both offer an agrarian lifestyle not found out in the deep sea upon which Odysseus wandered.[18] And the traveller in the folktale motif upon which the Homeric "inland journey" is based typically settles in a new climate and culture.[19] Odysseus therefore seemingly travels to an alternative homeland on this inland journey, though Tiresias transforms the tale type by indicating that Odysseus will return again to Ithaca.

After his first return to Ithaca, the disguised hero tells lying tales about "Odysseus" travelling on the mainland. According to the beggar, "Odysseus" is hosted by the Thesprotian king and visits to the oracle at Dodona.[20] Some have thought that this reflects, or rejects, an alternative tradition of Odysseus' return, whereby the hero seeks advice at Thesprotian Dodona concerning the best strategy of his homecoming.[21] Besides Dodona, another real world Thesprotian place may lie in the background of the Homeric epic: the *nekuomanteion* by the Thesprotian river of Acheron, which arguably is the inspiration for the Homeric *nekyia*, as Pausanias supposed.[22]

Travel by Odysseus on the mainland is also found in the *Telegony* of the Epic Cycle.[23] The summary by Proclus reports two journeys on the mainland after the

---

16 Hartmann 1917, 216–217.

17 Purves 2006, 2010, 65–96 is especially intriguing on the spatial aspects of the inland journey.

18 See Vidal-Naquet 1996.

19 Hansen 1977; Hansen 2002, 371–378; Hansen 2014.

20 14.314–359, 19.262–307.

21 See Danek 1998, 215–216; Malkin 1998, 129–130; Marks 2008, 89–92.

22 See Huxley 1958; Ogden 2001, xxiv-xxv, 43-64; and Paus. 1.17.4. See also the section on this issue on my website "In the Wake of Odysseus": http://homes.chass.utoronto.ca/~jburgess/rop/od.voyage.html

23 For recent discussions of the *Telegony*, see Martin 2011; West 2013, 295–306; Tsagalis 2015.

slaughter of the suitors. First Odysseus visits Elis to see after livestock.[24] After the Elis sojourn, Odysseus returns to Ithaca – note the back-and-forth sea and land interconnectivity at play here – before returning to the mainland for another journey. Odysseus now travels to Thesprotia, which becomes an alternative homeland when he stays with the queen and produces a child. In some respects Odysseus' mainland journey seems like a real-world version of the inland journey. Correspondence between the "inland journey" and the Thesprotian journey is implicitly implied in Apollodorus (*Epitome* 7.34), where it is specified that Odysseus in Thesprotia performs sacrifices enjoined by Tiresias in order to propitiate Poseidon. In the prophecy of *Odyssey* 11 these sacrifices are to happen when the hero reaches the inland place. Differently from the inland journey and similarly to the tale type upon which it is based, he settles in the new land – at least for a while. The eventual return home correlates with the return home from the inland journey mandated by Tiresias.

There were other legends connecting Odysseus with the mainland. For example, Apollodorus reports a story (*Epitome* 7.40) in which Odysseus is forced into exile after the slaughter of the suitors (a consequence which may be pre-Homeric, since it is at least considered a possibility in the *Odyssey*).[25] The judicial decision is made by Neoptolemus, now king of northwest Greece, who thereby seeks wider control over the region. As in the *Telegony*, Odysseus travels to the mainland. He starts up a new life in Aetolia with a new wife and son, though he never returns home. The narrative apparently stems from Aristotle's *Constitution of the Ithacans*, which indicates that the historical inhabitants of Ithaca mapped contemporary Ithacan and regional issues onto the mythological past. Plutarch provides further details of the Aristotelian account, with the variation that Odysseus goes into exile to Italy.[26] This is routinely modified so as to agree with Apollodorus, but the hero was often associated with the Italian peninsula, as was his son by Circe, Telegonus.[27]

---

24 For interpretation see Severyns 1962; Tsagalis 2008, 80–82; Tsagalis 2015, 380, 386–388; Steinrück 2008, 135–136; Marks 2010. For Peloponnesian traditions about Odysseus, see Nobili 2009.

25 See Burgess 2014b.

26 See Plut. *Quaest. Graec.* 14 (*Moralia* 294c-d); Arist. fr. 507 Rose. For other post-return adventures of Odysseus, cf. Apollod. *Epit.* 7.38–40; Parth. *Amat. Narr.* 3.1; *scholia* Lycoph. *Alex. ad* 806 = *FGrH* 115 F 354 (Theopomp.).

27 Phillips 1953 is the seminal study of this issue; see also Malkin 1998, 178–209; Debiasi 2004, 265–267, 270–271; West 2013, 302–303. The issue arises again below.

Death (apart) from the Sea

To interpret the *Telegony*'s account of Odysseus' death on Ithaca, we will want first to consider Tiresias's prophecy of the hero's death in *Odyssey* Book 11 (134 – 137):

> θάνατος δέ τοι ἐξ ἁλὸς αὐτῷ
> ἀβληχρὸς μάλα τοῖος ἐλεύσεται, ὅς κέ σε πέφνῃ
> γήρᾳ ὕπο λιπαρῷ ἀρημένον· ἀμφὶ δὲ λαοὶ
> ὄλβιοι ἔσσονται. τὰ δέ τοι νημερτέα εἴρω.
>
> And death to you indeed from the sea, gentle-like, will come to you, which will slay you weakened in your sleek old age. And the people about will be prosperous. I speak these things to you as true.

Tiresias predicts Odysseus' return to Ithaca after the inland journey and seems to foretell his eventual death in old age apart from the sea. But prophecies are famously misleading. The preposition ἐκ with a verb of movement, by comparative Homeric usage and by standard Greek usage, normally means "from", as in "originating from". So the words of Tiresias should mean "death will come to you out of the sea". However, many suppose that they mean "death will occur when you are away from the sea".[28] The misunderstanding is natural. Given the general circumstances of the prophecy (a return home, old age, the people of Ithaca prosperous), as well as the word order, Odysseus can be forgiven for thinking that he will be apart from the sea when he dies. That does not mean that we should. It is probable that the words of Tiresias misleadingly suggest a death removed from sea, but actually refer to death arriving out of the sea.[29]

---

28 A few Homeric passages (Od. 15.272, 19.7, Il. 14.129 – 130) seem to suggest an alternative meaning of the preposition as "apart from", with colloquial elision of implied motion. Interpretation of the passage was already debated in antiquity, with reference to Telegonus: see the scholia at West 2013, 301. For recent interpretations, see Nagler 1980; Ballabriga 1989, 294; Carrière 1992, 38-39; Danek 1998, 225 – 228; Cerri 2002, 155 – 156; Grossardt 2003, 214 – 215; Bostock 2007, 65 – 68; Gainsford 2012; West 2013, 307-308; Nagy 2013, 1.11, section 57; Burgess, 2014c. Bostock (2007, 65 – 67) thinks that the phrase syntactically expresses source of death but contextually references place of death, but with no intentional ambiguity.

29 Bostock (2007, 66) excellently delineates the correlations between Tiresias' prophecy and oracular responses, only to reject this approach because of "the tendency in oracular responses for the positive meaning to be open, the negative meaning cryptic, whereas in this case the opposite obtains". By this he means that a negative meaning ("from the sea") is syntactically primary, if unintended, with a positive meaning ("apart from the sea") secondary if contextually intended. My argument regards death arriving from the sea as negative and cryptic and death

An ancient audience aware of a traditional story of death arriving for Odysseus from the sea would recognize the real meaning, and also that Odysseus has been fooled by the ambiguity of the words.

Should we think there was a traditional story of Odysseus' death arriving from the sea? Death does come to Odysseus from the sea in the *Telegony.* Proclus reports that Telegonus unwittingly kills his father Odysseus. This occurs on Ithaca, but Telegonus has arrived from the sea. He also kills the hero with a spear tipped with the spine of a stingray, a sea creature.[30] The true meaning of the prophecy, "death will come to you from the sea", is thereby fulfilled. I am not implying that the *Telegony* predates the *Odyssey,* but rather that its story, or one very much like it, is pre-Homeric. West prefers to think that Tiresias alludes to the Aeschylean version of Odysseus' death, whereby a seabird drops barb-filled crap on the hero's head.[31] The rationalistic reasoning of this argument is unnecessary and inefficient, though it may be that there were multiform versions of death "from the sea". The hero's incorrect assumption that he is finally free from the dangers of the sea would always be the stinging point.

The sticking point for many is Tiresias' description of the death as "gentle" (ἀβληχρός) – how can this describe death by spear?[32] But since we know that ἐξ ἁλός can only be mistakenly interpreted as "away from the sea", we should be alert to the possibility that ἀβληχρός is misleading. In previous publications, I have suggested that the poisonous nature of the spine of the stingray, well-known in antiquity, might lead to a slow or "peaceful" death.[33] Alternative explanations exist. One is lexical. ἀβληχρός has a biform βληχρός; they both mean "weak, gentle", etc. Steve Reece persuasively explains the alpha-prefix of ἀβληχρός as resulting from oral/aural mis-division of words, with the attachment of a preceding word's final alpha to a preexisting βληχρός leading to the creation of the new form. Of interest to my argument is his admission that the prefix looks like an alpha-privative.[34] Thereby one can argue that the words of

---

apart from the sea as positive and open (as understood by Odysseus, at least); the passage is therefore consonant with oracular misdirection.

30 Apollod. *Epit.* 7.36. See Thompson 1947, 279–281 for the main ancient sources about the sting-ray and its role in the Telegonus story.

31 West 2013, 307–308.

32 Hartmann 1917, 74 n. 69, 221; a sentiment shared by many scholars.

33 Burgess 1995, 234 n. 70; Burgess 2001, 153–154; Burgess 2014a, 179; Burgess 2014b, 357 n. 8. Grossardt 2003, 212 notes the theme of poison in the Odyssey, which may contextualize my argument. However, Ael. *NA* 1.56, 2.36, 50 forcefully describes a wound by a sting-ray as immediately lethal; cf. Cicero's claim that Odysseus lamented much more from the pain of the wound in the Euryalus by Sophocles (fr. 461a Lloyd-Jones) than in the Niptra by Pacuvius.

34 Reece 2009, 122–132; see also Carrière 1992, 40–42; Cerri 2002, 156.

Tiresias seem to describe the death as gentle, but *really* mean that death will be "not-gentle". Presumably the true meaning is understood by Odysseus after the fact, as often with prophecies.

A third aspect of the prophecy, however, has generated further controversy. Tiresias adds that when Odysseus dies, "the people about will be fortunate" (ἀμφὶ δὲ λαοὶ ὄλβιοι ἔσσονται). An over-literal interpretation is that the hero's people are physically standing by the corpse; if they are "happy", the death must be gentle and natural.[35] The phrase should rather be understood to refer to the general state of the people ruled by Odysseus at the time of his death. The issue is important, given the just averted civil war at the end of the *Odyssey*, and the stories of exile for Odysseus as a consequence of his slaughter of the suitors. As elsewhere, the *Odyssey* here seeks to portray Odysseus as a good leader. A correlation between a ruler and his people's prosperity is a commonplace in early Greek epic,[36] and so Tiresias' remark about prosperous people would be welcome to the hero, and perhaps contribute to his readiness to misinterpret the prophecy as one of a peaceful death. By means of an elaborate and ingenious interpretation, Nagy sees in ὄλβιοι an allusion to a hero cult of Odysseus.[37] If so, Tiresias' remarks would also refer to the general state of the people, who are "blessed" because of a resulting cult status of Odysseus.

I am not necessarily convinced that ὄλβιοι here has cultic significance, though Homeric epic is aware of hero cult and eventually there was cult of Odysseus on Ithaca.[38] More importantly for my argument, there is no evidence that

---

35 Apparently, Tsagalis 2015, 393 n. 80 (discussing the interpretation by West): "How can his people surround him *in happiness* when Odysseus dies?" (by unnatural means). In the reconstruction by Tsagalis, an "older Thesprotian lay" told of Odysseus dying peacefully away from the sea at Thesprotia.

36 See *Od.* 19.108–114, Hes. *Op.* 225–247, with Haubold 2000.

37 Nagy 2013, 1.11, esp. section 40: "We need to keep in mind the non-local orientation of Homeric poetry as we consider the reference ... to people who are *olbioi*, "blessed", in the context of the death of Odysseus. Homeric poetry says only implicitly, not explicitly, that these people are made 'blessed' because they worship Odysseus as a cult hero ... This poetry refers only implicitly to existing practices of hero cult, without explicitly revealing the mysteries of the hero cult". See Bostock 2007, 65–66: "Nagler's argument (1980) that it [Tiresias' prophecy] refers to Odysseus' heroisation is too far-fetched to warrant discussion".

38 Hero cult as contemporaneous with Homeric epic: Burgess 2001, 167–169. The evidence of Ithacan cult worship of Odysseus, notably a votive inscription found at the Polis Bay cave, is Hellenistic. Some are optimistic, to various degrees, that the cult goes back to the time of the composition of the *Odyssey* (Malkin 1998, 100–110; Currie 2006, 52–53; Marks 2008, 97–100; Nagy 2013, 1.11, section 43; skeptical: Antonaccio 1995, 154; Boardman 2002, 68–71).

Odysseus was buried at Ithaca.[39] The ancients looked elsewhere for the location of the hero's grave; surviving testimony indicates burial on the Greek mainland, or possibly in Italy.[40] In the *Telegony*, the corpse of Odysseus is conveyed to Circe's island (Proclus), where apparently he is buried (so Hyginus *Fabulae* 127).[41] It may be that the Cyclic poem here refers to the Aeaea that was localized with Monte Circeo, where the Roman colony Circeii seems to have existed by the late sixth century (Polybius 3.22, Livy 1.56, 2.39). Already at *Theogony* 1011–1116 two of Circe's children, Agrios and Latinus, are vaguely connected with northwest Italy.[42]

A burial at Aeaea also occurs in Homer, but of Elpenor. In the underworld Odysseus is asked by the shade of the deceased oarsman to be buried with an oar to mark his tumulus, which Odysseus and his men later do (*Od.* 11.74–78, 12.9–15). As has often been noticed, Elpenor's reference to his oar occurs shortly before Tiresias speaks of Odysseus carrying an oar inland. Odysseus is directed by the seer to plant the oar in the ground, which corresponds to the planting of an oar on the mound of Elpenor. It almost seems as if Odysseus is to mark his own death on his inland journey. Nagy refers to the hero's planting of the oar as a "stylized image of his own tomb", and Purves notes the "structural similarity".[43] Arguably Odysseus' enactment of a quasi-tomb on the inland journey ref-

---

39 Nagy assumes that the burial is at Ithaca, as indicated by the second elision in the quotation in n. 37 above: "[a cult hero] whose corpse is buried in the earth that they cultivate, and that this 'blessing' is realized by way of physical contact with the earth containing the corpse of the hero … ".

40 Besides Apollod. *Epit.* 7.40, see Lycoph. *Alex.* 799–800, 805–806 with Schade 1999 and Hornblower 2015 ad loc.; see West 1984; West 2013, 295 n. 10.

41 Hartmann 1917, 53; Phillips 1953, 55; Wiseman 1995, 49; Braccesi 2010, 19; Debiasi 2004, 267; *contra:* West 2013, 305.

42 Jameson and Malkin 1998 argue convincingly that an eponymous Latinus is not anachronistic for the *Theogony.* Agrios ("Mr. Savage") reminds one of Homeric ethnography, for Odysseus encounters many *agrioi* in his travels (cf. Athena disguised as Mentes at *Od.* 1.197–199: ἀλλ' ἔτι που ζωὸς κατερύκεται εὐρέι πόντῳ / νήσῳ ἐν ἀμφιρύτῃ, χαλεποὶ δέ μιν ἄνδρες ἔχουσιν / ἄγριοι, οἵ που κεῖνον ἐρυκανόωσ' ἀέκοντα ["but yet I suppose alive he is detained on the wide sea, on a sea-girt island, and rough men have him, savages, who perhaps restrain that one against his will"]). Telegonus is mentioned as third child of Odysseus and Circe at *Theog.* 1014, often suspected as an interpolation (in an interpolation).

43 Nagy 2013, 1.11, section 57 ("There is no need to argue on this basis that the phrase *ex halos* somehow means 'away from the sea'. Rather, the double meaning of the *sema* or 'sign' for Odysseus … is formalized in the coincidence of opposites that shapes the whole myth: Odysseus finds the sign for his death from the sea precisely when he is farthest away from the sea"); Purves 2010, 81. See also Nagy 1990, 214–215; Nagy 2013, 1.11, section 47 (" … the ritual act of Odysseus when he sticks his own well-made oar into the ground … and sacrifices to Poseidon … points to

erences non-Homeric tales of the hero dying on the mainland.[44] But as tempting as this correlation is, Elpenor's request, occurring as it does shortly before Tiresias' prophecy, more likely foreshadows *Odysseus*' burial at Aeaea. In both the prophecy of Tiresias and the *Telegony* Odysseus returns to Ithaca and dies there. The *Odyssey* probably assumes that hero was not buried on Ithaca, since this was the prevalent view. The *Telegony* employs but need not have invented one version of this concept, burial of Odysseus at Aeaea. Perhaps Elpenor's request to be buried at Aeaea reflects an existing landmark linked to him,[45] but it may instead, or additionally, mirror a traditional belief that Odysseus was buried at Aeaea.

Conclusion

Since there were various narratives about the post-*nostos* adventures and death of Odysseus, multi-forms that were often incompatible with one other, it would be misguided to impose unity upon them. And given the lack of evidence, we can hardly feel confident when reconstructing potential Homeric allusions to non-Homeric material. For the purposes of this paper, it is most important to recognize that many stories of Odysseus feature a dichotomy of land and sea, and often a breakdown of the polarity. According to my analysis, the mandated inland journey is motivated by a need to appease the god of the sea, and Tiresias' apparent prediction of death for Odysseus apart from the sea actually references a death arriving from the sea. Elpenor is to be buried on land, but by the sea, on a headland on an island far away in the sea. This request may reflect the same type of land / sea burial for Odysseus, one manifestation of which occurred in the *Telegony*. However hypothetical my correlation of material found in the *Odyssey* and *Telegony* may be, a repeated binary of land and sea in these two epics is quite evident. And if we investigate into Odyssean myth and legend more widely, the same binary seems to obtain. For example, if one stacks paradigmatically the

---

the making of his own *sema* or 'tomb', corresponding to the *sema* or 'sign' given to him by Tiresias").

44 So Nagy 2013, 1.11, sections 45–50, 57–64, in an argument that links Tiresias' words with a burial of Odysseus in Arcadia.

45 Elpenor's tumulus at the cosmographical Aeaea in the *Odyssey* could hardly serve to publicize his fame among mortals (see Purves 2010, 84; Burgess 2014, 112). It has therefore sometimes been supposed that the Homeric passage aetiologically reflects a cape in the Black Sea with a tumulus (see West 2011, 296 with n. 50; West 2014, 215 with n. 115). A tumulus of Elpenor was eventually pointed out at Monte Circeo (notably, pseudo-Scylax 8; Theophr. *Hist. Pl.* 5.8.5).

inland journey, the *Telegony*'s Thesprotian episode, and testimony of the exile and/or death of Odysseus on the mainland together, these episodes naturally serve as multiforms filling one slot in the duality of land and sea.

As land that is amidst the sea, Ithaca has an ambiguous and often paradoxical status somewhere on the spectrum of land and sea. It has enough land to support an agrarian and pastoral economy, but it cannot support these activities on the scale of the mainland, with which it has economic and political connections. In myth Ithaca's importance is less than its ruler's *kleos* might suggest, and throughout history the island would remain on the periphery of more important power centers, east and west. For Odysseus it is the desired haven from the sea during his wandering at sea, yet it is vulnerable to incursion by the suitors, most of them from elsewhere. The *Telegony* plays with the ambiguity of Ithaca's status by revealing that the hero's death apart from the sea is actually a death caused by another incursion arriving by sea, this time by Telegonus. And the Cyclic poem then removes the hero's corpse from Ithaca, back to the uncharted marine world of the wanderings – or perhaps to the Italian world west across the sea, where Aeaea was localized. In ancient stories of Odysseus, the polarity of land and sea is prominent, and the spatial status of Ithaca as both land and sea, with all the ambiguity and paradox that this involves, is essential to understanding myth and literature about Odysseus.

Bibliography

Antonaccio, C. 1995. *An Archaeology of Ancestors.* Lanham, MD.

Ballabriga, A. 1989. La prophétie de Tirésias. *Mètis* 4: 291–304.

Bittlestone, R. 2005. *Odysseus Unbound. The Search for Homer's Ithaca.* Cambridge.

Boardman, J. 2002. *The Archaeology of Nostalgia.* London.

Bostock, R. 2007. *A Commentary on Homer: Odyssey 11* (unpublished Ph.D. Diss.). University of Exeter.

Braccesi, L. 2010. *Sulle rotte di Ulisse: l'invenzione della geografia omerica.* Rome.

Burgess, J.S. 1995. Achilles' Heel: The Death of Achilles in Ancient Myth. *ClAnt* 14: 217–243.

Burgess, J.S. 2001. *The Tradition of the Trojan War in Homer and the Epic Cycle.* Baltimore.

Burgess, J.S. 2014a. Ambiguous Homecomings: *Cold Mountain* and the *Odyssey.* In *Nostos: Odyssean Identities in Modern Cultures*, eds. H. Gardner and S. Murnaghan, 173–191. Columbus, OH.

Burgess, J.S. 2014b. Framing Odysseus. The Death of the Suitors. In *Crime and Punishment in Homeric and Archaic Epic*, eds. M. Christopoulos and M. Païzi-Apostolopoulou, 355–372. Ithaca.

Burgess, J.S. 2014c. The Death of Odysseus in the *Odyssey* and *Telegony.* In *Studies on the Greek Epic Cycle*, eds. G. Scafoglio and E. Lelle, 113–124. Rome and Pisa.

Burgess, J.S. 2015a. *Homer.* London.

Burgess, J.S. 2015b. Travelling to Ithaca. In *Politics, Mobility, and Identity in Travel Writing*, eds. G. Totten, J. Dubino, M. Cabañas and V. Salles-Reese, 143–154. New York.
Carrière, J.-C. 1992. La réponse de Tirésias: le denier voyage et la mort d'Ulysse selon l'Odysseé. In *Mélanges Pierre Lévêque*, eds. M.-M. Mactoux and E. Geny, 17–44. Paris.
Cerri, G. 2002. L'Odissea epicorica di Itaca. *MediterrAnt* 5: 149–184.
Clay, J.S. 2011. *Homer's Trojan Theater: Space, Vision, and Memory in the Iliad.* Cambridge.
Cresswell, T. 2004. *Place: A Short Introduction.* Malden, MA.
Currie, B. 2005. *Pindar and the Cult of Heroes.* Oxford.
Danek, G. 1998. *Epos und Zitat: Studien zu den Quellen der Odyssee.* Vienna.
Debiasi, A. 2004. *L'epic perduta. Eumelo, il Ciclo, l'occidente.* Rome.
Edwards, A.T. 1993. Homer's Ethical Geography: Country and City in the *Odyssey. TAPA* 123: 27–78.
Finley, M.I. 1954. *The World of Odysseus.* New York.
Gainsford, P. 2012. The Deaths of Beowulf and Odysseus: Narrative Time and Mythological Story-types. *C&M* 63: 247–278.
Grossardt, P. 2003. Zweite Reise und Tod des Odysseus. Mündliche Traditionen und literarische Gestaltungen. In *Ulisse nel tempo. La metafora infinita*, ed. S. Nicosia, 211–255. Venice.
Hansen, W.F. 1977. Odysseus' Last Journey. *QUCC* 24: 27–48.
Hansen, W.F. 2002. *Ariadne's Thread. A Guide to International Tales Found in Classical Literature.* Ithaca, NY.
Hansen, W.F. 2014. Odysseus and the Oar: A Comparative Approach to a Greek Legend. In *Approaches to Greek Myth* (2nd ed.), ed. L. Edmunds, 247–279. Baltimore.
Hartmann, A. 1917. *Untersuchungen über die Sagen vom Tod des Odysseus.* Munich.
Haubold, J. 2000. *Homer's People.* Cambridge.
Higgins, J. 2015. *From Fleet to Foot: An Analysis of Odysseus's Pedestrian Journeys in the Odyssey* (unpublished Ph.D. Diss.). University of Toronto.
Hornblower, S. 2015. *Lykophron. Alexandra.* Oxford.
Huxley, G.L. 1958. Odysseus and the Thesprotian Oracle of the Dead. *La Parola del Passato* 13: 245–248.
Jameson, M.H., and I. Malkin. 1998. Latinos and the Greeks. *Athenaeum* 86: 477–485.
Luce, J.V. 1998. *Celebrating Homer's Landscapes: Troy and Ithaca Revisited.* New Haven.
Malkin, I. 1998. *The Returns of Odysseus.* Berkeley.
Marks, J. 2008. *Zeus in the Odyssey.* Washington, DC.
Marks, J. 2010. Inset Narratives in the Epic Cycle. *Classics@* 6 http://chs.harvard.edu/CHS/article/display/3230.
Martin, R. 2011. *Telegony.* In *The Homer Encyclopedia*, ed. M. Finkelberg, 841–842. Malden, MA.
Merry, W. and J. Riddell. 1886. *Homer's Odyssey.* Vol. 1. Oxford.
Mueller, M. 2007. Penelope and the Poetics of Remembering. *Arethusa* 40: 337–362.
Mülder, D. 1931. Ithaka nach der Odyssee. *RhM* 80: 1–35.
Nagler, M.N. 1980. Entretiens avec Tirésias. *CW* 2: 89–106.
Nagy, G. 1990. *Greek Mythology and Poetics.* Ithaca, NY.
Nagy, G. 2013. *The Ancient Greek Hero in 24 Hours.* Cambridge, MA.
Nobili, C. 2009. L'Odissea e le tradizioni peloponnesiache. *Pasiphae* 3: 171–186.
Ogden, D. 2001. *Greek and Roman Necromancy.* Princeton, NJ.

Petrakis, V.P. 2006. History Versus the Homeric 'Iliad': A View from the Ionian Islands. *CW* 99: 371–396.
Phillips, E.D. 1953. Odysseus in Italy. *JHS* 73: 53–67.
Purves, A. 2006. Unmarked Space: Odysseus and the Inland Journey. *Arethusa* 39: 1–20.
Purves, A. 2010. *Space and Time in Ancient Greek Narrative.* Cambridge.
Reece, S. 2009. *Homer's Winged Words.* Leiden.
Schade, G. 1999. *Lykophrons Odyssee – Alexandra* 648–819. Berlin.
Severyns, A. 1962. Ulysee en Élide. *AC* 31: 15–24.
Skempis, M. and I. Ziogas (eds.) 2014. *Geography, Topography, Landscape: Configurations of Space in Greek and Roman Epic.* Berlin.
Snodgrass, A. 1974. A Historical Homeric Society? *JHS* 94: 114–125.
Souyoudzoglou-Haywood, C. 1999. *The Ionian Islands in the Bronze Age and Early Iron Age 3000–800 BC.* Liverpool.
Steinrück, M. 2008. *The Suitors in the Odyssey.* New York.
Stubbings, F.H. 1963. Ithaca. In *Companion to Homer*, eds. A.J.B. Wace and F.H. Stubbings, 398–421, New York.
Thompson, D'A.W. 1947. *A Glossary of Greek Fishes.* London.
Tsagalis, C. 2008. *The Oral Palimpsest: Exploring Intertextuality in the Homeric Epics.* Washington, DC and Cambridge, MA.
Tsagalis, C. 2012. *From Listeners to Viewers: Space in the Iliad.* Washington, DC and Cambridge, MA.
Tsagalis, C. 2015. Telegony. In *The Greek Epic Cycle and Its Ancient Reception*, eds. M. Fantuzzi and C. Tsagalis, 380–404. Cambridge.
Vidal-Naquet, P. 1996. Land and Sacrifice in the *Odyssey.* In *Reading the Odyssey*, ed. S. Schein, 33–54. Princeton.
Waterhouse, H. 1996. From Ithaca to the *Odyssey. ABSA* 91: 301–317.
West, M.L. 2011. *Hellenica.* Vol. 1. Oxford.
West, M.L. 2013. *The Epic Cycle. A Commentary on the Lost Troy Epics.* Oxford.
West, M.L. 2014. *The Making of the Odyssey.* Oxford.
West, S. 1984. Lycophron Italicised. *JHI* 104: 127–151.
Westbrook, R. 2005. Penelope's Dowry and Odysseus' Kingship. In *Women and Property in Ancient Near Eastern and Mediterranean Societies*, eds. M. Gagarin and R.W. Wallace, 3–23. Vienna.
Wiseman, T.P. 1995. *Remus: A Roman Myth.* Cambridge.

David Bouvier

The Correlation of Fountains and Altars in Archaic Greek Poetry

In the Ancient Greek epic or hymnic poetry, several examples may be found of an altar (βωμός) built close to a *krene*, a "source fountain", within a natural landscape. On the Helicon, the Muses dance "around the violet-dark *krene* (περὶ κρήνην ἰοειδέα) and the altar (βωμόν) of Cronus' mighty son"; at Aulis, the Achaean army gather "around a *krene* (ἀμφὶ περὶ κρήνην) accomplishing complete hecatombs to the immortals upon the holy altars (ἱεροὺς κατὰ βωμούς)"; on the island of Ithaca, approaching the town, one can see along the way, "the *krene* (κρήνη), sweet running and made of stone, where the townspeople go for their water; Ithakos has made this, and Neritos and Polyktor; and around it is a grove of black poplars, trees that grow by water, all in a circle, and there is cold water pouring down from the rock above; over it has been built an altar (βωμὸς ... τέτυκτο) of the nymphs, and there all the wayfarers make their sacrifice (ἐπιρρέζεσκον)"; in the *Homeric Hymn to Apollo*, the god wanted to build his temple and altar near a *krene*.[1]

Are the two elements, the altar and the *krene*, harmoniously compatible? How may the contrast between the cold water emerging from the earth and the hot blood poured on the altar for the gods be considered? There are numerous studies on both fountains and altars,[2] nevertheless their correlation has received little attention. Even when texts or images allude to their co-presence, commentators have usually focused their attention on one or the other element, considering their association as logical and non-problematic.[3] It may, however, be more complex than it seems, especially as it concerns the *thusia*.[4]

---

1 See, in order, Hes. *Theog.* 3–4; *Il.* 2.305–306; *Od.* 17.205–211; *Hom. Hymn Ap.* 247–253, 375. For Hesiod, I use the translation of G. Most, for the *Iliad* and the *Odyssey* that of R. Lattimore, and for the *Homeric Hymn to Apollo* that of M. West. Margaret Church completed the proofreading and English editing and I want to thank her very much for her precious help.

2 On sources and fountains, see Wycherley 1937; Dunkley 1935–1936; Ninck 1967; Tölle-Kastenbein 1985; 1990a; 1990b; Glaser 1983; 1987; 2000a; 2000b; Ballabriga 1986, 47–48; Buxton 1994, 109–113; Bouvier 2013; on altars, Yavis 1949 and Mare 1962. On the ritual use of water, see Moulinier 1952, 71. On water in Homer, Scott 1931.

3 See Burkert 1985, 85–86 and West 1997, 33 observing: "The oldest holy places are those fashioned by nature: trees and groves, springs, grottoes, rocks, and peaks"; and 34: "At Dodona there was the spring of Zeus Naios below the oak. Homer describes how the Achaeans at Aulis sacrificed on altars set around a spring under a plane-tree." Buxton 1994, 109 quotes

DOI 10.1515/9783110535150-004

Water is indispensable to human life. Villages, towns, sanctuaries had to be built close to sources of drinkable water.[5] Ancient Greeks used to build fountains there where sources gushed from the earth. The substantives *krene* and *krounos* both mean: "a natural spring" emerging or resurfacing forth from the earth, and "a fountain made available for use", an edifice purposefully made to pour water into a container.[6] Sources and fountains are easily confused since they are often the one and same thing. Coming from the depths of the earth, such as fire strikes from the sky when Zeus thunders,[7] the spring is a natural force to be tamed or canalized, like fire, in order to be used for social and religious life. The fountain and the hearth are proof that Man controls the natural forces of water and fire. This control is the condition and emblem of cultural and social life. As the principal ceremony coordinating the relations of men with the gods, the animal slaying significantly requires both: pure water (usually but not necessarily coming from the *krene*) and fire. The correlation of the *krene* and the altar (or of pure/lustral water and blood) is emblematic of a social life based on animal slaying. The fountain and altar are necessarily complementary, one requires the other. Is this complementarity obvious?

---

some examples of sanctuaries built near sources. On the 'concept of natural sanctuary', see Scheid 2007/2008, 626–635, who also proposes a definition of "spring".

4 Henrichs 2012 and Perceau and Wersinger made important remarks on the necessity to take into account the different conceptions of sacrifice in different poetical genres (comedy, tragedy, epic, etc.). I here approach only hexametric poetry, without consideration however for the different contexts of enunciation of a theogony, a hymn and a war epic.

5 Burkert 1985, 86 and Buxton 1994, 109: "Greek sanctuaries, like the Heraion near Argos and the temple of Apollo at Delphi, were founded near springs, since water was needed for visitors and sacrificial animals as well as for the ceremonial of sacrifice itself".

6 Κροῦνος and κρήνη are synonyms; the etymological connection is plausible (cf. *DELG*, *sub verbis*). See Ginouvès 1962, 21–28; Tölle-Kastenbein 1985; 1990a; 1990b; Glaser 2000a and 2000b. The opposition of πηγή (to indicate the natural source of a river: see *Il.* 21.312 and 22.147) and κρήνη (to signify Man's development and maintenance of a source for means of use) is confirmed by Thuc. 2.15.4–5; see also Thuc. 2.48.2; Pl. *Leg.* 758e; Dem. 3.29; 13.30. Wycherley (1937, 2–3) rightfully moderates this opposition (πηγή versus κρήνη) and underlines the instances where κρήνη designates a natural source (*Od.* 9.141). Along the same lines, see Tölle-Kastenbein 1985, 459. The term πῖδαξ (a possible synonym for κρήνη) is a hapax in Homeric poetry. The word is rare and means the water emerging, gushing from the earth. The verb πιδάω means "gush, forth". The compound πολυπῖδαξ is more frequent, and is used as an epithet of the Ida, the mountain with "many springs, many fountained": *Il.* 8.47; 14.157; 283, etc; see also Pl. *Leg.* 681e.

7 The source as a "root", see the difficult passages in Hes. *Theog.* 738 and 808, with the notes of West 1978, 364. More in Rudhardt 1971, 99; Ballabriga 1986, 47.

On many issues, the great interpretations of sacrifice in the twentieth century have recently been discussed and contested. In Burkert's *Homo Necans* and *Greek Religion* or in the collective book, directed by Marcel Detienne and Jean-Pierre Vernant, *The Cuisine of the Sacrifice*, the correlation of water and blood is mainly based on an opposition: the function of water is to purify, whereas blood means the animal's death: its killing. Burkert observes that, for the sanctuaries without a spring, nor a fountain, "the water must be fetched from further afield from an ever-flowing spring of from the always powerful sea" and carried in a hydria along with the procession to the altar.[8] Water was then poured from the jug over the hands of each participant in a ritual known as the washing of the hands (χέρνιψ), and then sprinkled on the animal "causing it to jerk its head, a gesture interpreted as the animal nodding its assent".[9] After other preliminaries, "comes the death blow. The women raise a piercing scream: weather in fear or triumph or both at once ... The blood flowing out is treated with special care; rather, it must hit the altar, the hearth, or the sacrificial pit. If the animal is small it is raised over the altar; otherwise the blood is caught in a bowl and sprinkled on the altar-stone. This object alone may, and must again and again, drip blood".[10] A special instrument, the *sphageion*, is used to receive the blood spurting from the slit throat in order, then, to pour it on the altar. Dripping with blood, the altar described by Burkert is like a blood fountain. This bloody altar is, for him, the emblem of Greek religion: "for this is the act of piety: bloodshed, slaughter – and eating". For Burkert, killing – and not eating – is "the basic experience of *sacred*": "*Homo religious* acts and attains self-awareness as *homo necans*".[11] The pure water of the spring is there to help the *homo necans* to kill: it is used to sprinkle the animal and cause it "to jerk its head ... nodding its assent". In his description, Burkert does not mention the necessity to clean the altar after the sacrifice: "the bloody altar is the strong image of his demonstration: "to stain the altar with blood (*haimassein*) is a pious duty".[12] The bloody altar is the animal killing justification. Contrary to the altar, the source must remain pure of blood.

---

8 Burkert 1985, 77–78.

9 Burkert 1985, 55; Dillon 2002, 56 adds that water was also used to purify the altar and the whetstone was used to sharpen the axe and sacrificial knife.

10 Burkert 1983, 5. Is the woman's cry, the *ololuge*, a scream of fear or triumph? Others have shown that the *ololuge* is not a lamentation cry but rather the expression of feminine excitement, without excluding enjoyment and ecstasy. See Graf 2012, 46; Perceau and Wersinger (forthcoming) 8.

11 Burkert 1983, in order 2, 3.

12 Burkert 1985, 56, for both quotations.

In the collective volume *The Cuisine of the Sacrifice*, the problem of a correlation between water and blood is posed in particular by J.-L. Durand's essay. Durand recognizes that the pure water has two main and complementary functions in the slaying ritual: a 'death dealing' meaning ("eaux mortifères") and a purification power.[13] Observing a series of vase depicting scenes related to ox sacrifice, he pays special attention to the presence, in some cases, of a *louterion*, a reservoir of pure water (drawn from a nearby source), occupying a place in the structure of the image analogous to that of the altar[14] and having a symmetrical function. Like the altar, the *louterion* means death for the animal, since it contains the "deadly waters" "les eaux mortifères" used to sprinkle the ox and obtain the gesture of assent. Durand insists on the distance separating the *louterion* from the altar: "The animals approach the basin in the same way that they move toward the altar; in the space of the rite as in that of the image, altar and basin, as signs of death, are separated like water and blood". This distance makes it necessary to carry the water from the *louterion* to the altar: water is first carried in the hydria, "but – adds Durand – the separation between the place of water and the place of blood is so great that a second relay is needed". Another instrument, a kind of basin, *chernips*, "is used to present water to the officiant, who puts his hands in it, sprinkles the animal, and thus puts it in contact with the drops signifying death".[15] From the *louterion* to the altar, as from the *krene* to the altar, the distance and separation are essential, and we must understand why. Durand suggests a possible answer that remains however incomplete: "the *louterion*, the reservoir of pure water is at a great distance from blood. It is there to wash the blood away in the post-sacrificial phase".[16] This reconstruction is not far removed from Burkert's conception that considers bloodshed as a mark of killing. Killing's blood is clearly a pollution. We know that in Classical Athens, a man accused of murder was forbidden to use the community's lustral water, and was prevented from libations.[17] The question is to know if sacrificial blood may be assimilated to murder's blood?

G. Ekroth's remarks may oppose Burkert's as Ekroth argues that "the blood at regular sacrifice was actually kept, prepared and eaten after a small quantity

13 Durand 1989b, 123–125; see also Durand 1989a; Durand and Lissarrague 1981.

14 Durand doesn't say if this sprinkle basin could correspond to the one found at the entrance of a temenos (the so called *perirrhanterion*); see Burkert 1985, 86.

15 Durand 1989, 124.

16 Durand 1989, 123. See also Van Straten 1995, 104: "Thus, when the animal's throat is cut, its blood will gush directly onto the altar, which, in fact, is already stained with the blood of previous sacrifices".

17 Naiden 2013, 107.

has been sprinkled on the altar".[18] Blood was not necessarily a murder's mark needing to be purified or cleaned as Burkert and Durand suggested. Little is said, in Greek texts, about the cleansing the place of sacrifice. It seems, in any case, logical to imagine that the *krene* needs to remain free of blood in order to provide men with pure water. The water drawn from the sea or from a *krene* had to be pure for washing the priest's hands, and for watering or sprinkling the animal to be sacrificed.[19] Without pure water, the sacrifice is obstructed.[20] Pure water is necessary to make the blood flow.

This however doesn't mean that water is absolutely opposed to blood. We know many examples of sacrifice made regularly to local river gods.[21] River gods may have a sacred district, altar, temple, and priest. In an accomplished paper on the rite of throwing a victim into the sea or a river, Renée Koch-Piettre studied different kinds and means of offerings, and she mentions many slaughtered animals thrown into the sea, rivers and even sources, letting the blood flow into waters, as if the water were thirsty for blood.[22]

This seems to also be the case already in the *Iliad*, even if the poem is never absolutely explicit. In *Iliad* 11.728, Nestor evokes the fine sacrifices he made (ῥέξαντες ... ἱερὰ καλά) to Zeus and other gods, offering a bull to the river Alpheus, and one to Poseidon (ταῦρον δ' Ἀλφειῷ, ταῦρον δὲ Ποσειδάωνι), but nothing is said regarding the way the bull is offered to the two gods. In 21.130, provoking the Scamandros, Achilles ironically evokes the bulls "offered" in vain (see: ἱερεύετε ταύρους) by the Trojans to the river god and "all the horses thrown down living into his waters (ζωοὺς δ' ἐν δίνῃσι καθίετε)". It is important to distinguish here the bloodless offerings of a living animal thrown into the waters and the blood sacrifice of the bulls. It is, however, not explicitly said if the bulls' carcasses are also lowered into the waters without being eaten, without the '*thusia*' in the sense that one portion was burned for a god and another consumed by the worshippers.[23] It is probable that the blood was poured into the Scamandros, but the poet doesn't clearly specify. Let us consider the last exam-

18 Ekroth 2005, 9.

19 See Parker 1983, 51, 150, 226–230, 293, 371; Burkert 1983, 4, 11, 125.

20 Durand 1979, 175. Several inscriptions recall the sanctions imposed on those who sullied the sacred water sources: *IG* II^2 1126.36; X II 5.569; Sokolowski, *LSS* 4; 50; *LSCG* 152; *SEG* XIII 521.180–202. See also Parker 1983, 291.

21 Graf 1998; Koch Piettre 2005. See also Ephorus *FGrH* 70 F 20.

22 Koch Piettre 2005, 77–100. An example of such a ritual may be found in Horace, *Carm.* 3.13. See Mader 2002, 51 who notes how "Horace's vivid picture of the blood sacrifice to the spring of Bandusia has left many readers feeling somewhat uneasy". Mader understands this picture as a metaphorical one.

23 Naiden 2013, 102.

ple in *Iliad* 23 when Achilles alludes to the holy sacrifice (ἱερὴν ἑκατόμβην) of fifty rams that old Peleus wanted him to "consecrate (ἱερεύσειν) to the Spercheios":

> Spercheios, it was in vain that Peleus my father vowed to you
> that there, when I had won home to the beloved land of my fathers,
> I would cut my hair for you and make you a grand and holy
> Sacrifice (ἱερὴν ἑκατόμβην) of fifty rams consecrate to the waters
> of your springs (ἱερεύσειν ἐς πηγάς), where is your holy ground and your
> smoking altar (ὅθι τοι τέμενος βωμός τε θυήεις).[24]

The expression ἱερεύσειν ἐς πηγάς is ambiguous and the hexameters 147–148 have been understood in two different ways, according to the value given to ἐς ("at" or "to") and to πηγή ("running water" or "source"). Either the fifty rams are consecrated "near the springs"; "there at the springs of the river, where is the grove and the altar fragrant with burnt-offerings",[25] or they are consecrated "to the waters of the springs, where is the holy ground and the smoking altar".[26] Referring to a sacrificial calendar of Mykonos, Koch Piettre mentions a sacrifice to the river Acheloos for whom some lambs were slaughtered on the altar and others "in to the river" (πρὸς τῷ β[ωμ]ῷ σ[φάτ]τετ[αι], τὰ [δὲ] [ἄ]λλα ἐς ποταμόν).[27] She is most probably correct to understand thus ἱερεύσειν ἐς πηγάς as a sacrifice into the stream. However, it is also possible to understand that the hecatomb was performed "near the sources". The mention of the precinct and altar (τέμενος βωμός τε θυήεις) is an important indication meaning that the victims were killed on the altar, on which was then roasted the pieces of flesh. In the *Odyssey*, no blood is poured into "the *krene* (κρήνη), sweet running and made of stone, where the townspeople go for their water", but we know that Odysseus roasted leg of lamb and goat on the altar for the Nymphs of the fountain (Νύμφαι κρηναῖαι).[28]

24 *Il.* 23.144–148.
25 Transl. by S. Butler 1898.
26 A translation also accepted by Richardson 1993, 65: "Sacrifices to river-gods were commonly lowered into the water itself. Horses were thrown into the sea by the Argives at a place where a freshwater spring rises". On the relation of horses with sources, see Bloch 1985, 127 on Poseidon Hippios: "Puissante divinité chthonienne, il se trouve lié au culte des sources et à celui du cheval"; and the pertinent remark of Buxton 1994, 113: "On Helikon, where Pegasos' hoof had struck ground and brought water at Hippokrene (*Horse Spring*), a poet meet those who inspired him".
27 *LSCG* 96. 34–37 and Koch Piettre 2005, 88–89.
28 *Od.* 17.240–246, 205–211.

Why can no description of sacrificial blood flowing into a river be found in Homeric poetry? Homeric *thusia* was not considered as a killing: it was a communication ritual with the gods. It would be a mistaken to assign a precise meaning to each gesture of the ritual. In ritual slaying, blood may suggest many values. *Thusia* was a practice not founded on a specific theory. Homeric heroes did not have to justify the slaying of living animals. Blood flowing into water is theoretically a paradoxical image (as the *slaying* is a paradoxical ritual): as a god, the river and its source deserve a blood sacrifice. There was no archaic theory of the sacrifice, but many images revealing the suggestive power of the ritual. What about blood flowing into a *krene* in the poem of war?[29]

When the Fountain is Polluted with Murder's Blood

In Book XVI, the Myrmidons are called to the battle.

> Achilles went meanwhile to the Myrmidons, and arrayed them
> all in their war gear along the shelters. And they, as wolves
> who eat flesh raw (ὠμοφάγοι), in whose hearts the battle fury is tireless,
> who have brought down a great horned stag in the mountains, and then feed
> on him, till the jowls of every wolf run blood (αἵματι φοινόν), and then go
> all in a pack to drink from a spring of dark-running water (ἀπὸ κρήνης μελανύδρου),
> lapping with their lean tongues the surface
> of the black water (μέλαν ὕδωρ) and belching up the murders' blood (ἐρευγόμενοι φόνον αἵματος).[30]

Richard Janko is correct to affirm that this is "one of Homer's best similes".[31] The bloodlust of the warriors anticipates the violence of the fight to come. Red blood (αἵματι φοινόν) and black water (μέλαν ὕδωρ) are opposed and mixed. The symmetry and the chiasmus of the construction (αἵματι φοινόν / ἀπὸ κρήνης μελανύδρου / μέλαν ὕδωρ / φόνον αἵματος) concurrently suggest an analogy and a contrast between water and blood. Often springs are described in the *Iliad* as

29 On the importance of metaphors to understand Homeric sacrifice, see Kitts 2005, 156–161.
30 *Il.* 16.155–162.
31 Janko 1992, 338.

pouring black water,[32] but it remains a mystery as to why the water emerging from the ground is black. Is the darkness linked to the depth of the source, to its temperature or even purity? More clearly, the red colour (φοινόν) of blood is associated here, through the alliteration, to murder (φόνον). Etymologically, the two words φοινόν ("red") and φόνον ("murder") have different roots: nevertheless, the chiasmus and the echoes indicate that the red gore is the result of murder. The Myrmidons are like wolves, eaters of raw flesh, hungry for war; the battle is also a feast of blood. But the simile does not say if the wolves reach the source to drink and sate their thirst or to wash and clean their jaws. After their passage, the source will be soaked in blood. Black or red? The chiasmus and the symmetrical repetition of words oppose and seem to reverse the black water and the red blood; in the *Iliad* the adjective 'black' (μέλας) qualifies water as well as blood.[33] In the simile, we even have the impression the wolves are not really drinking water but rather spewing blood into the source, pouring blood like a libation. Do the wolves really drink? If their slender tongues lap the black water, their jaws disgorge (ἐρευγόμενοι) blood. Here, the wolves' mouths are like the spouts of a living and disturbing fountain pouring blood.[34]

In their study published in *The Cuisine of Sacrifice*, M. Detienne and J. Svenbro have demonstrated the many affinities of wolves and 'sacrificers'. Wolves are not only hunters, they also have a natural vocation of butcher and cook, two qualities defining a worthy sacrifice. When it kills its prey, the wolf bleeds it (σφάζειν), and splits the victim's throat, acting, says Aristotle, with intelligence.[35] He also possesses the art of carving and apportioning the meat, making equal portions. This mastery of the butcher's art would make the wolf a perfect sacrificer in a city where social order is confirmed, through sacrifice (θυσία), by the equal distribution of the meat's parts. But the wolf's greediness makes this butchery specialist a paradoxical distributor who tries, in the end, to have always more than its fair share.[36] Worse still, acting as a ritual butcher, the wolf pays no respect to the meaning and justification of the sacrifice itself: the necessity to leave a portion for the gods. He is so voracious that "he gulps down into its

32 In the *Iliad*, we found four occurrences of the adjective μελάνυδρος always to qualify a κρήνη: *Il.* 9.14; 16.3; 16.160; 21.257. See also the only occurrence of the *Odyssey:* 20.158. See Pucci 2007, 38.

33 For water see also 2.825, 21.202, 23.806, and for blood: 4.149, 7.262, 10.298, 23.693.

34 Bouvier 2013.

35 Arist. *Hist. An.* 3.6 612b; see Detienne and Svenbro 1989, 154.

36 Detienne and Svenbro 1989, 157.

paunch even the long bones, the *meria* reserved for the gods in men's sacrifice".[37]

At the source, the wolves both simultaneously clean their mouths and spew the blood of the slain. They are like good sacrificers who, to honour a river, pour blood into its source. However, this blood is the result of a murder. Attention must be paid to the figure of hypallage; literally the expression ἐρευγόμενοι φόνον αἵματος means that the wolves are not pouring blood into the spring but "the murder of blood". What the wolves dedicate to the fountain is not "the blood of the slain" but "the slain of blood"!

Mixing blood and the pure water of a source: the wolves are an image of what the Myrmidons will become in a battle that will soon bathe with blood not only the plain of Troy but also the Scamander when Achilles will bring war close to the river and then into it, dying red its water with the blood of his enemies (ἐρυθαίνετο δ' αἵματι ὕδωρ[38]), leaving his victim, Asteropaios, soaking in black water (δίαινε δέ μιν μέλαν ὕδωρ, 21.202).

In Book XVI, the image of the wolves polluting the spring water is also an anticipation of the insults Achilles will address to the river Spercheios when, hurling the corpse of Lycaon into the water, he exclaims: "Lie there now among the fish, who will lick the blood away from your wound, and care nothing for you ... And there will not be any rescue for you from your silvery-whirled strong-running river, for all the numbers of bulls you dedicate (ἱερεύετε) to it and all the horses that you flung living into his waters".[39] There is an evident irony in Achilles' words: he throws the bloody corpse of Lycaon as an offering, mocking the very ritual of throwing animals into the river. Hearing Achilles' claim, the Scamander grew angry (χολώσατο). Even if sacrifices to river gods are special ones, Achilles has gone too far with his irony.

The simile of the wolves at the fountain not only describes the Myrmidons Achilles prepares to fight, but it also announces a violent war that will dissolve the boundary between murder (φόνος) and sacrifice. The scene of the simile is an expected one in the world of animals; it becomes a disturbing scene when the image refers to the world of men ready to plunge into "the bloody maw of battle" (19.313). If the altar soaked with blood is a norm of the ritual, the source must remain pure to provide the lustral water. Belching up the murderers' blood into the source's water, the wolves of the simile suggest a world ignoring the rules of sacrifice and reveal the danger of confusing the *krene* and the altar.

37 Detienne and Svenbro 1989, 154.
38 *Il.* 21.21. Cf. 10.484.
39 *Il.* 11.122–124, 130–132.

The Fountain and the Altar as Emblems of the Cosmic, Religious and Social Order

Hesiodic *Theogony* is at the same time the account and the performative confirmation of Zeus' final ordering of the cosmos. The poem itself appears to be the culminating point of this order: it is part of it and constitutes its keystone. The very possibility to sing the *Theogony* proves and reinforces the final and permanent stability of the *cosmos*.[40] It has a performative value. A chaotic situation would prevent any poet from singing anything. It is not that the poem reproduces the exact ordering of the cosmos but that its own order is the direct consequence and ultimate achievement of the cosmic order. If Chaos came into being at first (πρώτιστα) in the theogonic process, the poem narrating this process must start with the Muses.[41] The first divinities to be invoked and sung are not a primordial divinity neither the sovereign god but the Muses, daughters of Zeus and Mnemosyne, whose function and competence are to make possible the hymns that celebrate and charm their father. The cosmos is perfect when it echoes with the Muses' song.

As the direct result of the cosmic achievement and mirroring it, the Muses' performance in honour of Zeus is associated, in the very first lines of the poem, to an accurately described and highly significant landscape:

> From the Heliconian Muses let us begin to sing, who hold the great and holy mount of Helicon, and dance on soft feet about the deep-blue spring (violet dark fountain) and the altar of the almighty son of Cronos, and, when they have washed their tender bodies in Permessus or in the Horse's Spring (or Hippocrene) or Olmeius, make their fair, lovely dances upon highest Helicon and move with vigorous feet.[42]

The Heliconian Muses are not distinguished here from the Olympian ones. The epithet 'Heliconian', as observed by West, "only marks the place of their cult and the place they often haunt".[43] The Helicon indicates a point of departure from where the Muses will then leave to reach the foot of the Olympus continuing their song in the mountain inhabited by the gods. According to the place they

40 Hesiod never uses the word κόσμος with the meaning of "universe" as a whole. See West 1997, 137: "In early Greek there is no word for the universe as a whole".

41 Hes. *Theog.* 116.

42 Hes. *Theog.* 1–8.

43 West 1978, 152. When they met Hesiod under the holy Helicon, the same Muses are called "Muses of Olympus" (24). See also Nagy 1990, 58: "the local goddesses of Helikon are assimilated into the pan-Hellenic goddesses of Olympus".

visit, the Muses are Heliconian or Olympian. Do we need to recognize in the altar and fountain described by Hesiod a precise and existing place? Even if it were possible, we would still have to explain and emphasize the emblematic and poetical value, at the beginning of the *Theogony*, of the source's and altar's correlation.[44]

Ring-dances around altars or around sources are well attested.[45] M. West agrees with Sittl to propose that the dance around the fountain "was intended to ensure the continual flowing of water", an explanation P. Pucci is not ready to share. Even if he reminds how, in Pindar's *Isthmian* 6, the Muses were able to make the water of Dirke's spring gush forward, a flow the poet assimilated to the song itself, Pucci doesn't recognize in the *Theogony* any clear nor evident link between the Muses and a particular function of the water. This conclusion may be too radical. We certainly should pay more attention to repeated expressions in the proem assimilating the Muses' song or voice to a flow.[46] Four times, the poet repeats how the Muses "send forth their very beautiful voice": περικαλλέα ὄσσα ἱεῖσαι.[47] The same verb, ἵημι, is used to tell the gush of water from a spring or a river; Asteropaeus, one of the Trojans Achilles kills and abandons on the bank of the Scamandros, is proud to say that he is a descendant "from Axius, the water whereof flows the fairest over the face of the earth" (*Il.* 21.158: Ἀξιοῦ, ὃς κάλλιστον ὕδωρ ἐπὶ γαῖαν ἵησιν).[48]

In line 35, after the short description of his encounter with the goddesses and the mission he receives to celebrate them, Hesiod resumes the Muses' celebration, a celebration he hadn't really interrupted. This second beginning of the *Theogony* is a reaffirmation of the first; the poet recalls the necessity to start the celebration of Zeus' power with an invocation to the Muses themselves whose voice flows like a real river:

> Come then, let us begin from the Muses, who by singing for their father Zeus give pleasure to his great mind within Olympus, telling of what is and what will be and

---

44 Commentators separately considered the altar and the fountain, without paying attention to their correlation. This dissociation of both elements is evident in a comment of Strauss-Clay 2003, 54 that transforms the correlation into an alternative (the Muses "circle around a spring or the altar of Zeus").

45 Eur. *IA* 676; 1480; Poet. Lesb. fr. inc. 16; Callim. *Del.* 312 and West 1978, 152.

46 See the forthcoming work of C. Semenzato, *A l'écoute des Muses.*

47 Nagy 1979, 296–297. Nagy (1990, 47–48) recognizes in this expression a possible etymological explanation of Hesiod's name: "The very name Hesiodos at *Theogony* 22 means something like 'he who emits the voice: ἵημι and αὐδή'".

48 See also *Il.* 7.158; Hes. *Theog.* 10, 43, 45, 67.

> what was before, harmonizing in their sound (φωνῇ ὁμηρεῦσαι). Their tireless voice flows sweet from their mouths (τῶν δ' ἀκάματος ῥέει αὐδὴ / ἐκ στομάτων ἡδεῖα).[49]

The flow of the voice is more than a metaphor here. "Tireless" (ἀκάματος) like the fire of the sun,[50] the voice of the Muses is assimilated to a natural element: a permanent flow that fulfils a specific and essential function in the world and rejoices not only Zeus himself but also the places around him. Thus, the house of Zeus "rejoices (γελᾷ) at the goddesses' lily-like voice as it spreads out (σκινδναμένη) and snowy (νιφόεντος) Olympus' peak resounds (ἠχεῖ), and the mansions of the immortals".[51] The world blossoms and reaches its plenitude when it resounds with the music of the Muses. For Zeus, the perfect world is the one in which the song of his celebration flows constantly like a river.

To what extent can we establish any kind of relation between the fountain around which the Muses dance and the tireless voice flowing from their mouths? Could the explanation of a dance aiming to ensure the continual flowing of a river also mean the hope of the continual flowing of a song celebrating Zeus? We must pay attention to the structure of the poem and to the importance Hesiod attaches to the right beginning. The poem starts with a verb precisely indicating the act of beginning and the divinity it must begin with. Narrating his encounter with the goddesses, Hesiod recalls how they ordered him to sing of themselves first and last and how he should proceed: "Come then, let us begin from the Muses ... Their tireless voice flows sweet from their mouths".[52] The religious necessity to start a song with the Muses emanates from the fact that the goddesses are the very source of poetry and eloquence. Hesiod explicitly recounts that the Muses "pour sweet dew upon the tongue of the king they behold (τῷ μὲν ἐπὶ γλώσσῃ γλυκερὴν χείουσιν ἐέρσην), and his words flow soothingly from his mouth (τοῦ δ' ἔπε' ἐκ στόματος ῥεῖ μείλιχα)"; and the singer is blessed "whomever the Muses loved, for the speech flows sweet from his mouth (γλυκερή οἱ ἀπὸ στόματος ῥέει αὐδή)".[53] The flow of words and speeches poured from the king's and singer's mouths is but the continuation of the tireless flow of the Muses' voice.[54]

---

49 Hes. *Theog.* 36–40. See also 84 and 97.

50 See Pucci 2007, 78.

51 Hes. *Theog.* 40–46. See also 36–37. For the use of the verb σκίδναμαι to mean the spread of water, see also *Od.* 7.130–131: ἐν δὲ δύω κρῆναι ἡ μέν τ' ἀνὰ κῆπον ἅπαντα / σκίδναται.

52 Hes. *Theog.* 36–40.

53 Hes. *Theog.* 83–84 and 96–97.

54 Compare τῶν δ' ἀκάματος ῥέει αὐδὴ / ἐκ στομάτων ἡδεῖα (39–40), τοῦ δ' ἔπε' ἐκ στόματος ῥεῖ μείλιχα (84) and γλυκερή οἱ ἀπὸ στόματος ῥέει αὐδή (97).

It is certainly not a coincidence if at the beginning of a poem recounting the importance of initially celebrating the right divinity, we find an image of the Muses, sources of poetry and eloquence, dancing around a *krene* that is both, as we said, a fountain and a source, the place where the water emerging from the depths of the earth starts to flow down the mountain. Sources are always the image of a beginning, a point of departure. Hesiod doesn't speak of a god who had to kill a *drakon* near this source on the Helicon: a well-known motif.[55] But the *krene* of the proem is significantly associated to an altar specifically dedicated to Zeus. Is the source that emerges from the depth of the earth a kind of cosmic root? Hesiod doesn't suggest it, but the correlation with the altar is sufficient to make it very clear that the water of the source is now controlled and tamed, useful for any sacrifice on the altar.

The altar is an important emblem of Zeus' power in the *Theogony.* As J.-P. Vernant's analysis has established, both the stories of Prometheus' trick and of the fabrication of woman constitutes, inside the *Theogony*, a whole and coherent ensemble and certainly not "an unsatisfactory synthesis of heterogeneous elements".[56] In the middle of the poem, the Titan Prometheus is an enigmatic figure whose real intention, when he tries to deceive Zeus, is never explicitly revealed. Between the sovereign god and the Titan, we assist in a duel of ruses and counter-ruses, a battle, could we say, of intelligence. The instauration of the sacrifice procedure will be one of the results of this battle.

The conflict starts at the time when a separation or a crisis occurs between gods and men who were until then living together. Prometheus wants to provoke Zeus' power through favouring mankind. Does he want to usurp the supremacy or help men? If this remains unclear, the intention to test Zeus' perceptiveness and cleverness is obvious. Prometheus chooses to divide the parts of an ox, making two very unequal portions: the good pieces, flesh, on one side but covered by the unappetizing skin and stuffed in the stomach, and inedible bones, on the other side, but hidden beneath appetising and gleaming fat. Zeus is invited by the Titan to choose the part he wants. But Zeus is the god "who knows eternal counsels" (ἄφθιτα μήδεα εἰδώς: line 550); he immediately recognizes the deception and decides to accept the wickedly appetizing part. Doing so, he thwarts the intention Prometheus had to favour men and obtain benefits for them. It is difficult to understand how Prometheus wished to better the human condition. The first fragment of the *Catalogue of women* recalls a time when "feasts were in common and in common the seats for the immortal gods and mortal human be-

55 On this important motive see Ogden 2013, 165–178.

56 As says Strauss Clay 2003, 100, summarizing Vernant's analysis.

ings".[57] What was the common food for gods and men? As J.-P. Vernant explained well, the consequence of Prometheus' ruse is the definition itself of the new and definitive status of the human condition, with its new mode of existence and carnivorous alimentation characterised by the necessity of sacrifice. Men will become a race completely separated from their ancient table companions. The corruptible meat of the ox stuffed in the stomach will become the emblem of men's mortality and of their insatiable hunger. On the contrary, the incorruptible white bones constitute the immortal gods' portion, who receive them in the form of smoke.

We know that the story doesn't end there. Zeus harshly and astutely responds to Prometheus' trick by depriving the mortals of fire, condemning them to live wild, to eat raw food like wolves. However before narrating the second episode, Hesiod anticipates its conclusion. He explains how just after the discovery of the white bones under the white appetizing fat Zeus "became enraged in his breath and how wrath came upon his spirit" (554). At this point, Hesiod comments: "And ever since then the tribes of human beings upon the earth burn white upon smoking altars for the immortals (θυηέντων ἐπὶ βωμῶν)" (556–557). The conclusion is anticipated but it gives the impression that Zeus already decided the end of the whole conflict with his rival and its consequence for men. He knew he would deprive mortal beings of fire but Prometheus refuses to abandon them and steals fire to offer it to mankind, thus making possible the practice and good procedure of sacrifice.

Could we consider the ox of Prometheus' ruse as a sacrificial victim? Most commentators do so. Certainly, the story explains the origin of the sacrificial practice. But Hesiod does not mention the killing of the ox, nor the place where it happened. Even if Prometheus' ruse offers the model to adopt for the division of the victim's parts, it remains to be known if he killed the animal on an altar or not. If no altar were there, it can't legitimately be considered a sacrifice, even if the practice of sacrifice *(thusia)* originates from it.

Anticipated in line 556, the image of the smoking altars men will need after Prometheus' trick is a definitive sign of Zeus' victory over the Titan and of the new condition for mankind. The altar proves the end of the conflict with Prometheus and the definitive status accorded to men. Exactly as the fountain proves the control of water, the altar means for men the good use of fire.

Just after the short narration of his encounter with the Muses, Hesiod returns to the first purpose of his poem with a famous and enigmatic line, to say that he will not spend more time speaking about secondary themes: "But what is this to

---

57 Hes. fr. 1 M.-W. = Most 1.

me, around an oak or a rock?" (περὶ δρῦν ἢ περὶ πέτρην). Even if an oak and rocks are often associated one with the other in Greek poetry, the exact meaning of the expression has always been unclear, for the Ancients as for us. If the origin of the proverb remains mysterious, it is commonly accepted however that the general idea indicates the moment to abandon a less important argument for a more important one.[58] In this case, it is rather surprising that no one really pays attention to the possible parallel with lines 3 and 4. Semantically and also phonetically (περὶ κρήνην echoes περὶ πέτρην), the proverbial and enigmatic expression recalls the indications of the fountain and the altar from where the musical performance of the Muses starts. In iconography, springs, when they are natural, are typically represented by trees or by conical piles of stones.[59] The altar is nothing but a stone worked by men and dedicated to a divinity. The tree and the rock are natural elements to which the edified fountain and the altar correspond on a cultural level. Having to return to his main argument, Hesiod asks himself: "why do I have to speak around an oak or a rock when I have to celebrate the Muses that dance around the fountain and the altar?" In the *Theogony* this fountain and altar are emblematic of Zeus' control of the world and are the perfect place to start the celebration of the first of the gods himself.

Dancing around the fountain and the altar "with their soft feet (πόσσ' ἀπαλοῖσιν)", the Muses mark out an area that unifies and coordinates both elements. They are not making a sacrifice but their dance establishes a harmonious link between two places allowing for social life and sacrifice. Their dance illustrates the good and harmonious relationship of the fountain and altar, water and fire, the right balance of the world governed by Zeus. If we admit that the context of performance of the *Theogony* was ritual and sacrificial, it would make great sense.[60]

However, if the correlation of the *krene* and the altar appears to be an ideal *locus* at the beginning of the *Theogony*, we must not forget that their association is not an immediate one. As learned from the *Homeric Hymn to Apollo*, the source often has its own guardian that must be killed in order to access the water.[61] If the correlation of the source and the altar makes the sacrifice and the killing of the ox possible, the ritualization of the murder never definitively cancels the opposition of pure water and blood and the threat to see the fountain or the altar becoming the theatre of a more dangerous form of violence.

58 Longo 1956.

59 Ogden 2013, 167.

60 As supposed by Perceau and Wersinger 2014, 131–132.

61 See Ogden 2013, 165–178. A developed version of this study will pay more attention the *Homeric Hymn to Apollo*.

To conclude let us consider the description of what was for the Achaeans a crucial moment. In Book II of the *Iliad*, after nine years of war, the Achaean army is discouraged. Odysseus has just restored order among the panicked troops; he has silenced Thersites and tries to awaken Achaean's ardour, reminding them of the initial prophecy of Calchas and of the omen that appeared to them nine years before. They were at Aulis, waiting to embark to Troy and ready to sacrifice to the gods:

> And we around a spring (περὶ κρήνην) and upon the sacred altars (ἱεροὺς κατὰ βωμοὺς)
> were accomplishing complete hecatombs to the immortals
> under a fair plane tree whence ran the shining of water.
> There appeared a great sign; a snake, his back blood-mottled (δαφοινός),
> a thing of horror, cast into the light by the very Olympian,
> wound its way from under the altar and made toward the plane tree.[62]

The analepsis brings us back nine years before: to a moment so crucial that its memory is still a vivid one, as – observes Odysseus – if it were yesterday. The sacrifice was on this day a decisive one, its success was the condition to guarantee the whole expedition of the Achaeans. Not a word in the *Iliad* about Iphigenia.[63]

Descriptions of landscapes are very rare in the *Iliad*, and when there is one, it must be closely examined. Odysseus is a fine orator. He knows that the precision of his description will contribute to stress the authenticity of his reminiscence. We don't know if the gods have accepted the sacrifice, upon this day. The apparition of the serpent sent by Zeus is a sign of more importance than the result of sacrifice itself. On the highest branch of the tree the serpent was reaching for, lay a brood of eight young sparrows peeping with their mother. The serpent ate the eight chicks:

> While the mother flew about lamenting her little ones; but the serpent threw his coils about her and caught her by the wing as she was screaming. Then, when he had eaten both the sparrow and her young, the god who had sent him made him become a sign; for the son of scheming Cronos turned him into stone, and we stood there wondering at that which had come to pass.[64]

---

62 *Il.* 2.305–310.

63 Nothing is said about Iphigenia in the *Iliad* but it was certainly known by the listeners of the poem from a very ancient time. See Bouvier 2002, 305–308; Naiden 2013, 115.

64 *Il.* 2.215–220.

It belongs to Calchas, the interpreter of omens, to reassure the Achaeans explaining that the eight sparrows and their mother, nine in total, correspond to the nine years that they will have to fight at Troy, before taking the town on the tenth year. Calchas' interpretation is right but the picture of the omen says much more: it clearly announces the war and the violence to come during the sack of the town. Near the spring (περὶ κρήνην), the serpent that darts from under the altar recalls an ancient guardian of the spring. His back δαφοινός ("blood-reeking") seems "maculated with blood". At Aulis, spring and altar were the theatre of a terrific omen announcing a war in which men will sometimes be like beasts drinking blood. The *drakon* transformed into a stone near the tree is a disturbing element. Stones are used to build altars and serpents with a back "δαφοινός" seem thirsty of blood.[65]

In Greek poetical imaginary, the correlation of the fountain and the altar is a suggestive one that says much about sacrifice. But the inquiry is far from being concluded.

Bibliography

Bloch, R. 1985. Quelques remarques sur Poséidon, Neptune et Nethuns. In *D'Héraklès à Poséidon*, ed. R. Bloch, 125–140. Geneva and Paris.

Bouvier, D. 2002. *Le sceptre et la lyre: L'Iliade ou les héros de la mémoire.* Grenoble.

Bouvier, D. 2010. Les jardins de l'Odyssée, lieux de l'ambiguïté et de la mémoire. In *Jardins antiques*, ed. K. Reber, 21–47. Gollion.

Bouvier, D. 2013. Les fontaines à tête de lion. Quand un motif sculptural invite à relire l'*Iliade*. In *Il paesaggio e il sacro: l'evoluzione dello spazio di culto in Grecia: interpretazioni e rappresentazioni*, ed. E. Villari, 59–76. Genua.

Burkert, W. 1983 (1972). *Homo Necans: The Anthropology of Ancient Greek Sacrificial Ritual and Myth.* Berkeley. Los Angeles and London.

Burkert, W. 1985 (1977). *Greek Religion.* Cambridge, MA.

Buxton, R. 1994. *Greece Imaginery.* Paris.

Detienne, M. and J. Svenbro. 1989. The Feast of the Wolves, or the Impossible City. In Detienne and Vernant 1989, 148–163.

Detienne, M. and J.-P. Vernant (eds.). 1989. *The Cuisine of Sacrifice Among the Greeks.* Chicago and London.

Dillon, M. 2002. *Girls and Women in Classical Greek Religion.* London.

Dunkley, B. 1935–1936. Greek Fountain-Buildings Before 300 BC. *ABSA* 36: 142–212.

---

65 Ogden 2013, 165: "Rivers, with their inherently serpentine courses, and *drakontes* offered ready metaphors for each other: Hesiodic poetry already compares, from the one side, the constellation of Draco to a flowing river and, from the other, the river Cephisus to a *drakon* as it winds through Orchomenos".

Durand, J.-L. 1989a. Greek Animals: Toward a Topology of Edible Bodies. In Detienne and Vernant 1989, 87–118.
Durand, J.-L. 1989b. Ritual as Instrumentality. In Detienne and Vernant 1989, 119–128.
Durand, J.-L. and F. Lissarrague. 1981. Un lieu d'image? L'espace du loutérion. In *Arts et légendes d'espace*, ed. C. Jacob, 125–148. Paris.
Ekroth, G. 2005. Blood on the Altars? On the Treatment of Blood at Greek Sacrifices and the Iconographical Evidence. *AK* 48: 9–29.
Faraone, C.A. and F.S. Naiden (eds.). 2012. *Greek and Roman Animal Sacrifice: Ancient Victims, Modern Observers*. Cambridge and New York.
Ginouvès, R. 1962. *Balaneutikè: recherches sur le bain dans l'Antiquité grecque*. Paris.
Glaser, F. 1983. *Antike Brunnenbauten (krēnai) in Griechenland*. Vienna.
Glaser, F. 1987. Brunnen und Nymphäen. In *Die Wasserversorgung antiker Städte*, ed. W. Eck, 103–131. Mainz am Rhein.
Glaser, F. 2000a. Fountains and Nymphaea. In *Handbook of Ancient Water Technology*, ed. Ö. Wikander, 413–452. Leiden.
Glaser, F. 2000b. Water Landscaping. In *Handbook of Ancient Water Technology*, ed. Ö. Wikander, 453–466. Leiden.
Graf, F. 1998. Flussgötter. In *Der Neue Pauly* IV, 576–578.
Graf, F. 2012. One Generation after Burkert and Girard: Where are the Great Theories? In Faraone and Naiden 2012, 32–51.
Henrichs, A. 2012. Animal Sacrifice in Greek Tragedy. In Faraone and Naiden 2012, 180–194.
Janko, R. 1992. *The Iliad: A Commentary*, vol. iv: Books 13–16. Cambridge.
Kitts, M. 2005. *Sanctified Violence in Homeric Society: Oath-making Rituals and Narratives in the "Iliad"*. Cambridge and New York.
Koch Piettre, R. 2005. Précipitations sacrificielles en Grèce ancienne. In *La cuisine et l'autel: les sacrifices en questions dans les sociétés de la Méditerranée ancienne*, eds. S. Georgoudi, R. Koch Piettre and F. Schmidt, 77–100. Turnhout.
Lincoln, B. 2012. From Bergaigne to Meuli. How Animal Sacrifice Became a Hot Topic. In Faraone and Naiden 2012, 13–31.
Longo, V. 1956. Un'interpretazione di Omero. In *Antidōron Hugoni Henrico Paoli Oblatum: miscellanea philologica*, ed. A.F. Messina, 183–203. Genua.
Mader, G. 2002. That St(r)ain Again: Blood, Water, and Generic Allusion in Horace's Bandusia Ode. *AJPh* 123: 51–59.
Mare, W.H. 1962. *A Study of the Greek Bomos in Classical Greek Literature* (Ph.D. Diss., Dissertation Abstracts 23: 1011–1012). University of Pennsylvania.
Moulinier, L. 1952. *Le pur et l'impur dans la pensée des Grecs d'Homère à Aristote*. Paris.
Nagy, G. 1990. *Greek Mythology and Poetics*. Ithaca and London.
Naiden, F.S. 2013. *Smoke Signals for the Gods: Ancient Greek Sacrifice from the Archaic through Roman Periods*. Oxford.
Ninck, M. 1967 (Leipzig 1921). *Die Bedeutung des Wassers im Kult und Leben der Alten*. Darmstadt.
Ogden, D. 2013. *Drakôn. Dragon Myth and Serpent Cult in the Greek and Roman Worlds*. Oxford.
Parker, R. 1983. *Miasma: Pollution and Purification in Early Greek Religion*. Oxford.

Perceau, S. and A.G. Wersinger. 2014. Retour sur le prétendu "sacrifice" grec. In *Polutropia: d'Homère à nos jours. Mélanges offerts à Danièle Aubriot*, eds. S. Perceau and O. Szerwiniack, 281–300. Paris.

Perceau, S. and A.G. Wersinger-Taylor (forthcoming). Ritual Differences: Cutting up and Dividing (Gender, Genus, and Genres). What about Blind Spots in Anthropology? (paper delivered at Stanford University in October 2012, at a meeting on "Language, Ritual, Performance: Evaluating Theoretical Approaches to Greek Culture" [paper accessible on www.academia.edu]).

Pucci, P. 2007. *Inno alle Muse. Esiodo, Teogonia, 1–115*. Pisa and Rome.

Richardson, N.J. 1993. *The Iliad: A Commentary*, vol. vi: Books 21–24. Cambridge.

Rudhardt, J. 1971. *Le thème de l'eau primordiale dans la mythologie grecque*. Berne.

Scheid, J. 2007/2008. Le culte des eaux et des sources dans le monde romain. *Annuaire du Collège de France* 108: 622–637.

Scott, J.A. 1931. Homer's Attitude Towards Water. *CJ* 26.9: 695–698.

Semenzato, C. (forthcoming, 2017). *A l'écoute des Muses*. Berlin and Boston.

Strauss Clay, J. 2003. *Hesiod's Cosmos*. Cambridge.

Tölle-Kastenbein, R. 1985. Der Begriff κρήνη. *AA* 1985/3: 453–470.

Tölle-Kastenbein, R. 1990a. *Antike Wasserkultur*. Munich.

Tölle-Kastenbein, R. 1990b. *Archeologia dell'acqua: la cultura idraulica nel mondo classico*. Milan.

Van Straten, F.T. 1995. *Hiera Kala: Images of Animal Sacrifice in Archaic and Classical Greece*. Leiden and New York.

Vernant, J.-P. 1989. At Man's Table: Hesiod's Foundation Myth of Sacrifice. In Detienne and Vernant 1989, 21–86.

West, M.L. 1978 (1966). *Hesiod, Theogony*. Oxford.

West, M.L. 1997. *The East Face of Helicon: West Asiatic Elements in Early Poetry and Myth*. Oxford.

Wycherley, R.E. 1937. ΠΗΓΗ and ΚΡΗΝΗ. *CR* 51: 2–3.

Yavis, C.G. 1949. *Greek Altars: Origins and Typology, Including the Minoan–Mycenaean Offertory Apparatuses*. Saint-Louis.

Eleni Peraki-Kyriakidou

Iris as Messenger and Her Journey: Speech in Space and Time

Fama and its personification in ancient myth and literature represent the dissemination of speech; in recent years there has been much discussion on the subject.[1] However, there is another mythological figure with similar capacities who also makes her presence felt as early as in the archaic epic and this is none other than Iris.[2]

Fame or Φήμη derives from φημί[3] and Iris from εἴρω – both meaning "to speak", "to say" (LSJ s.v. B). Like Fama, Iris does not bear a complete myth of her own.[4] Unlike Fama, however, Iris has – at least in the archaic epic – a more concrete or specific role to play: on most occasions, that is, she is an intermediary among gods or between gods and humans when carrying a message.

One way of approaching Iris' function in epic is through etymology. Whether explicit or implicit, it has a significant role to play in the ancient texts. While the explicit is quite obvious, the implicit is the one resulting through cognate words.[5] Very often, especially in Homer, a proper name, in one way or another, seems to be etymologically related with its attribute or other lexical elements in the sentence.[6]

1 Seminal is Hardie 2012; see also Syson 2013 and Kyriakidis 2016.

2 Laird (2003, 158) talks about the "visual characterization of speech as something reaching down from heaven to earth".

3 Already in Pl. *Cra.* 408b: καὶ ἥ γε Ἶρις ἀπὸ τοῦ εἴρειν ἔοικεν κεκλημένη, ὅτι ἄγγελος ἦν ("Iris also seems to have got her name from εἴρειν, because she is a messenger"; transl. H.N. Fowler, Loeb).

4 By 'complete' I mean a full-blown story, a myth, that is, with a beginning, middle and end. However, there are some traits of a plot, e.g. in Alcaeus. See Gow 1952, 346: Alc. fr. 13: ... δεινότατον θέων, / <τὸν> γέννατ' εὐπέδιλλος Ἶρις / χρυσοκόμαι Ζεφύρῳ μίγεισα ("Iris with the beautiful sandals having slept with Zephyr with the golden hair bore the most powerful of all the gods" [i.e. the God of Love]), apud Plutarch *Mor.* [*Amatorius*] 765e. Cf. Parthenius [Lightfoot, fr. 2]). As is obvious, Iris does not appear as 'messenger' in all cases: e.g. Hom. *Il.* 5.353–354; Theoc. *Id.* 17.133–134: ἓν δὲ λέχος στόρνυσιν ἰαύειν Ζηνὶ καὶ Ἥρῃ / χεῖρας φοιβήσασα μύροις ἔτι παρθένος Ἶρις ("and single is the couch that Iris, virgin still, her hands made pure with perfumes, strews for the sleep of Zeus and Hera"; transl. Gow 1952). On Iris' virginity see Gow 1952 on 17.134; Lightfoot 1999, 141.

5 Peraki-Kyriakidou 2002; Peraki-Kyriakidou 2003; Tsitsibakou-Vasalos 2007.

6 Peraki-Kyriakidou 2003: the proper name as 'point of concentration'.

DOI 10.1515/9783110535150-005

With his insistence on etymologizing a proper name (explicitly or implicitly through verbs or epithets or even adverbs and phrases) the poet retains the reader's attention on that specific name by disclosing qualities and characteristics of it. Quite often the sum of different etymological approaches to a proper name seems to create a kind of sub-narrative. In this way, the proper name turns into a focal point,[7] around which the narrative is woven. As a consequence, etymologizing contributes considerably to the lengthening of the reading time of a passage.

Before looking into the *Iliad* (Iris does not appear in the *Odyssey*), let us read Iris' genealogy in Hesiod's *Theogony* 265 – 269:

> Θαύμας δ' Ὠκεανοῖο βαθυρρείταο θύγατρα
> ἠγάγετ' Ἠλέκτρην· ἡ δ' ὠκεῖαν τέκεν Ἶριν
> ἠυκόμους θ' Ἁρπυίας, Ἀελλώ τ' Ὠκυπέτην τε,
> αἵ ῥ' ἀνέμων πνοιῇσι καὶ οἰωνοῖς ἅμ' ἕπονται
> ὠκείῃς πτερύγεσσι· μεταχρόνιαι γὰρ ἴαλλον.

> And Thaumas wedded Electra, the daughter of deep-flowing Ocean, and she bore swift Iris and the Harpies with the beautiful hair, Aello (Storm) and Ocypete (Swift-flier) who on their swift wings keep pace with the blasts of the winds and the birds; for they were flying high up in the sky.

Iris is the daughter of Thaumas and Electra. As has been noted, her genealogy[8] with the significant names of her parents contributes to the perception of her visual aspect and this perhaps explains – according to the ancient scholia – the purpose of that genealogy: schol. on Hesiod's *Theog.* 266a2.2: Ἶρις θαύματός ἐστιν ἀξία ("Iris equals a miracle"). Further down, sight combines with speech (266a2.3 – 266a2.4): Ἶρις δέ ἐστιν ὁ προφορικὸς λόγος· ὁρῶντες γὰρ καὶ θαυμάζοντες προφερόμεθα λόγους ("Iris is the oral speech; for seeing and admiring we utter words").[9] However, in Hesiod, beyond Iris' visual qualities, her kinetic

7 See previous note.

8 *BNP:* "Her genealogy characterizes her: she herself is thought to be fast, and in Greek physics, the rainbow can produce winds. In mythological accounts, she is to a very great extent detached from her element, and has the function of a messenger of the gods (Hom. *Il.* 2.786, etc.) which she shares post-Homerically with Hermes in such a way that she is assigned in particular to Hera. Only Roman poetry sometimes playfully recalls the underlying natural element (Ov. *Met.* 11.589)".

9 In Plato, the philosophical perception of the genealogy of the goddess is prominent; *Tht.* 155d: Θεόδωρος γάρ, ὦ φίλε, φαίνεται οὐ κακῶς τοπάζειν περὶ τῆς φύσεώς σου. μάλα γὰρ φιλοσόφου τοῦτο τὸ πάθος, τὸ θαυμάζειν· οὐ γὰρ ἄλλη ἀρχὴ φιλοσοφίας ἢ αὕτη, καὶ ἔοικεν ὁ τὴν Ἶριν Θαύμαντος ἔκγονον φήσας οὐ κακῶς γενεαλογεῖν. ἀλλὰ πότερον μανθάνεις ἤδη δι' ὃ ταῦτα τοιαῦτ' ἐστὶν ἐξ ὧν τὸν Πρωταγόραν φαμὲν λέγειν, ἢ οὔπω.

characteristics are also vividly acknowledged. She is described as ὠκεῖα ("quick", "swift", 266), whereas her two sisters, Ἀελλώ ("Storm")[10] and Ὠκυπέτη ("Swift-flying"), the Harpies, with their swift wings (ὠκείῃς πτερύγεσσι) also share with Iris the characteristic of mobility and swiftness.[11]

Iris, therefore, with the aforementioned characteristics, seems also to allegorize movement in space and time. The ancient scholia, in fact, render the adjective μεταχρόνιαι of Hesiod as "movement in the sky":[12] (269.1–2) μεταχρόνιαι γὰρ ἴαλλον: ἀντὶ τοῦ ἔτρεχον, ἐπέτοντο· καὶ τὸν οὐρανὸν γὰρ χρόνον καλοῦσι ("they were flying high up in the sky: instead of *they were running*, *they were flying*; for they call the sky as *time*"). This scholiastic phrase is very interesting as it acknowledges the transmutation of time into space.

Iris in the *Iliad:* Reflection and Immediacy

In Homer, as in Hesiod, Iris' mobility is of major importance; however, her visual aspects are played down. In the *Iliad* her only visual attribute is the adjective χρυσόπτερος ("gold-winged") which appears twice (9.398, 11.185). In contrast, all the other descriptive attributes concern her movement and the conveyance of a message. The implicit etymologizing of the goddess' name in the archaic epic seems to play a crucial role as it enhances Iris' two major roles in the *Iliad,* that of the traveller and of the conveyor of a message.

As early as in Plato's *Cratylus,* her name – as we saw above – was etymologized from εἴρω = "to speak", "to say", an etymology related to her capacity as a

(Soc.): I see, my dear Theaetetus, that Theodorus had a true insight into your nature when he said that you were a philosopher, for wonder is the feeling of a philosopher, and philosophy begins in wonder. He was not a bad genealogist who said that Iris (the messenger of heaven) is the child of Thaumas (wonder). But do you begin to see what is the explanation of this perplexity on the hypothesis which we attribute to Protagoras?
(transl. by B. Jowett: http://classics.mit.edu/Plato/theatu.html)

10 "This may be regarded as a short form of Ἀελλόπους ...": West 1966 on Hes. *Theog.* 267. Iris' relation to the winds is blown to a whole scene at *Il.* 23.198–216, although her journey is not described at all.

11 West 1966 on Hes. *Theog.* 266.

12 Iris traverses all space; but she rarely visits Styx: παῦρα δὲ Θαύμαντος θυγάτηρ πόδας ὠκέα Ἶρις / ἀγγελίῃ πωλεῖται ἐπ' εὐρέα νῶτα θαλάσσης. / ὁππότ' ἔρις καὶ νεῖκος ἐν ἀθανάτοισιν ὄρηται ... ("only rarely does Thaumas' daughter, Iris, with the swift feet, come to her [i.e. Styx] with a message over the wide back of the sea. But when strife and quarrel arise among the immortal gods ...", Hes. *Theog.* 780–782 (cf. Ap. Rhod. *Argon.* 2.291–295 when she takes an oath on the waters of the Styx).

messenger. Parallel to this, her mobility and the impetus in her movement and swiftness obviously relate her name to the verb ὄρνυμαι (e. g. *Il.* 24.77: ὦρτο), and only obliquely to the air, ἀήρ,[13] the privileged area of Hera, as it appears in a wide variety of texts: e. g. Plato, *Cra.* 404c.1–3: ὁ νομοθέτης τὸν ἀέρα Ἥραν ὠνόμασεν.[14] As we saw in Hesiod, Iris is the sister of Ἀελλώ[15] and Ὠκυπέτη, while in the *Iliad* is ποδήνεμος ("wind-swift", *Il.* 5.353), and ἀελλόπος and ὠκεῖα.

Two examples would suffice, one from Book 2 and one from the last Book (24) of the *Iliad: Il.* 2.786–787: Τρωσὶν δ' ἄγγελος ἦλθε ποδήνεμος ὠκέα Ἶρις / πὰρ Διὸς αἰγιόχοιο σὺν ἀγγελίῃ ἀλεγεινῇ ("and swift and wind-footed Iris came to the Trojans as a messenger from Zeus who bears the aegis, with a grievous message"); and at 24.77: ὣς ἔφατ', ὦρτο δὲ Ἶρις ἀελλόπος ἀγγελέουσα ("so he spoke, and storm-footed Iris hasted bearing a message").

Nearly every word of the above passages has an implicit relation with the name of the goddess and her two basic qualities, that of mobility in space and the conveyance of speech. With the increase of the reading time allocated to Iris through her attributes, however, the reader is helped to realize and become aware of these functions of the goddess. Thus, her name becomes the metonymy of speech itself and its (quick) transference from one place (usually that of the gods) to another (that of humans or gods). The relation of her name with

---

13 e. g. Anaximenes, test. 7.25; Epicurus *Ep. ad Pyth.* 109.9–10; Isid. *Orig.* 13.10.1: *dicitur iris quasi aeris, id est quod per aera ad terras descendat* ("she is called Iris, as if of aer, because she comes down to earth through the air").

14 See also Eust. *Comment. ad Hom. Iliadem* 4.153.13: Ἰστέον δὲ ὅτι πᾶσαν Ἶριν Ἥρα πέμπει, ἤγουν ἀήρ ("it should be known that Hera on all occasions has as a messenger Iris, that is aer"); Olympiodorus, in Arist. *Mete. comment.* 63.25: Τότε γὰρ γίνεται ἶρις καὶ ἅλως, ὅταν ἀὴρ πυκνωθῇ ("when the air becomes dense, it is then that iris [rainbow] becomes round"). In addition, the texts point to another etymological relation, that with the verb ὁράω [see schol. in Hes. *Theog.* 266a2.3–266a2.4]; she is also related with ἔρις (since she is connected with Hera who is directly associated with ἔρις in the scholia): *EtM* 475.45: Σημαίνει καὶ ... ἔριν (cf. Hsch. *Lexicon* epsilon 5821: ἔριδας· τὰς ἐν οὐρανῷ ἴριδας. Ἀττικῶς). From a fragment of Alcaeus Iris' relation with love [ἔρως] becomes evident: see above, n. 4 (*EtM* 470*.268–269: τούτου χάριν καὶ ὁ Ἀλκμαῖος (*sic*) Ζεφύρου καὶ Ἴριδος τὸν Ἔρωτά φησι. See also *EtM* 475.37–475.42: Ἶρις: σημαίνει τὴν ἄγγελον τῶν θεῶν· καὶ γίνεται παρὰ τὸ εἴρειν καὶ λέγειν τὰ μέλλοντα· σημαίνει καὶ τὴν νεφελώδη ζώνην, τὸ τόξον τὸ ἐν τῷ οὐρανῷ φαινόμενον· καὶ αὐτὸ ἀπὸ τοῦ εἴρω· παρὰ τὸ σημαίνειν καὶ προλέγειν τὸ μέλλον ἔσεσθαι, χειμῶνα ἢ εὐδίαν ("Iris signifies the messenger of the gods; and her name derives from the verb εἴρειν, to foretell the future; it also means a strip of cloud, the arch that appears in the sky [i. e. the rainbow]; the latter also derives from the verb εἴρω as it is related to the verb 'to signify' and 'to predict' either the bad or the good weather"). See also above, n. 3.

15 See above, n. 10.

aer in Homer seems to point mainly to her mobility and swiftness and not the space of the *aer* itself.

Further to the extension of narrative and reading[16] time through etymologizing, in the *Iliad* another – different – technique is also attested. One case, which is not unique in the epic (another example is at 24.146 – 158, 175 – 187),[17] is characteristic. It is from the long episode of Book 8 where Zeus (8.397), enraged with Hera and Athena who insist on helping the Greeks, tries to prevent them. He calls Iris (398) and recites to her the message she has to convey (399 – 408). She then travels from Mount Ida to Olympus, finds the goddesses, gives them the message of Zeus and returns to Ida (8.397 – 425):

Ζεὺς δὲ πατὴρ Ἴδηθεν ἐπεὶ ἴδε χώσατ' ἄρ' αἰνῶς,
Ἶριν δ' ὄτρυνε χρυσόπτερον ἀγγελέουσαν·
"βάσκ' ἴθι Ἶρι ταχεῖα, πάλιν τρέπε μηδ' ἔα ἄντην
ἔρχεσθ'· οὐ γὰρ καλὰ συνοισόμεθα πτόλεμον δέ.
ὧδε γὰρ ἐξερέω, τὸ δὲ καὶ τετελεσμένον ἔσται· {415}
γυιώσω μέν σφωϊν ὑφ' ἅρμασιν ὠκέας ἵππους, [416]
αὐτὰς δ' ἐκ δίφρου βαλέω κατά θ' ἅρματα ἄξω· [417]
οὐδέ κεν ἐς δεκάτους περιτελλομένους ἐνιαυτοὺς [418]
ἕλκε' ἀπαλθήσεσθον, ἅ κεν μάρπτῃσι κεραυνός [419]
ὄφρα ἰδῇ γλαυκῶπις ὅτ' ἂν ᾧ πατρὶ μάχηται. [420]
Ἥρῃ δ' οὔ τι τόσον νεμεσίζομαι οὐδὲ χολοῦμαι· [421]
αἰεὶ γάρ μοι ἔωθεν ἐνικλᾶν ὅττί κεν εἴπω." [422]
Ὣς ἔφατ', ὦρτο δὲ Ἶρις ἀελλόπος ἀγγελέουσα,
βῆ δ' ἐξ Ἰδαίων ὀρέων ἐς μακρὸν Ὄλυμπον.
πρώτῃσιν δὲ πύλῃσι πολυπτύχου Οὐλύμποιο
ἀντομένη κατέρυκε, Διὸς δέ σφ' ἔννεπε μῦθον·
"πῇ μέματον; τί σφῶϊν ἐνὶ φρεσὶ μαίνεται ἦτορ;
οὐκ ἐάᾳ Κρονίδης ἐπαμυνέμεν Ἀργείοισιν.
ὧδε γὰρ ἠπείλησε Κρόνου πάϊς, ᾗ τελέει περ,
γυιώσειν μὲν σφῶϊν ὑφ' ἅρμασιν ὠκέας ἵππους,
αὐτὰς δ' ἐκ δίφρου βαλέειν κατά θ' ἅρματα ἄξειν·
οὐδέ κεν ἐς δεκάτους περιτελλομένους ἐνιαυτοὺς
ἕλκε' ἀπαλθήσεσθον, ἅ κεν μάρπτῃσι κεραυνός·
ὄφρα ἰδῇς γλαυκῶπι ὅτ' ἂν σῷ πατρὶ μάχηαι.
Ἥρῃ δ' οὔ τι τόσον νεμεσίζεται οὐδὲ χολοῦται·
αἰεὶ γάρ οἱ ἔωθεν ἐνικλᾶν ὅττι κεν εἴπῃ·
ἀλλὰ σύ γ' αἰνοτάτη κύον ἀδεὲς εἰ ἐτεόν γε

16 In studying the Homeric epic in this paper we try to see what was its impact in later periods, in Virgil's *Aeneid* and in Ovidian *Metamorphoses*. The orality of the epic is not within the interests of this paper.
17 Also at: *Il.* 11.187 – 194, 202 – 209; 15.160 – 162, 176 – 178.

τολμήσεις Διὸς ἄντα πελώριον ἔγχος ἀεῖραι."
Ἡ μὲν ἄρ᾽ ὣς εἰποῦσ᾽ ἀπέβη πόδας ὠκέα Ἶρις.

But when father Zeus saw them from Ida he waxed wondrous wrath, and sent forth golden-winged Iris to bear a message: "Up, go, swift Iris; turn them back and suffer them not to come face to face with me, seeing it will be in no happy wise that we shall join in combat. For thus will I speak and verily this thing shall be brought to pass. I will maim their swift horses beneath the chariot, and themselves will I hurl from out the car, and will break in pieces the chariot; nor in the space of ten circling years shall they heal them of the wounds wherewith the thunderbolt shall smite them; that she of the flashing eyes may know what it is to strive against her own father. But against Hera have I not so great indignation nor wrath, seeing she is ever wont to thwart me in whatsoe'er I have decreed." So spake he, and storm-footed Iris hasted to bear his message, and went forth from the mountains of Ida to high Olympus. And even at the entering-in of the gate of many-folded Olympus she met them and stayed them, and declared to them the saying of Zeus: "Whither are ye twain hastening? Why is it that the hearts are mad within your breasts? The son of Cronos suffereth not that ye give succour to the Argives. For on this wise he threateneth, even as he will bring it to pass: he will maim your swift horses beneath your chariot, and yourselves will he hurl from out the car, and will break in pieces the chariot; nor in the space of ten circling years shall ye heal you of the wounds wherewith the thunderbolt shall smite you; that thou mayest know, thou of the flashing eyes, what it is to strive against thine own father. But against Hera hath he not so great indignation nor wrath, seeing she is ever wont to thwart him in whatsoe'er he hath decreed. But most dread art thou, thou bold and shameless thing, if in good sooth thou wilt dare to raise thy mighty spear against Zeus". When she had thus spoken swift-footed Iris departed. (transl. A.T. Murray, Loeb)

From the above 28 lines, lines 399–408 are the words of Zeus with the message that Iris has to carry. Lines 409–412 refer to the journey of the goddess and her meeting with Hera and Athena. Then follows as a textual mirroring[18] (or as an echoing) the repetition of the message. Iris repeats the message of Zeus almost *verbatim*, making only the necessary changes for the *oratio obliqua*. As a mirror reflection (or an echo) the message is doubled; this results in the increase of the reading and narrative time and the emphasis thus given to it. Besides, this message is not common: it is the message which equates Greeks and Trojans and tries to bring an equilibrium between the opposing forces.

If we accept that a myth usually develops around some concepts – the pillars of the narrative – we shall not be hampered in understanding, as we have said above, that in the specific textual context Iris is nothing more and nothing less than a personification of the very word of the god who sees everything and everywhere and whose will and word reach far: this mythological example precise-

18 Textual mirroring is not a rare phenomenon in antiquity: see Kyriakidis 2007, mainly 52–66.

ly justifies the epithet εὐρύοπα, ὁ ("far-sounding" or / and "wide-eyed") for Zeus.[19]

In lines 409 – 412 of the above passage (between the message and its textual reflection) we have the report of Iris' departure from Ida, her arrival at Olympus and her meeting with the goddesses. However, the journey itself through the air is not described at all and the space in between Ida and Olympus remains as if non-existent for the reader of the *Iliad*.[20] In this way Iris, the word of the god that is, brings into contact two different places and parts and becomes an agent of transition, functioning as a unifying character, similar to Fama's unifying character.[21] At the same time, this lack of description may correspond to the needs of a formulaic brevity and – if I am allowed to use a phrase from Fitter – it can "exert influence on narrative action".[22]

On another occasion, Iris' journey from Mount Ida to Troy occupies larger narrative space but again the space itself in the air between the two different places is not described. This is when Iris bears a message of Zeus to Poseidon (*Il.* 15.168 – 173):

> ὣς ἔφατ', οὐδ' ἀπίθησε ποδήνεμος ὠκέα Ἶρις,
> βῆ δὲ κατ' Ἰδαίων ὀρέων εἰς Ἴλιον ἱρήν.
> ὡς δ' ὅτ' ἂν ἐκ νεφέων πτῆται νιφὰς ἠὲ χάλαζα
> ψυχρὴ ὑπὸ ῥιπῆς αἰθρηγενέος Βορέαο,
> ὣς κραιπνῶς μεμαυῖα διέπτατο ὠκέα Ἶρις,
> ἀγχοῦ δ' ἱσταμένη προσέφη κλυτὸν ἐννοσίγαιον.

So he spoke, and wind-footed, swift Iris obeyed, and went down from the mountains of Ida to sacred Ilios. And as when from the clouds there flies snow or hail, driven by the blast of the North Wind that is born in the bright heaven, so sped in eagerness swift Iris; and drawing near she spoke to the glorious Shaker of Earth.

---

19 Cf. Apollonius Soph. *Lex. Homericum* 79.19 – 21: εὐρύοπα, ἐπίθετον Διός, ἤτοι τὸν μεγάλως ἐφορῶντα, ἢ τὸν μεγάλους ἤχους καὶ ψόφους ἀποτελοῦντα, ἢ τὸν μεγαλόφθαλμον. ἐπίθετον Διός ("εὐρύοπα is an attribute of Zeus, meaning either he who oversees far and wide or he who causes strong sounds and noises, or the wide-eyed").

20 See Lateiner's observation (2005, 413) for the *Iliad:* "Homer compresses space into claustrophobic dimensions". In *Hom. Hymn to Delian Apollo* 107 – 108 once more there is no description of the space she traverses, as it carries no narrative value for the poet: αὐτὰρ ἐπεὶ τό γ' ἄκουσε ποδήνεμος ὠκέα Ἶρις / βῆ ῥα θέειν, ταχέως δὲ διήνυσε πᾶν τὸ μεσηγύ ("when swift Iris, fleet of foot as the wind, had heard all this, she set to run; and quickly finishing all the distance [she came …]"; transl. H. Evelyn-White, Loeb). The dactyls of these lines represent Iris' fast journey; cf., however, *Il.* 24.78 – 79 where there is some kind of description of the space of the goddess' journey; see also 24.95 – 97 on her journey back to the sky together with Thetis.

21 Hardie 2012, 95, 171, 481; Clément-Tarantino 2016.

22 Fitter 1995, 49.

The simile seems to refer to the journey of the goddess; however, in essence, as Janko notes, it rather alludes to the "chilling threat" of Zeus.[23]

With the few instances we have seen above, the reader understands that, first, Homer has very little to say about Iris' rich visual aspect (the appearance, that is, of the rainbow, as it is commonly perceived[24]); secondly, he also avoids describing the goddesses' journey in space and time; his attention is clearly focused on the message she has to convey by doubling on occasion the textual space it occupies (and the reading time it requires). This kind of mirroring is the textual transmutation of the natural phenomenon of the rainbow's reflection, which was already observed in antiquity: e. g. Diog. Laert. *Fragmenta logica et physica* 692.1–2: ἶριν δὲ εἶναι αὐγὰς ἀφ' ὑγρῶν νεφῶν ἀνακεκλασμένας ("the rainbow is the light of the sun reflected upon the humid clouds").

Iris in Apollonius' *Argonautica:* Preparing the Roman Reception

Centuries later, Apollonius' *Argonautica* would be the Hellenistic epic with many similarities but also with significant differences from the Homeric *Iliad*. Iris is still a 'Homeric' figure; however, some details of characterization seem to change.

The goddess appears in Books 2 and 4. In Book 2 Iris watches the sons of Boreas, Zetes and Calais chasing the Harpies and descends from heaven (*Arg.* 2.278–294):

ὡς δ' ὅτ' ἐνὶ κνημοῖσι κύνες δεδαημένοι ἄγρης
ἢ αἶγας κεραοὺς ἠὲ πρόκας ἰχνεύοντες
θείωσιν, τυτθὸν δὲ τιταινόμενοι μετόπισθεν
ἄκρης ἐν γενύεσσι μάτην ἀράβησαν ὀδόντας·
ὣς Ζήτης Κάλαΐς τε μάλα σχεδὸν ἀίσσοντες
τάων ἀκροτάτῃσιν ἐπέχραον ἤλιθα χερσίν.
καί νύ κε δή σφ' ἀέκητι θεῶν διεδηλήσαντο
πολλὸν ἑκὰς νήσοισιν ἔπι Πλωτῇσι κιχόντες,
εἰ μὴ ἄρ' ὠκέα Ἶρις ἴδεν, κατὰ δ' αἰθέρος ἆλτο
οὐρανόθεν, καὶ τοῖα παραιφαμένη κατέρυκεν·
"Οὐ θέμις, ὦ υἱεῖς Βορέω, ξιφέεσσιν ἐλάσσαι

23 Janko 1992 on 15.170–171.

24 See *Il.* 11.27; in another case Iris is substituted by Athena who, enveloped in a lurid cloud (πορφυρέῃ νεφέλῃ), is compared to the rainbow (πορφυρέην ἶριν, *Il.* 17.547/551) and comes down to urge on the Danaans (ὀρνύμεναι Δαναούς, 546). The etymological relation between ὀρνύμεναι and ἶρις is obvious again (see also above, n. 12).

> Ἀρπυίας, μεγάλοιο Διὸς κύνας· ὅρκια δ' αὐτή
> δώσω ἐγὼν ὡς οὔ οἱ ἔτι χρίμψουσιν ἰοῦσαι
> Ὣς φαμένη λοιβὴν Στυγὸς ὤμοσεν, ἥ τε θεοῖσιν
> ῥιγίστη πάντεσσιν ὀπιδνοτάτη τε τέτυκται,
> μὴ μὲν Ἀγηνορίδαο δόμοις ἔτι τάσδε πελάσσαι
> εἰσαῦτις Φινῆος, ἐπεὶ καὶ μόρσιμον ἦεν.

> And as when upon the mountain-side, hounds, cunning in the chase, run in the track of horned goats or deer, and as they stain a little behind gnash their teeth upon the edge of their jaws in vain; so Zetes and Calais rushing very near just grazed the Harpies in vain with their finger-tips. And assuredly they would have torn them to pieces, despite heaven's will, when they had overtaken them far off at the Floating Islands, had not swift Iris seen them and leapt down from the sky from heaven above, and checked them with these words: "It is not lawful, O sons of Boreas, to strike with your swords the Harpies, the hounds of mighty Zeus; but I myself will give you a pledge, that hereafter they shall not draw near to Phineus." With these words she took an oath by the waters of Styx, which to all the gods is most dread and most awful, that the Harpies would never thereafter again approach the home of Phineus, son of Agenor, for so it was fated. (transl. R.C. Seaton, Loeb)

In this instance, it is not Zeus (or some other god) who sends her but she comes of her own accord – representing of course Zeus' will.[25] As in Homer, the journey from heaven is not described here either. Once again, the reading and the narrative time of the journey is contracted (κατὰ δ' αἰθέρος ἅλτο / οὐρανόθεν, 286–287); a few lines below the return trip of the goddess will be briefly mentioned but with a vivid difference: whereas in the case of the descent the poet relies only on the attribute ὠκέα (286), in the case of ascent he refers to Iris' speedy wings, again alluding to the speedy journey (ἡ δ' ἀνόρουσεν Οὔλυμπόνδε θοῇσι μεταχρονίη πτερύγεσσιν, *Arg.* 2.299–300). In this phrase the Hesiodean epithet μεταχρόνιος has returned. It is the attribute which – as we saw above – renders in time what actually takes place in space.

Iris herself in this instance is not carrying any divine message and consequently her words are not a repetition. It is she who promises to the Boreads that she will not allow the Harpies[26] to come close to Phineus. The Homeric technique of repetition has been transposed: it is Iris herself who reiterates the content of her speech as an oath (293–294) without, however, the phenomenon of

---

25 See Hunter's remark (1993, 78): "Apollonius greatly reduces the prominence of the divine. Gone are the easy appearances of gods to mortals and the conversations between them ... The Argonauts' only direct contact is with minor divinities ... Iris' intervention to prevent the Boreads from killing the Harpies forms a partial exception."

26 On the relation of Iris' to the Harpies see Hunter 1993, 81–82.

(textual) mirroring. Whereas the context of the two instances is the same, the wording is totally disparate.

Similarly, in Book 4 of the *Argonautica*, Iris is ordered by Hera to watch over the Argonauts as they leave Circe (4.753–756);[27] Iris obeys and then Hera bids her to carry a message to 1) Thetis, 2) Hephaestus, and 3) Aeolus asking him to let the west wind blow till the heroes reach the island of Alcinous.[28] Iris' description once again is limited to the element of her speedy wings (*Arg.* 4.757–758): Ἶρι φίλη, νῦν, εἴ ποτ' ἐμὰς ἐτέλεσσας ἐφετμάς, / εἰ δ' ἄγε λαιψηρῇσι μετοιχομένη πτερύγεσσιν ("dear Iris, now come, if ever thou hast fulfilled my bidding, hie thee away on light pinions"; transl. R.C. Seaton, Loeb).

A few lines further down, the emphasis on her wings is similar; again, as in Book 2, the journey in space is not portrayed (*Arg.* 4.770–772):[29] Ὣς ἔφατ'. αὐτίκα δ' Ἶρις ἀπ' Οὐλύμποιο θοροῦσα / τέμνε, τανυσσαμένη κοῦφα πτερά· δῦ δ' ἐνὶ πόντῳ / Αἰγαίῳ ("so she spoke, and straightway Iris leapt down from Olympus and cleft her way, with light wings outspread. And she plunged in the Aegean Sea"; transl. R.C. Seaton, Loeb).

The end of her mission receives a mere mention;[30] the return to Olympus is not contained at all and the reference to her repose signifies the end of the scene (*Arg.* 4.778–779): ὄφρα δὲ καὶ τῷ / ἀγγελίην φαμένη θοὰ γούνατα παῦεν ὁδοῖο ("and when she had given her message to him also and rested her swift knees from her course"; transl. R.C. Seaton, Loeb).

In the *Argonautica* Iris' journey again is not described; the reference is only to her quick movement in space; in one case, however, there is a more pictorial rendering (τανυσσαμένη κοῦφα πτερά, 4.771). Beyond this, there is no visual trait of the goddess. Further to this, the textual and narrative space which could be ensured by the *verbatim* repetition of the message in Homer, has no room in the Hellenistic epic.

Iris in Virgil's *Aeneid:* A Shift of Interest

As we turn to the Roman epic, things tend to change. In the *Aeneid* there are three major references to Iris and in these cases – contrary to the *Iliad* – she

27 Hunter 1993, 96 sees as intertext Callimachus' *Hymn to Delos* 4.61–69.

28 On Iris' manifold mission, see Hunter 1993, 96 who refers the reader to *Il.* 24.7476, where the message has two recipients, Thetis and Priam.

29 Hunter 1993, 100.

30 Livrea 1973 on 4.770.

functions exclusively as Juno's envoy (in *Iliad* 8 Juno was the recipient of the message).[31] In *Aeneid* 4 Iris comes to release the spirit of dying Dido; in Book 5 in serving Juno's intentions, she comes down among the Trojan women, transformed to old Beroe[32] and incites them to burn the ship. The third instance is in Book 9 when she visits Turnus as Juno's envoy in order to guide his thoughts.

Juno's words, which Iris has to transmit, are never repeated *verbatim*. In all three instances, however, from the words of Iris, the scheme of Juno is disclosed. The Homeric technique of mirroring or of echoing does not appear in the *Aeneid*. Only in Book 4 do we have – at least in reported speech – the words of Juno (694–699). But even there, between these verses and the words of Iris, there is only one tying element: the verb *resolveret* (4.695) and the verb *solvo*, the last word of Iris (703). The scene is completed with the death of Dido without any reference to the return-trip of the goddess to the sky.

In the Roman epic there is a clear shift of the poetic interest to the goddess' visual characterization and not to her speediness. At *Aen.* 9.5 the *Thaumantias* goddess is described *roseo ... ore*. In Book 4 the colours come from her saffron-hued wings[33] (*croceis ... pennis*, 700) and in her journey she creates a thousand colours (*trahens*, 701). The sky, therefore, is illuminated with coloured stripes, a description directly linked to the nature of the rainbow (*Aen.* 4.693–705):

> *Tum Iuno omnipotens longum miserata dolorem*
> *difficilisque obitus Irim demisit Olympo*
> *quae luctantem animam nexosque resolveret artus.*
> *nam quia nec fato merita nec morte peribat,*
> *sed misera ante diem subitoque accensa furore,*
> *nondum illi flauum Proserpina vertice crinem*
> *abstulerat Stygioque caput damnaverat Orco.*
> *ergo Iris croceis per caelum roscida pennis*
> *mille trahens varios aduerso sole colores*
> *devolat et supra caput astitit. "hunc ego Diti*
> *sacrum iussa fero teque isto corpore solvo":*
> *sic ait et dextra crinem secat, omnis et una*
> *dilapsus calor atque in ventos vita recessit.*

But Juno who has all power took pity on the long anguish of her difficult death, and sent Iris down from Olympus to release the wrestling spirit from the twined limbs. For since she

---

31 Laird 1999, 264: "There are five episodes in the poem in which a messenger (Iris or Mercury) is sent down to earth under the specific direction of Jupiter or Juno" (Iris: 4.694–595, 5.606 = 9.2)

32 Laird 1999, 172 n. 53 (Iris acts a part of the scene).

33 An imagery reminiscent of the Homeric Dawn who spreads over the earth with her saffron-robe: *Il.* 8.1 (~ 24.695): Ἠὼς μὲν κροκόπεπλος ἐκίδνατο πᾶσαν ἐπ' αἶαν; cf. also 19.1 and 23.227.

> perished neither by destiny nor by a death deserved, but tragically, before her day, in the mad heat of a sudden passion, Proserpine had not yet taken a golden lock from her head, to assign her life to Stygian Orcus. So therefore Iris, saffron-winged, sparkling like dew and trailing a thousand colours as she caught the light of the sun, flew down across the sky. She hovered over Dido's head: "By command I take this lock as an offering to Pluto; and I release you from the body which was yours." Speaking so, she held out a hand and cut the lock. At once, all the warmth fell away, and the life passed into the moving air (transl. D. West).

Here, instead of the textual mirroring (and the echoing effect) of the Homeric message, we have the description of the reflection of light (ἀνάκλασιν ... φωτός, "reflection of light" – the Aristotelian term for ἶρις [Arist. *Analyticorum posteriorum paraphrasis* 5.1.60.3 – 5.1.60.9]), which renders a thousand colours. From the Roman text some of the Homeric attributes, like ἀγγελέουσα, ἀελλόπος, or ποδήνεμος and ὠκεῖα, are missing, elements which stressed the role of Iris as a messenger and swift-traveller.

This time, the poet keeps the reader's attention with a couple of lines (700 – 701) on the journey itself through the sky. Although there is again no description of any geographical location over which Iris flies, the emphasis on her course through the sky attributes to the journey a 'transcendental' character and in this way a superhuman nature (or, better, an extra-human nature) to the message Iris conveys.

We have seen that the name of Iris is related to the air.[34] But here there is a meaningful shift in this relation: While in Homer this relation had nothing to do with the space Iris travels through (as the main care of the poet was to render the goddess' speed), in the case of the Roman epic it has to do with her substance as a whole, as well as with that of Hera/Juno who sends her. Since the sky in its broader sense includes the air [see for example, Eust. *Comm. ad Hom. Iliad.* 1.170.16 οὐρανὸς δὲ ὁ ἀήρ], then we note that there is a rather allegorical shift in the etymological relation between *Iris* and *aer*. Iris travels in the sky and in the air, the space of Hera / Juno. Juno and Iris seem to share the same space or be the space itself. Besides, at 9.803 Iris is characterized as *aeria* (*aeriam caelo nam Juppiter Irim / demisit germanae haud mollia iussa ferentem*, Jupiter had sent from the sky down to earth the 'aerial' Iris, bringing stern orders to the sister 9.803 – 804). The space now matters. The interest turns not only to the purpose of the journey, the message, that is, the goddess has to carry, or the act she has to perform, but also to the travelling space.

34 See above, n. 14.

The space of the air, with the humidity reflecting the light of the sun, takes thousands of colours and in this way the space of the journey becomes visible acquiring its own functional existence.[35] The emphasis on the colours adorning it, fully justify the attribute *Thaumantias*, Virgil uses at *Aen.* 9.5 for Iris.

The above shift of interest in the *Aeneid* gives Iris a different function than the one she had in Homer. To begin with, contrary to the *Iliad*, her mission is directed to the earth and only to the humans, while the authority of the message is never questioned. In order to achieve the resoluteness and the irreversibility of the divine decision, Iris must go across the space of the sky. The space with the colours of the rainbow now is lit and the reader with the eyes of his/her imagination sees the rainbow painting the sky. This imagery gives substance to the journey which gives the impression of being long; the distance separating humans from the divine seems great as is the sky. The vastness of the air-space and the sense of the duration of the journey are the two conditions which load the mission with extra weight. Indeed, what Iris has to say in the epic is always crucial. Instances like the ones at *Iliad* 15 where Iris is sent by Zeus to Poseidon (15.158–161) with the order to stop the war and the latter to dispute the power of the former to dictate what he has to do (15.185–199), are not found in the Virgilian scenes of Iris. For Virgil, it is the course through the heavenly space which renders each of her missions decisive. The Homeric immediacy of the message with what it entails, is non-existent.

Iris in Ovid' *Metamorphoses:* Poetics and Metapoetics

In the *Metamorphoses* the term (messenger, 1.270) is found already in the first time the poet employs the name of the goddess, although in that case she carries no message. Ovid will also use the epithet *Thaumantias* three times (4.480, 11.647, 14.485); it is the adjective which Virgil had used for Iris at *Aen.* 9.5, thus pointing to her visual aspects. Two are the other major instances of Iris' appearance: in Book 11 when Juno sends her to the kingdom of Sleep to convince Ceyx to disclose through a dream to his wife Alcyone that he is dead, and at the end of Book 14 when she is sent by Juno to assist Hersilia to see her dead husband, Romulus.

35 The notion of space in the *Aeneid* is of greater importance than in the *Iliad* (see above, n. 20). In this respect the Virgilian epic is perhaps charged with more of the Odyssean element; see Lateiner 2005, 413: "In the Odyssey, contrariwise, twenty years of time is often compressed and space often expanded".

As in Virgil, the Iris of the *Metamorphoses* is full of colours again (*Met.* 11.589 – 591):

induitur velamina mille colorum
Iris et arcuato caelum curvamine signans
tecta petit iussi sub nube latentia regis.

and Iris put on her cloak of a thousand colours and, trailing across the sky in a rainbow curve, she sought – as ordered – the cloud-concealed halls of the king of sleep. (transl. F.J. Miller, slightly adapted, Loeb)

The colours, however, in the *Metamorphoses* are not her own, as in the *Aeneid*; they are of the *velamina* ("garments", 11.589) and in a few lines below of her *vestis* ("clothing" / "garment", 11.617). She is *varios induta colores* ("clothed in robes of many colours", *Met.* 1.270 – 271 / *induitur, Met.* 11.589). And when she enters the kingdom of Sleep: *vestis fulgore reluxit / sacra domus* ("the sacred house glowed with the brightness of her garments"; *Met.* 11.617– 618).

Ovid blends the Homeric with the Virgilian features of the Iris-imagery: for Juno, Iris is not simply her *nuntia*, but her *fidissima nuntia vocis* (my most faithful messenger of my word, 11.585).[36] This means that when she dictates to her the message at lines 586 – 588 Iris is expected to repeat it as it is. However, a proper repetition and a proper textual mirroring never take place. In this way, lines 627– 629 only partly justify the phrase *fidissima nuntia vocis* (11.585).[37] Nevertheless, we do have a repetition – of a sort – of the message which Juno herself dictates:[38]

In these lines there is a crucial difference from both Homer and Virgil: The message is not delivered to its final recipient. Iris passes it to Somnus who passes it to Morpheus to pass it to Ceyx. A. Laird realizes that "the complicated structure of this messenger scene still bears on the nature of epic communication" and considers that Iris "herself ... embodies that epic form".[39] A major issue, therefore, is at stake: if we accept that the Ovidian epic is fraught with countless epic reminiscences and reformations of the past literature, then Iris, a minor epic character, seems to assume the major role of an epic voice[40] which will be reproduced and adapted perpetually. In other words, as a messenger of the divine and as a poetic voice she will be developed over and over again, as the poet had de-

36 Laird 2003, 165.
37 Laird (1999, 282) calls attention to the similar way Jupiter earlier calls Mercury: *fide minister* (2.837).
38 Laird 1999, 282: "After some preliminary liturgical courtesy, the essence of Iris' delivery is indeed very faithful to Juno's words." See also Laird 1999, 283.
39 Laird 2003, 164; see also Laird 1999, 283.
40 Laird 1999, 284.

clared elsewhere that his epic would always keep growing in the future (*crescens*, *Tristia* 1.7.22).[41]

As we have said, Iris' colours in the Ovidian *Metamorphoses* are not her own. The goddess, the metonymy of speech, travels 'invested' with colours in the sky, which could also allude to 'variations of speech' or 'kinds of style'. As we know, both words *color* and the verb *induo* are related to speech and oration.[42] What we have here is the visualization of the original word which is 'invested' with countless different influences affecting in its course the final form. In this description it is not only the word of the poet that is affected by external influences [*induta*]; more than this, it is the poet himself who also leaves his trace in the sky as this *caelum ... signans* (11.590) clearly suggests: *Iris et arcuato caelum curvamine signans* ("Iris ... trailing across the sky in a rainbow curve"; transl. F.J. Miller, Loeb).

The metapoetic reading is facilitated greatly by the above line: The word *signans*[43] may function in a way as a substitute for the Virgilian *trahens* (*Aen.* 4.701); it plays a much more significant role, however, as it clearly shows the impact of the word on its course. The space of the journey becomes the space for the reception and the transmutation of the word.

When Iris is sent by Juno to Hersilia in Book 14, her duty is to encourage her to stop crying for Romulus' loss and to lead her to *collis Quirinalis* (14.836); in this way the deification of Romulus is announced.[44] The message of Juno will not be repeated by Iris as would have been repeated in Homer. Ovid, however, acknowledges the off-stage repetition of the message: *Hersilien iussis conpellat vocibus Iris* (Iris addressed Hersilia with the words which she was ordered [to convey]). The poet of the *Metamorphoses* does not describe the goddess herself when travelling in the sky as Virgil did in Book 4 of the *Aeneid*. It is the space and the *arcus* of the rainbow which are described as in an ekphrasis: *in terram pictos delapsa per arcus* (coming down to earth through the painted bows, 14.838).[45] Art and metapoetics are met in order to point to the Ovidian course in poetry and art.

41 Kyriakidis 2013, 354, 359–362.

42 Cf. for *color:* Cic. *De Orat.* 3.96.1: *ornatur igitur oratio genere primum et quasi colore quodam et suco suo* ("well, then, the embellishment of oratory is achieved in the first place by general style and by a sort of inherent colour and flavour"; transl. H. Racham, Loeb); and for *induere:* Quint. *Inst.* 4.1.28: *liceat ... fictam orationem induere personis* ("we may ... place fictitious speeches in the mouths of our characters"; transl. H.E. Butler, Loeb).

43 See Reed 2013, on 6.63–65; Hardie 2015, on 14.829–851, 830.

44 Laird 1999, 286–287.

45 A similar vocabulary is used at *Pont.* 4.4.11–18, where Fama is described in a similar fashion (*lapsa*), as 'flying through the air': Michalopoulos 2016.

Bibliography

Boys-Stones, G.R. (ed.) 2003. *Metaphor, Allegory and the Classical Tradition.* Oxford.
Clément-Tarantino, S. 2016. Wanderings of Fama and 'Fame's Narratives' in the Aeneid. In Kyriakidis 2016, 55–70.
Ferri, R. (ed.) 2011. *The Latin of Roman Lexicography.* Pisa and Rome.
Fitter, C. 1995. *Poetry, Space, Landscape. Toward a New Theory.* Cambridge.
Gow, A.S.F. 1952. *Theocritus, Edited with a Translation and Commentary*, 2 vols. Cambridge.
Hardie, P. 2012. *Rumour and Renown. Representations of Fama in Western Literature.* Cambridge.
Hardie, P. 2015. *Ovidio Metamorfosi, vol. VI, libri XIII–XV.* Rome.
Hunter, R. 1993. *The Argonautica of Apollonius. Literary Studies.* Cambridge.
Janko, R. 1992. *The Iliad: A Commentary*, vol. iv: Books 13–16. Cambridge.
Kyriakidis, S. 2007. *Catalogues of Proper Names in Latin Epic Poetry: Lucretius, Virgil, Ovid* (Pierides I). Newcastle upon Tyne.
Kyriakidis, S. 2013. The Poet's Afterlife: Ovid between Epic and Elegy. In Papanghelis, Harrison and Frangoulidis 2013, 351–366.
Kyriakidis, S. (ed.) 2016. *Libera Fama. An Endless Journey* (Pierides VI). Newcastle upon Tyne.
Laird, A. 1999. *Powers of Expression, Expressions of Power. Speech Presentation and Latin Literature.* Oxford.
Laird, A. 2003. Figures of Allegory from Homer to Latin Epic. In Boys-Stones 2013, 151–188.
Lateiner, C. 2005. Proxemic and Chronemic in Homeric Epic: Time and Space in Heroic Social Interaction. *CW* 98: 413–421.
Lightfoot, J.L. 1999. *Parthenius of Nicaea. Extant Works Edited with Introduction and Commentary.* Oxford.
Livrea, E. 1973. *Apollonii Rhodii Argonauticon Liber IV.* Florence.
Maltby, A. 1991. *A Lexicon of Ancient Latin Etymologies.* Leeds.
Maltby, A. 2011. Servius on Stylistic Register in his Virgil Commentaries. In Ferri 2011, 63–73.
Michalopoulos, A. 2016. Famaque cum domino fugit ab urbe suo: Aspects of Fama in Ovid's Exile Poetry. In Kyriakidis 2016, 94–110.
Nifadopoulos, C. (ed.) 2003. *Etymologia. Studies in Ancient Etymology* (Proceedings of the Cambridge Conference on Ancient Etymology, 25–27 September 2000). Münster.
Papanghelis, T., S. Harrison and S. Frangoulidis (eds.) 2013. *Generic Interfaces in Latin Literature: Encounters, Interactions and Transformations* (Trends in Classics 20). Berlin and Boston.
Peraki-Kyriakidou, H. 2002. Aspects of Ancient Etymologizing. *CQ* 52: 478–493.
Peraki-Kyriakidou, H. 2003. Homer's Etymologising in the Aeneid. Simile and the Point of Concentration. In Nifadopoulos 2003, 478–493.
Reed, J.D. 2013. *Ovidio. Metamorfosi*, volume v. libri X-XII (transl. from English by A. Barchiesi). Rome.
Syson, A. 2013. *Fama and Fiction in Vergil's Aeneid.* Ohlo.
Tsitsibakou-Vasalos, E. 2007. *Ancient Poetic Etymology: The Pelopids: Fathers and Sons* (Palingenesia 89). Munich.
West, M.L. 1966. *Hesiod. Theogony, with Prolegomena and Commentary.* Oxford.

Stratis Kyriakidis*

The Patronymics *Pelides* and *Aenides*: Past, Present and Future in Homeric and Virgilian Genealogical Catalogues

The use of patronymics is particularly common in the Homeric epics and they appear in every part of the narrative.[1] Here I shall mainly confine myself to examining the use of patronymics in catalogues of proper names and their frame. The reason for my focusing on this issue is that – in my view – the presence of a patronymic in a catalogue of proper names, where it 'claims' in a way its own space, acquires a particular spatial and temporal function compared to when it is found to substitute for, or accompany, a proper name in other parts of the narrative.

As a catalogue of proper names I consider a piece of text which is at least two lines long and contains a minimum of three proper names; an essential prerequisite is that all names share a common denominator which would justify their listing together in catalogue form.[2] However, the presence of patronymics, as we shall see, may challenge to a degree the definition as to what a catalogue of proper names comprises.[3]

Before commencing, I would like to define what we mean by the term 'patronymic': the ancient definition of the term by Dionysius Thrax (*Ars Grammatica* 1.1.25–26: πατρωνυμικὸν μὲν οὖν ἐστὶ τὸ κυρίως ἀπὸ πατρὸς ἐσχηματισμένον, καταχρηστικῶς δὲ καὶ τὸ ἀπὸ προγόνων, οἷον Πηλείδης, Αἰακίδης ὁ Ἀχιλλεύς: "a patronymic is formed mainly from the name of the father and, by an excessive use of language, from that of the ancestors, as Pelides, Aeacides for Achilles") shows that a patronymic is the adjective which relates someone to his/her father, but also on occasions to his/her grandfather or another ancestor. This definition is similarly repeated by Servius: *patronymica non a patre tantum, sed a parentibus tracta* (Gramm. Lat. 438.3, Keil).[4]

---

* I am grateful to Mary Plastira, Robert Maltby and Stephanos Matthaios for sharing their knowledge.

1 Patronymics are multi-functional. See Tsitsibakou-Vasalos' closure of her book (2007, 226).

2 Kyriakidis 2007, xiii; see also Gassner 1972, 64; Mainberger 2002; Sammons 2010, 9.

3 See Reitz 2013, 230: "Though a definition ... is helpful as a starting point, the catalogue format becomes most interesting when the texts challenge the definitions".

4 Cf. e.g. Choeroboscus schol. in *Theodosii Canones*, Gramm Gr. (1889) IV/1, 157, 21–22, Hilgard, Leipzig: ὅτι τὰ πατρωνυμικὰ εἰς τὸ υἱὸς καὶ ἔγγονος διαλύονται. Donatus *Ars Grammatica* II.2

DOI 10.1515/9783110535150-006

In the *Iliad* (but not exclusively), where each catalogue of proper names often constitutes a narrative unit which defines its limits clearly, the presence of a patronymic[5] may suggest the indirect insertion of a proper name and on certain occasions – especially in genealogical catalogues – it has the dynamics of functioning as a proper addition to it.[6] This practice may condense the catalogue space to the degree of creating a register of even one verse, affecting in this way the above definition of what a catalogue is. One such example comes from the *Iliad* at 5.159–160, with the generations of Dardanus, Priam and his sons appearing in one verse: ἔνθ' υἷας Πριάμοιο δύω λάβε Δαρδανίδαο ("then he [i.e. Diomedes] took two sons of Priam, the descendant of Dardanus").

In later times the phenomenon will be repeated in the *Aeneid* 1.617–618: *tune ille Aeneas quem Dardanio Anchisae / alma Venus Phrygii genuit Simoentis ad undam?* ("Are you that Aeneas whom caring Venus bore to Dardanius Anchises by the river Simois?").

But what is the function of such an addition and what does it aim at? The recital of a genealogical catalogue is often considered as a kind of suspension of the narrative. In my view, rather the opposite is the case, since it is a vertical[7] time-component which enriches temporally the narrative.[8] When a patronymic couples with the proper name, although it condenses the textual space that the catalogue covers, at the same time it augments[9] the catalogue's time span significantly since a whole phase of a particular myth – albeit time-contracted – is encapsulated within the catalogue.[10] With its presence, that is, a patronymic blows up temporally that part of the catalogue to which it belongs, by giving

---

(Gramm. Lat. [1864] IV.373.23, Keil): *alia patronymica, ut Atrides Pelides: haec et ab avis et a matribus saepe fiunt*; also *Explan. in Don.* (Gramm. Lat. II.537.7, Keil): *De anonomicis et patronomicis. sunt patronomica, ut Tydides Pelides, quae a vocabulo paterno trahuntur.*

5 It has been considered in the past that the presence of a patronymic may serve metrical needs of the text. As H. Paul Brown 2006, 26 has shown, however, following D. Shive 1987, there are instances where a patronymic – in particular that of Achilles' – may occupy exactly the same metrical position as that of the hero's name. This means that the use of a patronymic instead of the proper name itself is not haphazard nor is it governed exclusively by metrical reasons: "The distinction of form could have some social or discourse-specific, pragmatic explanation"; see also Table 2 on p. 27.

6 This indirect addition happens either when the patronymic accompanies the proper name (e.g. Πηληϊάδεω Ἀχιλῆος, *Il.* 1.1), or it substitutes for it (σὺ Πηλεΐδη, *Il.* 1.146).

7 Bettini 1991, 167–169: "This image of the past as something that stands 'on high' brings to mind what is perhaps our most interesting vertical model of time: the genealogical tree, or stemma".

8 From a different perspective, this is similar to the function of a simile.

9 Lateiner 2005, 415: "Length of speech may indicate importance".

10 On the condensed texts see Horster and Reitz 2010.

room to an extra generation in a dense form, through the minimization of a whole period. It is also considered that the introduction of such a word may on occasion remind the reader of the "fecund totality of the entire [story] tradition", as John Miles Foley had suggested (in Beck's paper).[11]

The Homeric Πηλείδης: The Unifying Aspect of a Patronymic

Beginning with Homer, an initial note should be made: the time allotted to the reading of a patronymic, which actually consists of a single word, does not permit easily the recollection of a myth in full. What is retrieved from memory depends entirely on the interaction between the patronymic and the character's name within the frame of the surrounding context. It is this interaction with the context which finally determines the kind of the reception of a patronymic.

In *Iliad* 21 Achilles is ready to attack Asteropaeus, Pelegon's son, and he challenges him on the grounds of genealogy: τίς πόθεν εἰς ἀνδρῶν ὅ μευ ἔτλης ἀντίος ἐλθεῖν; ("who are you, and where are you from, that you dare to come against me?", *Il.* 21.150). To this arrogant question Asteropaeus replies with a rhetorical question (21.153): Πηλεΐδη μεγάθυμε τί ἦ γενεὴν ἐρεείνεις; ("magnanimous son of Peleus, do you really ask me about my descent?"), thus introducing the catalogue of his genealogy by reciting it himself. It is useful at this point to look into Brown's finding (2006, 27) that "Akhilleus is addressed by his given name by characters who can be fairly designated as friends or allies; on the other hand, he is addressed with his patronymic by Trojan enemies ... That is to say, in the case of Akhilleus, the poet seems to be making a social distinction, since he is not making a metrical one".[12] These observations are relevant for the *Iliad* and they may perhaps help us in interpreting certain instances which *prima facie* do not belong in the above categorization (see e.g. *Il.* 1.144–147, when Agamemnon addresses Achilles as Πηλεΐδη).[13] In the *Odyssey*, howev-

11 Beck 1998/1999, 122.

12 This case, however, is not the rule for others, as, for instance, for Agamemnon (Brown 2006, 30) "who is never addressed except with his patronymic or with the title of his office ... This fact suggests that for Agamemnon, within the context of the *Iliad*, status, particularly in relation to ancestry and office, is of great importance for defining his social persona." Brown further argues that the method of his work concerns the "growing tendency among scholars and readers of Homer to see context as important in reading traditional characters in traditional poetry (among these, Beck, Higbie, Foley, Janko, Kahane and Nagy)": Brown 2006, 42 and n. 79.

13 See below, n. 27.

er, things are somehow different, as we shall see below when the spirit of Agamemnon addresses that of Achilles (again) with his patronymic.[14] At any rate, what becomes clear from Brown's analysis is that the use of patronymics – and in particular that of Achilles' – is not made haphazardly. But let us see the Iliadic text (21.152–153, 157–160):

> τὸν δ' αὖ Πηλεγόνος προσεφώνεε φαίδιμος υἱός· 152
> Πηλεΐδη μεγάθυμε τί ἦ γενεὴν ἐρεείνεις;
> εἴμ' ἐκ Παιονίης ἐριβώλου τηλόθ' ἐούσης
> ...
> αὐτὰρ ἐμοὶ γενεὴ ἐξ Ἀξιοῦ εὐρὺ ῥέοντος 157
> Ἀξιοῦ, ὃς κάλλιστον ὕδωρ ἐπὶ γαῖαν ἵησιν,
> ὃς τέκε Πηλεγόνα κλυτὸν ἔγχεϊ· τὸν δ' ἐμέ φασι
> γείνασθαι.

> Then the glorious son of Pelegon addressed him: "magnanimous son of Peleus, do you really ask me about my descent? I come from fertile Paeonia, a far away place ... But my origin is from wide-flowing Axios, Axios, who lets flow the fairest water on earth, who begot Pelegon renowned for his spear, and they say he begot me.

Just before the self-recited genealogy of Asteropaeus, it is the poet who anticipates the hero also presenting the same genealogy (*Il.* 21.139–143):[15]

> τόφρα δὲ Πηλέος υἱὸς ἔχων δυλιχόσκιον ἔγχος
> Ἀστεροπαίῳ ἐπᾶλτο κατακτάμεναι μενεαίνων
> υἱέϊ Πηλεγόνος· τὸν δ' Ἀξιὸς εὐρυρέεθρος
> γείνατο καὶ Περίβοια Ἀκεσσαμενοῖο θυγατρῶν
> πρεσβυτάτη· τῇ γάρ ῥα μίγη ποταμὸς βαθυδίνης.

> Meanwhile the son of Peleus holding his spear with the long casting shadow lept on Asteropaeus, the son of Pelegon, eager to slay him; Pelegon was born to broad-flowing Axius and Periboea, the eldest daughter of Acessamenus; for the deep-eddying river coupled with her.

In the above two passages both heroes – Achilles and Asteropaeus – are characterized by the name of their father or their patronymics: Πηλέος υἱός (139), Πηλεΐδης (153)[16] for Achilles; υἱέϊ Πηλεγόνος (141) and Πηλεγόνος ... υἱός (152) for Asteropaeus. As regards the name *Pelegon,* in Strabo we read that: οἱ γὰρ Παίονες Πελαγόνες ἐκαλοῦντο ("for the Paeonians were called Pelagones", *Geog.* fr. 7.39,

---

14 *Od.* 24.36.

15 On the structure of these catalogues see Kyriakidis (forthcoming) 3, 6–7.

16 For Shive 1987, 110 the phrase Πηλεΐδη μεγάθυμε is just a "vocative formula".

Jones, Loeb). The proper name *Pelegon* seems to derive from the ethnic *Pelagon.*[17] The proximity, however, in the text of Πηλεγόνος ... υἱός (for Asteropaeus, 152) to Πηλεΐδη (for Achilles, 153) and of Πηλέος υἱός (139) to υἱέϊ Πηλεγόνος (141) seems to activate an etymological play between the name *Pelegon* and *Peleus*' son: it is as though *Pelegon* stands for ὁ γόνος τοῦ Πηλέος, a sort of Πηλέος υἱός for the opponent of Achilles. To my reading, the structure of the passage does not seem to be haphazard: the(se) patronymics seem to relate, or even connect – in fact they equate the two rivals – Achilles and Asteropaeus, before the combat starts.[18] The technique seems to allude to the common fate awaiting both of the heroes. It is an implied comment of the poet effected through the involvement of agents from the previous generation and external to the scene. In this case (and it is not the only one) past seems to be temporally neutralized and functions transcendentally, forewarning the common future.

Perhaps, Achilles was haunted by worries ever since his mother warned him (*Il.* 18.95–96) about his imminent death[19] in the event Hector was killed, an ominous prospect with which the hero seems to have reconciled himself after the loss of his dear comrade, Patroclus (18.90–93, 98–126). It is a telling detail that having left Asteropaeus dying, Achilles recited his own divine lineage as if to shake off the threat of such a fate (*Il.* 21.184–191).[20]

In dealing, however, with Achilles' patronymic in the above catalogue – the most prominent patronymic in the *Iliad* – , I came to wonder about its function in the very first line of the epic and, indeed, beyond the narrow limits of a catalogue: unlike the *Odyssey* and the *Aeneid,* in the *Iliad* the name of the central hero –who "holds within himself all the heroic virtues"[21] – is prominently disclosed together with his patronymic at the opening of the epic: Μῆνιν ἄειδε θεὰ Πηληϊάδεω Ἀχιλῆος. For Benardete, Achilles "is marked off from all other men because of his father; as an only son, without brothers, he was entirely Peleus' heir (24.538–540)".[22] Besides, taking into account also the patronymic *Aea-*

---

17 Richardson 1993, on 21.141.

18 The emphasis is here given by another common element: in verse 159 Pelegon's mastery of the spear is praised (ὃς τέκε Πηλεγόνα κλυτὸν ἔγχεϊ) and a little further (162) Achilles' spear made of Pelian ash is mentioned (Πηλιάδα μελίην, 162): see Tsitsibakou-Vasalos 2007, 58–59 and the next note.

19 Cf. also *Od.* 24.29.

20 Tsitsibakou-Vasalos (2007, 58–59) also realizes the etymological relation between Peleus and Pelegon.

21 Benardete 1963, 1.

22 Benardete 1963, 12.

cides, the same scholar[23] argues: “In Achilles’ patronymic is summed up part of his own greatness. He is partly the work of generations”.

Indeed various interpretations may be given for the appearance of the patronymic in the first programmatic line. I will repeat here James Armstrong’s view (1993) which, I think, is significant and to the point:[24] “It seems to me now an unusually apt use of a traditional locution. The poet chose to sing of a ‘wrath’ and ‘a goddess’ and ‘Achilles,’ but there is more. There is Peleus and that name touches the central and primary event which led to the Trojan War, namely, the marriage of Peleus and Thetis (Achilles’ mother) which gave occasion to the Apple of Discord and the Judgment of Paris (see *Il.* 24.35–36). Overarching the whole poem and often explaining the action are the forces unleashed by the Judgment of Paris, his grievous violation of guest friendship and Helen’s departure for Troy”.

The first word of the epic, μῆνις, Achilles’ wrath, seems to connect on the one hand allusively the epic with its past since it was the wrath of the goddess Eris, who – uninvited to the wedding of Peleus and Thetis – triggered the sequel of events leading to the Trojan war, and on the other the wrath of Peleus’ son, Achilles, for Agamemnon over Briseis. According to Kirk,[25] wrath is the ‘dominant theme’ established as the narrative of Book 1 develops and on which the plot of the *Iliad* is internally organized. In the pre-Iliadic wrath-plot, Peleus’ participation obviously is minimal; within the epic, however, that wrath will turn into that of Peleus’ own son. Nevertheless, it is within this epic that a place of honour for Achilles’ father is reserved in the words of Nestor (*Il.* 7.124–128):

ὢ πόποι ἦ μέγα πένθος Ἀχαιΐδα γαῖαν ἱκάνει.
ἦ κε μέγ’ οἰμώξειε γέρων ἱππηλάτα Πηλεὺς
ἐσθλὸς Μυρμιδόνων βουληφόρος ἠδ’ ἀγορητής,
ὅς ποτέ μ’ εἰρόμενος μέγ’ ἐγήθεεν ᾧ ἐνὶ οἴκῳ
πάντων Ἀργείων ἐρέων γενεήν τε τόκον τε.

Alas! A great sorrow has come on the land of Achaea. Truly the old horseman, Peleus, the noble counselor and orator of the Myrmidons, would lament, he who once was so pleased asking me in his own house of the origin and birth of all the Argives.

Disappointed by the reaction of the Achaeans to the challenge of Hector for a man-to-man combat, Nestor rebukes his comrades and chooses to open his speech with a reference to Peleus, the “noble counselor and orator of the Myrmi-

23 Benardete 1963, 12.
24 Armstrong 1993, 61–62.
25 Kirk 1985, 46.

dons" (7.126)[26] who would have lamented for this behavior of the Achaeans. In Book 9, it is also Peleus– as Odysseus recalls – who had wisely advised his son: σὺ δὲ μεγαλήτορα θυμὸν / ἴσχειν ἐν στήθεσσι ("but you control your great anger in your chest"; *Il.* 9.255–256). Peleus' words and advice obliquely enhance his son's inactivity due to his wrath.

The mention of Peleus' behavior in this critical moment of the plot may have further implications: according to the scholiast on *Il.* 7.125, Nestor's reference to Peleus anticipates, as it were, the embassy to Achilles in Book 9 (schol. in *Il.* 7.125.10 †προμνηστεύει† οὖν τὰς Λιτάς, bT, Erbse). Indeed, the mention of Achilles' father in this instance, instead of Achilles himself, may be an oblique way of referring to the hero whose hurt pride kept him away from the rest of the Achaeans; it may also be an imperceptible flattery of Achilles (schol. in *Il.* 7.125.12–13: καὶ τὸν Ἀχιλλέα ὑποθωπεύει λεληθότως, bT, Erbse), according to the same scholiast. The poet then may well exploit Achilles' absence and through Peleus' mention could keep the gathered Achaeans and the reader aware of the fact that Achilles is not far off. Notwithstanding Achilles' absence, the hero's participation remains a central issue and Peleus' mention in this part prepares the next stage of the narrative.

The above cases suggest that reference to the father and the employment of a patronymic may have a drastic effect in the present-time of the narrative by conditioning its course. This is effected again not exactly because it refers to the past but because Peleus – denuded from any blemishes or any defects related to parts of the myth referring to him – becomes an everlasting "model" of behavior and an agent in moving the narrative forward.

Turning back to the function of the patronymic *Pelides* in the Iliadic catalogues, we have to note that Achilles himself is very rarely included in a catalogue. One such instance is when the hero recites his own divine origin to which we referred earlier (*Il.* 21.184–191):[27]

---

26 According to Phoenix (*Il.* 9.479–484; also 23.89–90), Peleus was also very hospitable.

27 Further to his lineage, Achilles' name (either with, or replaced by his patronymic) will be involved in a catalogue form at least two more times in the Homeric works; once in the beginning of the *Iliad* (1.144–147) and in the beginning of the last book of the *Odyssey* (24.19–23).

Il. 1.144–147:

εἷς δέ τις ἀρχὸς ἀνὴρ βουληφόρος ἔστω,
ἢ Αἴας ἢ Ἰδομενεὺς ἢ δῖος Ὀδυσσεὺς
ἠὲ σὺ Πηλεΐδη πάντων ἐκπαγλότατ' ἀνδρῶν,
ὄφρ' ἥμιν ἑκάεργον ἱλάσσεαι ἱερὰ ῥέξας.

And let one man who is a counselor be the leader, either Aias, or Idomeneus or noble Odysseus or you, son of Peleus, most violent of all men, so that you may appease him who works from a distance by offering sacrifice.

χαλεπόν τοι ἐρισθενέος Κρονίωνος
παισὶν ἐριζέμεναι ποταμοῖό περ ἐκγεγαῶτι.
φῆσθα σὺ μὲν ποταμοῦ γένος ἔμμεναι εὐρὺ ῥέοντος,
αὐτὰρ ἐγὼ γενεὴν μεγάλου Διὸς εὔχομαι εἶναι.
τίκτέ μ᾽ ἀνὴρ πολλοῖσιν ἀνάσσων Μυρμιδόνεσσι
Πηλεὺς Αἰακίδης· ὃ δ᾽ ἄρ᾽ Αἰακὸς ἐκ Διὸς ἦεν.
τῶ κρείσσων μὲν Ζεὺς ποταμῶν ἁλιμυρηέντων,
κρείσσων αὖτε Διὸς γενεὴ ποταμοῖο τέτυκται.

It is hard even for one begotten of a river to rival with the children of the mighty son of Cronos. You said that you are the offspring of the wide flowing river, but I am proud to be of the stock of great Zeus. The father who begot me, Peleus, son of Aeacus, rules among the Myrmidons and Aeacus comes from Zeus. In this way, as Zeus is mightier than the rivers flowing into the sea, so the race of Zeus is mightier than that of a river.

It is only after his combat with Asteropaeus, moments before the latter's death, that Achilles recites his own lineage in order to compare it with that of his victim: as his stock ultimately originates from Zeus, he feels superior to his dying opponent. Achilles soliloquising is eager to prove – I wonder to whom at that stage if not to himself[28] – the superiority of his lineage and through it to claim the survival of his name, at least. He obviously needs to be under the illusion that his superior descent will ensure him eternity which could not be secured for Asteropaeus. In this passage, beyond the patronymic of Zeus, Κρονίωνος (184) which gives temporal depth to Achilles' generation, the god's name is repeated four more times (184, 187, 190, 191). And to add prestige to himself, the hero places the first person singular pronoun ἐγώ – together with the verbal phrase εὔχομαι εἶναι – in the same verse with the name of the supreme god (187) and he then states his line of descent in two more verses (188–189). Besides, by using the patronymic Αἰακίδης (189), next to the name of Αἰακός he adds prominence and renown not only to his father, Peleus, but also to himself, owing to his grandfather's piety and illustrious past. Aeacus was considered as the εὐσεβέστατος πάντων.[29] "Achilles is not only the son of Peleus but the grandson of Aeacus; and yet to be called 'Aeacides' when he is actually 'Pe-

It is when Agamemnon in his quarrel with Achilles over the spoils proposes to sacrifice to Apollo and place Chreseis on board a ship. We note that the motive of wrath associated with Achilles from the very first line of the epic is here developing as part of the narrative with Achilles being addressed as the son of Peleus. In this instance if Brown 2006 is correct in his analysis, then Agamemnon's address of Achilles as Πηλεΐδη, expresses his spite.

28 Kyriakidis (forthcoming) 7.

29 Apollod. *Bibl.* 3.12.6: ἦν δὲ εὐσεβέστατος πάντων Αἰακός ... τιμᾶται δὲ καὶ παρὰ Πλούτωνι τελευτήσας Αἰακός, καὶ τὰς κλεῖς τοῦ Ἅιδου φυλάττει ("Aeacus was the most pious of all men ... After his death Aeacus was honoured by Pluto and watches over the keys of Hades").

leides' means that he has inherited something that was common to all his first ancestors."[30] His going back to the family past is what will restore his confidence in the future. Both past and the 'future' he claims serve the hero's present.

The name of Achilles together with his patronymic will appear again in a catalogue at the end of the *Odyssey*,[31] when Hermes leads the spirits of the suitors to the Underworld where they meet the spirit of Achilles and of other important heroes of the *Iliad*. Then follow the spirits of Agamemnon and those who were slain with him (*Od.* 24.15–23):

εὗρον δὲ ψυχὴν Πηληϊάδεω Ἀχιλῆος
καὶ Πατροκλῆος καὶ ἀμύμονος Ἀντιλόχοιο
Αἴαντός θ', ὃς ἄριστος ἔην εἶδός τε δέμας
τε τῶν ἄλλων Δαναῶν μετ' ἀμύμονα Πηλεΐωνα.
ὣς οἱ μὲν περὶ κεῖνον ὁμίλεον· ἀγχίμολον δὲ
ἤλυθ' ἔπι ψυχὴ Ἀγαμέμνονος Ἀτρεΐδαο[32]
ἀχνυμένη· περὶ δ' ἄλλαι ἀγηγέραθ', ὅσσοι ἅμ' αὐτῷ
οἴκῳ ἐν Αἰγίσθοιο θάνον καὶ πότμον ἐπέσπον.
τὸν προτέρη ψυχὴ προσεφώνεε Πηλεΐωνος.

They met the soul of Achilles, son of Peleus, and of Patroclus and of noble Antilochus and of Aias who in his form and bodily stature excelled of all the Danaans second to the son of Peleus. So these were gathering about him [i.e. Achilles], and the grieving soul of Agamemnon, son of Atreus, came near and other souls gathered together around him, those who had been slain with him in the house of Aegisthus and encountered their fate. The soul of the son of Peleus spoke to him first.

It is the catalogue which leads to the colloquy of the spirits of the two heroes, Achilles' and Agamemnon's, whose names are the only ones in the catalogue appearing with their patronymics (15, 18, 20, 23). They are the ones, therefore, who acquire temporal depth because of their patronymics and this singles them out among the catalogue names. Immediately after the catalogue, the spirits of the two Greek heroes, Achilles and Agamemnon, address each other with their pat-

---

30 Benardete 1963, 12.

31 Whitehead 1984, 119 sees in this second Homeric *Nekyia* the belated connection between the *Iliad* and the *Odyssey*: "In this episode, in which the shades of Achilles and Agamemnon meet those of the suitors, and the stories of the burial of Achilles and the revenge of Odysseus are juxtaposed, the Odyssey self-consciously comments on its function as a moral lesson or example; but beyond this, its implications extend back to the Iliad, and it can be seen as supplying a delayed epilogue to the earlier poem". In this way he answers in essence the issue of whether these verses bear any value whatsoever to the text and sees that what is said here for Achilles forms a 'conclusion' of the *Iliad*.

32 On the importance of the verse-placement of this patronymic, see Bassett 1926, 124–125.

ronymic or father's name respectively (Ἀτρεΐδη, *Od.* 24.24; ὄλβιε Πηλέος υἱέ, θεοῖσ' ἐπιείκελ' Ἀχιλλεῦ, 24.36), a locution which the poet himself,[33] as we have said, has also observed. Achilles first accepts the inescapable fate of all humans (μοῖρ' ὀλοή, τὴν οὔ τις ἀλεύεται, ὅς κε γένηται, "grim fate that not one avoids of those born"; *Od.* 24.29) and takes pity on Agamemnon, who could not have a glorious death under the walls of Troy and thus leave a noble name to his son: ἠδέ κε καὶ σῷ παιδὶ μέγα κλέος ἤρα' ὀπίσσω ("and you would have also won great renown in the future for your son"; *Od.* 24.33). For Agamemnon, Achilles is fortunate for he was slain in battle (an event that takes place beyond the Iliadic narrative). This means that, according to Agamemnon, it is Achilles' glorious death that brought renown to his own name: ὣς σὺ μὲν οὐδὲ θανὼν ὄνομ' ὤλεσας, ἀλλά τοι αἰεὶ / πάντας ἐπ' ἀνθρώπους κλέος ἔσσεται ἐσθλόν, Ἀχιλλεῦ ("so, even in death you did not lose your name, Achilles, but your noble fame will be among all men"; *Od.* 24.93–94). In this way the renown is recognized as being carried into the future either through a descendent who will bear the patronymic of his famous ancestor, or by fame itself which will travel freely as a *libera fama* according to the Ovidian description (*Met.* 15.853).

As a matter of fact, one of the basic reasons that the name of Peleus survived in myth and literature was because he was the father of Achilles,[34] the hero whom the Greeks aspired to and the Romans eagerly wished to surpass. Achilles – Peleus' son – did not need a descendant for his name to remain in memory. This is something that Agamemnon had implied in their last colloquy, a detail that Virgil also realized and applied to his own purposes.

Pelides and *Aenides* in the *Aeneid*: Transgression and Prospect

Centuries of literary production separate the Homeric epic from the *Aeneid* and yet the Roman epic *par excellence* has followed many of the techniques introduced by the former. The differences between them are enormous (although the similarities are also striking). It is worth mentioning one general remark here: in the catalogues of the *Aeneid* (not only the genealogical ones) the presence of patronymics is not very common; there is, however, a noteworthy pres-

33 It is interesting that at the opening of the last book of the *Odyssey* Achilles appears in the Underworld with the appellation Πηληϊάδεω Ἀχιλῆος (*Il.* 1.1) with which the *Iliad* opens. See Whitehead 1984, 121; Heubeck 1992 on *Od.* 24.15.

34 On this see Ov. *Met.* 15.856; Kyriakidis (forthcoming) 17.

ence of epithets of origin. This is to a great extent due to the nature of each epic work. Like the *Odyssey*, a main feature of the *Aeneid* is the journey. Obviously, the journey entails development in space and time. Although the *Iliad* has some remarkable openings in respect of the narrative development in space (on occasion in small scale, as e.g. the use of epithets of origin, or in a larger narrative scale by way of digressions) it nonetheless remains rather more sparing in the inclusion of many place names.[35]

Even though in the *Aeneid* patronymics are seldom found in catalogues of proper names, their employment is crucial in the interpretation of the narrative. We turn again to the patronymic *Pelides* which appears in a catalogue of Book 2 in Aeneas' narration to Dido of the night Troy was taken. The wooden horse enters the city and a number of Greeks climb down from it; one of them was Ulixes. All this was effected by his cunning scheme. For Aeneas, however, the internal narrator, Ulixes is not characterized as πολύτροπος; he is simply *dirus* (fearful, 2.261); he neither bears a patronymic nor is there a place of origin referred. It is also interesting to note that Sthenelus participates in this catalogue, the hero who in the *Iliad* (4.370–405) had claimed that his generation is better than the one of their fathers (*Aen.* 2.259–265):

illos patefactus ad auras
reddit equus laetique cavo se robore promunt
Thessandrus Sthenelusque duces et dirus Vlixes,
demissum lapsi per funem, Acamasque Thoasque
Pelidesque Neoptolemus primusque Machaon
et Menelaus et ipse doli fabricator Epeos.
invadunt urbem somno vinoque sepultam

The horse opens and delivers from its wooden cavity the Greeks to the air and cheerfully come out the leaders Thessandrus and Sthenelus and fearful Ulixes sliding by the rope they had let down, and Acamas and Thoas and Neoptolemus, the descendant of Peleus, and first of all Machaon and Menelaus and Epeos himself the maker of this artful deception. They invade a city buried in sleep and wine.

The bearer of the patronymic *Pelides* in line 263 is not Achilles, but his son, Neoptolemus. Neoptolemus was almost non-existent in the *Iliad*. He was growing up at Skyros and Achilles did not even know whether he was alive (*Il.* 19.326–327). From other sources (e.g. *Od.* 11.506–537; Soph. *Phil.* 343–356; Apollod. *Epit.* 5.11;[36] Paus. 10.26.4, etc.) we learn that he was particularly connected with Odysseus who had fetched him from there. Aeneas, therefore, the inter-

35 Bettini 1991, 138–139.

36 See Frazer's note (1921, 224).

nal narrator, by attributing the patronymic *Pelides* to Neoptolemus obviously bypasses Achilles, the central hero of the *Iliad*. Instead, that is, of introducing a distinct temporal stage in the lineage, that of Peleus – as patronymics often do – the patronymic *Pelides* in this case functions rather conversely. It is as though Aeneas – and the poet for that matter – attempts to disregard the most important phase of the Trojan war, that of the *Iliad*. Besides, in using the name of 'Neoptolemos' – instead of his earlier name, 'Pyrrhus' – Aeneas turns his attention to the period after the end of Achilles' involvement in the war. It is the etymology of the name 'Neoptolemus' which also contributes to this perception, as – although it may mean 'a young warrior'[37] – it may also allude to the new stage of the war, after Achilles. This is the first time the name 'Neoptolemus' is employed instead of that of 'Pyrrhus', as used elsewhere.

Further down in the *Aeneid* we have the association once again of the same patronymic but with Achilles this time (2.548).[38] According to Aeneas' narration, Priam had witnessed the death of his own son, Polites, by the hands of Pyrrhus-Neoptolemus. Enraged, the old king speaks to the young Greek and doubts that he is a true son of Achilles as the Greek hero had treated him with respect when as a suppliant he had asked for the dead body of Hector (*Aen.* 2.540 – 543):

At non ille, satum quo te mentiris, Achilles
talis in hoste fuit Priamo; sed iura fidemque
supplicis erubuit corpusque exsangue sepulcro
reddidit Hectoreum meque in mea regna remisit.

But Achilles whom you falsely claim to be your father, was not so cruel to his enemy but he respected the suppliant's rights and my trust in him and gave me back the lifeless body of my Hector for burial and sent me back to my kingdom.

In Aeneas' narration in the *Aeneid* while Priam recalls the past, Pyrrhus looks to what comes afterwards[39] answering jeeringly to king Priam (*Aen.* 2.547 – 549):

---

37 Austin 1964 ad loc. O'Hara 1996, 132 – 133 and the sources therein; Paschalis 1997, 89 – 90; According to Pausanias, the name of 'Neoptolemus' was given by Phoenix "because his father Achilles had gotten his start in warfare while still young" (10.26.5: ὅτι Ἀχιλλεὺς ἡλικίᾳ ἔτι νέος πολεμεῖν ἤρξατο, transl. Peradotto 1990, 136); Berlin 1998, 15. Servius, however, states (ad loc.) that it was Neoptolemus, and not Achilles, who entered warfare at a young age.

38 Achilles in the Roman epic is also called *Aeacides*, as in *Il.* 11.805, 16.854 (see above, n. 4); Benardete 1963, 12. He was also called *Larisaeus* (*Aen.* 2.197; 11.404) with the latter attribute stating the place of his origin.

39 Berlin 1998, 19.

cui Pyrrhus: 'referes ergo haec et nuntius ibis
Pelidae genitori. illi mea tristia facta
degeneremque Neoptolemum narrare memento.

To him Pyrrhus answered: "well, then, you will go as a messenger to my father, the son of Peleus, and say these things. Remember to tell him about my foul deeds and the degenerate Neoptolemus.

Pyrrhus refers to his father with his patronymic,[40] avoiding the name of Achilles itself; it is as though he wishes – among other things – to allude to an existing distance between an absent father and his growing son or even perhaps to dissociate Achilles from this new and final stage of the Trojan war.

Within the space of so few lines we have an intriguing text where things are happening in front of our eyes. Textually, we have a narrativized catalogue which deals with the three generations of the House of Peleus: Peleus himself, Achilles and his son Pyrrhus-Neoptolemus. But this is only one aspect of the narrative. When Priam contrasts the worthy father to the unworthy son (2.540 – 543)[41] Virgil emphatically puts the father's name, Achilles, at the end of the hexameter (540) and his short speech is a summary of the way the Greek hero treated the suppliant king in the Homeric epic.[42] With all this name-playing, a peculiarly odd situation is formed in which the generation of Achilles is disjointed – if not shattered– by Achilles' own son and with it the 'best of the Achaeans' is doomed to oblivion. This thought is backed by Neoptolemus' self characterization as *degener* which in the context of the passage, although it may suggest, according at least to OLD (s.v. 2a) "one less admirable than one's forebears", may also mean "fallen away from one's origin" (Cassel's *Latin English Dictionary*, s.v.), "one who is not connected with his / her stock". For Pyrrhus, *Pelides* seems to operate again as a transgression of the chief exponent of the Iliadic heroic model and his generation.

Quite plausibly, the above verses will be associated in the reader's mind with another passage of the *Aeneid* (9.641– 656) and may lead to a comparison of the two:

'macte nova virtute, puer, sic itur ad astra,
dis genite et geniture deos. iure omnia bella
gente sub Assaraci fato ventura resident,
nec te Troia capit.' simul haec effatus ab alto

40 Berlin 1998, 15: "the juxtaposition of *degenerem* (549) and the patronymic, which occupy the same position in successive verses underscores the irony of the inversion within the allusion".
41 Austin 1964, ad loc.
42 Berlin 1998, 15.

aethere se mittit, spirantis dimovet auras
Ascaniumque petit; forma tum vertitur oris
antiquum in Buten. hic Dardanio Anchisae
armiger ante fuit fidusque ad limina custos;
tum comitem Ascanio pater addidit. ibat Apollo
omnia longaevo similis vocemque coloremque
et crinis albos et saeva sonoribus arma,
atque his ardentem dictis adfatur Iulum:
'sit satis, Aenide, telis impune Numanum
oppetiisse tuis. primam hanc tibi magnus Apollo
concedit laudem et paribus non invidet armis;
cetera parce, puer, bello.'

'Well done my boy on your new courage! This is the way to the stars, son of the gods and future father of gods. All wars to come by fate will justly cease under an offspring of Assaracus, Troy cannot hold you!' As soon as Apollo said these words he plunged down from the sky and parting the gusty winds, searched for Ascanius; he then takes the form and features of old Butes. He was once the armour bearer to the Dardanian Anchises and the trusted guard of his door when Aeneas made him Ascanius' aide. Apollo came similar to the old man in every detail – voice, complexion, white hair, weapons clanging grimly – and with these words he counsels Iulus, now glowing in excitement: 'Let it be enough, son of Aeneas, that Numanus fell to your shafts and you are safe. Great Apollo has granted you this first feat of glory and does not envy your arrows equal to his own. For the rest, abstain, my boy, from the war'.

These lines are part of the words that Apollo addressed to Ascanius-Iulus after the killing of Numanus Remulus in Book 9. Although the passage looks like a catalogue, it is not; I would say that the catalogue has been totally narrativized. Nevertheless, all the names of Aeneas' lineage are present. Whereas Neoptolemus was never called *Achillides*,[43] in the above passage of the *Aeneid* we have the unique presence of the patronymic *Aenides* (653) for Ascanius which contextually gives a clear future dimension to the succession of Aeneas. The *Aeneid* for the Romans was the epic of their beginnings, which should be furthered so that there is a continuity. This continuity is seen in Ascanius, who deservedly replaces his absent father. Though absent, Aeneas has left his mark by the use of *pater* (649) and the patronymic *Aenides* (653)[44] with which Ascanius is uniquely ad-

43 See Tsitsibakou-Vasalos 2007, 226 for the 'artful' suppression on the part of the poet [sc. Homer] of patronymics of the sort of *Odysse-ides, Odyssei-ion*. See also De Jong 1993, 302 n. 27, who notes that there is no patronymic Ὀδυσσιάδης or some other similar patronymic in the Homeric epics. On the lack of a patronymic Ἀχιλλείδης see Kyriakidis (forthcoming). Ovid will employ this patronymic for Pyrrhus at Ov. *Her.* 8.3.

44 Servius, ad loc.: *Aenide patronymicon hoc non venit ab eo, quod est 'hic Aeneas': nam 'hic Aeneades' et 'o Aeneade' faceret.* For Hardie 1994 "this form of the patronymic, as if from **Ae-*

dressed in the same way that his grand-father Anchises[45] is connected with the earlier members of the race through the patronymic *Dardanius.*[46] As Coleman argues: "Usually only the heroes in the *Aeneid* are called by a patronymic. The boy is growing to heroic proportions".[47]

In the *Iliad* the patronymic Πηλείδης – the most potential of Greek patronymics in the Homeric epics – refers to the central character of the *Iliad* and it thus concerns the narrative presence. In the *Aeneid* the corresponding patronymic, *Anchisiades* (e.g. 5.407; 6.126, 348; 8.521, etc.) similarly refers to the central hero of the Roman epic, Aeneas, who is particularly closely connected with his father. His course in the narrative until his *catabasis* in the middle of the epic was characterized by a strong sense of devotion to his father (6.695–696). Its climax is his urge to meet his father in the Underworld. Anchises' prophecy on the future of his son's race has as a major goal to admonish Aeneas to stick to his mission and to this purpose he discloses to his son the glorious prospects of his race and the pageant of heroes who will succeed him (*Aen.* 6.756–759):

> *'Nunc age, Dardaniam prolem quae deinde sequitur*
> *gloria, qui maneant Itala de gente nepotes,*
> *inlustris animas nostrumque in nomen ituras,*
> *expediam dictis, et te tua fata docebo.'*

> 'Come then, I shall reveal to you what glory will follow in time to come the sons of Dardanus and the descendants who are in store from the Italian stock, noble spirits and future heirs of our name and fame and I shall tell you your fate'.

In the last book of the *Odyssey* Achilles is addressed by the spirit of Agamemnon with his patronymic, Πηλείδης who envies him for his future renown; in the *Aeneid* the hero himself is also addressed with his patronymic, as *dux Anchisiade* (6.348), by Palinurus. Both cases are from meetings that take place in the space of the Underworld. However, unlike the Greek epic where Achilles and Agamemnon are concerned with their own renown, in the *Aeneid* Anchises with his prophecy gives the sense of succession, continuity and glory of the race. The same sense is again reaffirmed by the creation of the unique patronymic *Aenides* which concerns the son and successor of the central hero, Aeneas,

---

neus (cf. 7.484 *Tyrrhidae),* is found only here; *Aeneada* is perhaps avoided as that form is used indiscriminately to refer to all the Trojans".

45 Anchises has been called *Dardanius* by Dido herself when she first met with Aeneas at 1.617–618. There we have a condensed catalogue of three generations within one verse (see above).

46 Kyriakidis (forthcoming).

47 Coleman 1942, 146.

and is, therefore, oriented to the future of the race and beyond the narrative of the Roman epic. The effort to create the sense of continuity and a promising prospect for the wandering Trojans is quite evident in the poetic discourse of the Roman epic which also attempts to surpass the Trojans' traumatic past.

Bibliography

Armstrong, J.I. 1993. A Note on the Proemium of the Iliad: 'Sing, Goddess, the Wrath of the Son of Peleus, Achilles ...'. *CW* 87: 61–62.

Austin, R.G. 1994. *Aeneidos Liber Secundus, with a Commentary.* Oxford.

Bassett, S.E. 1926. The So-called Emphatic Position of the Runover Word in the Homeric Hexameter. *TAPhA* 57: 116–148.

Beck, D. 1998/1999. Speech Introductions and the Character Development of Telemachus. *CJ* 94: 121–141.

Benardete, S. 1963. Achilles and the *Iliad. Hermes* 91: 1–16.

Berlin, N. 1998. War and Remembrance: Aeneid 12.554–560 and Aeneas's Memory of Troy. *AJPh* 119: 11–41.

Bettini, M. 1991. *Kinship, Time, Images of the Soul* (transl. by J. van Sickle). Baltimore and London.

Brown, H.P. 2006. Addressing Agamemnon: A Plot Study of Politeness and Pragmatics in the *Iliad. TAPhA* 136: 1–46.

Coleman, R.E. 1942. Puer Ascanius. *CJ* 38: 142–147.

De Jong, J.F. 1993. Studies in Homeric Denomination. *Mnemosyne* 46: 289–306.

Erbse H. 1969/1983. *Scholia Graeca in Homeri Iliadem* (Scholia vetera). 6 vols. Berlin.

Frazer, G.J. (transl.) 1921. *Apollodorus, The Library* (Loeb Classical Library). London and New York.

Gassner, J. 1972. *Kataloge im römischen Epos. Virgil – Ovid – Lucan* (Ph.D. Diss.). Munich.

Hardie, P. 1994. *Virgil. Aeneid Book IX.* Cambridge.

Heubeck, A. 1992: see Russo, Fernandez-Galiano and Heubeck 1992.

Hilgard, A. 1889/1894. *Prolegomena et scholia in Theodosii Alexandrini canones.* In *Grammatici Graeci*, I-IV. Leipzig. (reprint Hildesheim 1965)

Horster, M. and C. Reitz (eds.) 2010. *Condensing Texts – Condensed Texts* (Palingenesia 98). Stuttgart.

Jones, H.L. 1967. *The Geographies of Strabo.* vol. III (Loeb Classical Library). Cambridge, MA.

Kirk, G.S. 1985, *The Iliad: A Commentary*, vol. i: Books 1–4. Cambridge.

Kyriakidis, S. 2007. *Catalogues of Proper Names in Latin Epic Poetry* (Pierides I). Newcastle upon Tyne.

Kyriakidis, S. (forthcoming). Looking Backwards to Posterity: Catalogues of Ancestry from Homer to Ovid. In Lämmle, Scheidegger-Lämmle and Wesselmann (forthcoming).

Lämmle, R., C. Scheidegger-Lämmle and K. Wesselmann (eds.) (forthcoming). *Lists and Catalogues in Ancient Texts: Towards a Poetics of Enumeration* (Trends in Classics). Berlin and New York.

Lateiner, C. 2005. Proxemic and Chronemic in Homeric Epic: Time and Space in Heroic Social Interaction. *CW* 98: 413–421.

Mainberger, S. 2002. *Die Kunst des Aufzählens: Elemente zu einer Poetik des Enumerativen* (Quelle und Forschungen zur Literatur und Kulturgeschichte). Berlin and Boston.
Manuwald, G. and A. Voigt (eds.) 2013. *Flavian Epic Interactions* (Trends in Classics, Suppl. 21). Berlin and Boston.
O'Hara, J.J. 1996. *True Names: Vergil and the Alexandrian Tradition of Etymological Wordplay.* Ann Arbor.
Paschalis, M. 1997. *Virgil's Aeneid: Semantic Relations and Proper Names.* Oxford.
Peradotto, J. 1990. *Man in the Middle Voice. Name and Narration in the Odyssey.* Princeton, NJ.
Reitz, C. 2013. Does Mass Matter? The Epic Catalogue of Troops as Narrative and Metapoetic Device. In Manuwald and Voigt 2013, 229–243.
Richardson, B. (ed.) 2002. *Narrative Dynamics. Essays on Time, Plot, Closure, and Frames.* Ohio.
Richardson, N.J. 1993. *The Iliad: A Commentary*, vol. vi: Books 21–24. Cambridge.
Russo, J., M. Fernandez-Galiano and A. Heubeck 1992. *A Commentary on Homer's Odyssey.* vol. iii: Books 17–24. Oxford.
Sammons, B. 2010. *The Art and Rhetoric of the Homeric Catalogue.* Oxford.
Shive, D.M. 1987. *Naming Achilles.* Oxford.
Tsitsibakou-Vasalos, E. 2007. *Ancient Poetic Etymology: The Pelopids: Fathers and Sons* (Palingenesia 89). Munich.
Whitehead, O. 1984. The Funeral of Achilles: An Epilogue to the *Iliad* in Book 24 of the *Odyssey. G&R* 31: 119–125.

Part II **Drama**

Anton Bierl

The Bacchic-Chor(a)ic Chronotope: Dionysus, *Chora* and Chorality in the Fifth Stasimon of Sophocles' *Antigone*

Introduction: Time and Space in Greek Tragedy

Greek drama is the ideal test case for the study of time and space in Greek myth. As its literary, poetic, or, better said, performative frame is linked to a specific place and occasion, i.e. Athens at Dionysian festivals, the genre occupies that realm between literacy and orality. The tragic poet composes a script designed for a single performance at a specific time in front of a known Athenian audience in the theatre of Dionysus. The chorus assigned by the polis to the poet plays a decisive factor, representing a group surrounding and yielding space to the professional actor. To a certain extent they express a "together with" sensation that shapes the space with reference to other figures and creates the deeper intellectual and emotional matrix that mediates between the actors and the audience.[1] The entire theatrical setting of the theatre of Dionysus involves a complicated structure of spatial enclosures. The audience of ten to fifteen thousand people, more than a third of the male citizenship of the entire polis, grouped on the north-western slopes of the Acropolis encloses the orchestra where the chorus sings and dances, encircling the *skene* that constitutes the lower back stage where the actors perform – it all embraced by the majestic landscape of the steep Acropolis at the audience's back and the Attic landscape behind the *skene* reaching to the sea at the horizon.[2] A festive time of exception, a brief moment, characterizes time. The outer frame also influences the play itself so that we can speak of an oscillating exchange between inner-dramatic and extra-dramatic perspectives. In the dramatic plays, time is subordinated to space: the fictive time situated in the mythic past, the plots taken from episodes in the mythic megatext emerging from the oral principles of variation and selection, a great network weaved together over a long process.[3] Plot variations are allowed as

1 Haß and Tatari 2014, esp. 81; Nancy 2014, esp. 108; Haß 2014. On the Greek chorus, see Bierl 2001, esp. 11–104 (Engl. 2009, 1–82).

2 On the theatrical space in general, see e.g. Wiles 2003; Kramer and Dünne 2009.

3 See Segal 1983, esp. 176; Segal 1986, esp. 52–53.

DOI 10.1515/9783110535150-007

long as the basic structure remains recognizable. Weaving with mythic fabric, the tragedians perpetuate this collective work while also prompting the audience to reflect upon contemporary political events. The stories can zoom in and out of Athens and her actual situation, the chorus as a polyphonic instance serving as a shifting and mediating tool in this regard.[4] All things considered, the tragic plot takes place during a relatively short period of time. Since practically everyone knows the mythic outcome from the beginning, little emphasis is placed on narrative progress, step-by-step analysis or suspense produced by temporal evolution.[5] From the start the actors and the chorus act as if everything were already pre-determined; usually the outcome is already revealed in the prologue or shortly thereafter.[6] The chorus can glide back to preceding times, jump to the future or zoom-in into the present of the actual performance, creating special effects of presence.

Space, the other basic Kantian category, dominates. The play is situated in a mythic locale. As has been pointed out, Thebes represents the place of the Other,[7] where all categories and distinctions collapse. Many tragedies are located there while Athens, the place of the performance, is widely avoided. We can define an axis of decreasing Otherness: Thebes, or some distant barbarian region, housing the Other; Argos, Delphi, and Sparta representing a middle ground, and Athens functioning as a mirror of Athenian identity – even in mythical times it is the least othered.[8] Dionysus, the honoured god intrinsically linked to drama, can be understood as the deity of the Other.[9] Typically his place of birth is Thebes, but we also have indications for Thrace, Phrygia and Lydia. Yet he can also facilitate, via zooming, a perspective of the here and now, of the actual performance in his honour at the Athenian theatre. Furthermore, the chorus can project itself onto numerous places: it can intone other landscapes, reaching near and far on a horizontal axis, traveling from the Underworld to the sky on the vertical axis. Moreover mythic time and space interact with ritual reenacting myth in cul-

---

4 Sourvinou-Inwood 1989, esp. 136, 138, 144; on the fluidity of choral voices and the chorus as shifter, see Bierl 2001, esp. Introduction; see Index Chor / 'Fluktuation (Ambiguität) der Instanzen und Rollen' (Engl. 2009); see also Gagné and Hopman 2013b, 1–28.

5 Nancy 2014.

6 Haß 2014, 143.

7 Zeitlin 1990.

8 Zeitlin 1990; Zeitlin 1993. On Dionysus in tragedy from Thebes to Athens, see also Bierl 1991, 45–110.

9 Gernet 1953, 393 was the first to interpret Dionysus as "the Other". See also Vernant 1965, 358; Vernant 1981, 18; Vernant 1983, 42–43; Vernant 1985, 246; Vernant 1986, 291–292. See Bierl 1991, 15–16 and Gödde 2011, 85–88.

tic terms while myth itself creates a scenario of distorted forms of violence and anti-culture.

The Concept of *Chronotope*

Time and space always interfere with each other in specific configurations. For chronological and spatial interconnected structures in literature the Russian formalist and classicist Mikhail Bakhtin (1895–1975) coined the term, and theory of, chronotopes. His concept, on the one hand, is rather ahistorical and abstract, taken from mathematics and the theory of relativity, yet, on the other hand, he attempts to link certain chronotopes to real-life experiences. For some genres and epochs, the chronotopes boil down to objects of motifs. His analysis of the chronotope of the ancient novel became well known. He argues that the primary dimension is space, whereas the temporal axis moves almost against the zero-point. The adventure time that perpetuates itself in the plot of the erotic novels has very little effect on history or processes drawn from everyday life, least of all on the biological time of the heroes, who do not undergo any kind of maturation or mental development.[10] Even in the field of the ancient novel, critics rehabilitated the temporal axis. Be it as it may, the tragic chronotope obviously favors space over time as well.

The purely literary approach was recently complemented through the exploration of several chronotopes in competition with each other. Richard Seaford discusses chronotopes in tragedy as matrices of social behaviour. In the vein of Émile Durkheim, he sees them in terms of models for a cognitive perception of the world and as social constructions based on socio-economic and cultural achievements. According to Seaford, various chronotopes operate in the tragic text as if in a battleground. Like in the Hegelian or Marxist evolutionary model of three steps, one stage evolves into the next one to reach the final solution. The reciprocal chronotope of Homeric time comes under attack after the invention of money, leading to the unlimited or monetised chronotope. From this internal conflict, a reintroduction of communal ritual in the new aetiological chronotope emerges. Tragedy mirrors the socio-historical evolution in the progression from unlimitedness to limitedness played out in the institutionalization of a final ritual.[11]

10 Bakhtin 1981, 84–258; on the ancient novel, see Bierl 2006, esp. 73 (Engl. 2014).
11 Seaford 2012, esp. 1–10, 11–121. On *Antigone*, see 327–332, on the fifth stasimon, 331.

Deeply influenced by his model and teacher, Lucas Murrey draws upon Seaford when emphasizing the role of mysteries in tragedy. In a book on Hölderlin, whose translations of Greek tragedy inspired his new hymnic poetry, he develops an internal three-step evolution based on mystic ritual of three sub-chronotopes from the so-called "Dionysiac chronotope". Disoriented by an overpowering and absolute limitlessness, the initiate overcomes the crisis of death to experience complete unity. Thus, like in a *rite de passage*, the unlimited chronotope, which he extends to visual and abstract thought, facilitates a near death experience from which the ordered, limited chronotope emerges, one equal to Seaford's ritualistic and aetiological chronotope. Moreover, he connects the unlimited, monetised sub-chronotope to the visualized chronotope. In this framework he explores how the visual and the audible are captured in tragedy and modern reception.[12] The Dionysian, as Murrey points out, reveals itself in the distorted utterances of fear and horror as well as in the vocal cry of liberation and the specific visual experiences acting out this moment of limited unity.[13]

Departing from my previous work on Dionysus, I will introduce a Bacchic-chor(a)ic chronotope,[14] which will not be based on socio-cultural construction and internal development, but instead it will focus on the concrete moment of eruptive energy as an expression of chorality. It is thus punctual instead of extensive in the sense of a rite of passage or an internal conflict of different stages. Thus, in a way, it is not constructive diachrony but pure, aesthetic synchrony.

Dionysus as God of the Other, Ambivalence, Mediation, Performance and Presence

Before we can further pin down the chronotope we must first endeavor to define the god behind it. Dionysus' dazzling ambiguity fascinates us more than ever in these (post)postmodern times. Like a kaleidoscope he constantly oscillates between manifold manifestations. Therefore the elusive god resists clear-cut and simple definitions.[15] As a multi-faceted, ambiguous and transgressive deity,

12 Murrey 2015; on the "Dionysiac chronotope", see esp. 9–23. On the fifth stasimon of *Antigone*, see 55.

13 Murrey 2015, 37–60 (with examples taken from Greek tragedy).

14 Bierl 1991; Bierl 2011a; Bierl 2011b; Bierl 2013a.

15 The following passage and few other parts are based on Bierl 2012. On the role of Dionysus in all three dramatic genres, see Bierl 2011a. On the following description of the god, see Bierl 2011a, 315–316 and Bierl 1991, 13–20. On some basic ideas about Dionysus as the personification of the ecstatic performance, see Ford 2011, esp. 347–355; Bierl 2012; Bierl 2013b, esp. 36. On

full of energy and abounding vitality, Dionysus is always on the move and in constant flux. He notoriously exposes others to transformations within a range of categorical oppositions. Dionysus occupies the "in-between" between oppositions, simultaneously present on both sides, which, understood as energetic forces in dynamic reciprocation, tend to fuse under his influence. He is not only the violent, ecstatic and destructive power, but also the central urban deity stabilizing society. It is well known that myth and ritual complement each other. Myth encompasses scenarios of inversion and violence, while cult embraces phenomena of group cohesion like festivity, enjoyment and happiness. In the realm of Dionysus, however, myth and cult cannot be neatly separated. Ritual reenacts myth and myth mirrors ritual. Taking everything into account, it makes the most sense to assert Dionysus as the figure of the Other, also in the Lacanian sense, since he epitomizes difference more than any other Greek deity.[16] Most of all, he tends to show his liminal presence when displaying his whole array of signification. We constantly encounter him as arriving from afar or even from the realm of the dead, manifesting himself in his manifold forms as an epiphanic god *par excellence* (ἐπιφανέστατος θεός).[17] This attitude is counterbalanced by scenarios of resistance which he must overcome through miracles and other spectacular manifestations.

Among his main features and areas of responsibility we can name: 1. wine and inebriation; 2. wild nature, vegetation and animality; 3. madness and ecstasy; 4. the Underworld and death; 5. mysteries and afterlife; 6. sex and eroticism; 7. dance, music and performance; 8. mask and costume; 9. fiction, imagination, vision and miracle.[18] The last three items establish his role as the god of theatre. A community assembled in a θέατρον watches the spectacle or show (θέα) of a procession bringing the deity into the city,[19] a celebration of the arrival of "the

---

Dionysus in tragedy, see Bierl 1991; on the chorus, also often in a Dionysiac context, see Bierl 2001 (Engl. 2009). On Dionysus in general, see Henrichs 1982; Henrichs 1996a; Schlesier 1997. On Dionysus as a different god, see the volume by Schlesier 2011.

16 See Gernet 1953, 393 and above n. 9.

17 See the inscriptions of Antiochia *CIG* III 3979 und *CIG* 1948 (named together with the chthonic epithet Eubouleus; see the gold leaf from Thurii fr. 491.2; see also fr. 488.2, 489.2 and 490.2 Bernabé = 5–7 Graf and Johnston); on Dionysus' particular presence and tendency to show himself in an epiphany, see Otto 1933, esp. 70–80; Henrichs 2008, 19; Henrichs 2011.

18 See Henrichs 1982, esp. 139; Henrichs 1996a, esp. 479; Henrichs 2008, esp. 23; Schlesier 1997 (esp. C: "Wirkungsbereich"; Engl.: Dionysus. Brill's New Pauly. Brill Online, 2013. Reference. Universitaetsbibliothek Basel. 29 January 2013 <http://referenceworks.brillonline.com/entries/brill-s-new-pauly/dionysus-e320270>).

19 See Kavoulaki 1996 (she does not include Soph. *Ant.* 1115–1152).

coming god".[20] Dionysus can function as a θεατής (spectator), masked actor or even the leader of his ritual chorus that celebrates him with song and dance, we also find him among his entourage of mythic female maenads and male satyrs occupied with the same performative activities. The procession, θεωρία or πομπή, serves as a matrix for drama that evolves from further choral expansions. These χοροί were performed after the reception of the god in the city centre and eventually evolved into theatre performed with the polis as audience. As the coming god Dionysus bridges the fixed boundaries of time and space, old and new, past, present and future, absence and presence, inside and outside, here and then, death and live, beneath and above. Dionysian chorality helps create a performance of blurring perspectives mediating between myth and ritual, dramatic action and cultic framing as well as time and space with regards to the mimetic-representational, the imaginary and the pragmatic here and now of the Athenian theatre of Dionysus. In his epiphanic quality of *deus praesentissimus*,[21] manifesting himself through performative means, Dionysus becomes the emblem of the new thinking of presence in mediality upon which we have embarked following the excesses of poststructuralism.[22]

Upon this background the Bacchic choric chronotope takes shape. Intense Dionysian songs featuring this chronotope happen often at decisive turning-points, moments of *metabole* or even *peripeteia*, moments of choral self-referentiality often paired with choral projection.[23] Theatre reflects theatre in a metatheatrical manner.[24] The chronotope is positioned at transitions, at the boundaries, those in-between points of liminality where myth and ritual fuse and the categories of time and space blur.

At these moments of epiphany and choral manifestation, tragedy becomes aesthetic form and pure performance.[25] Medium, so to speak, becomes the message:[26] A sudden, miraculous expression of a holy time and place in synaesthe-

20 See Otto 1933, esp. 74–80; on the "kommende Gott" as a Romantic concept (G.F. Creuzer, F. Hölderlin [see *Brot und Wein*, line 54]), see Frank 1982 and Henrichs 1984, 216–219.

21 See Ov. *Met.* 3.658–659: *nec enim praesentior illo / est deus*. See Henrichs 2011, esp. 105.

22 On the new era of medial presence, see Kiening 2007a and the contributions collected in Kiening 2007b.

23 On Dionysiac choral self-referentiality, see Segal 1982, 242–47; Bierl 1991, e.g. 35–36, 83–84, 99, 106–107, 129, 155, 164, 190–191, 224, 242–243 (where Dionysus is associated with self-referential and metatheatrical utterances) and Henrichs 1994/1995. On choral projections, see Henrichs 1994/1995, esp. 68, 73, 75, 78, 88, 90.

24 Bierl 1991, 111–119 and Bierl 2001, 43–44 (Engl. 2009, 29–30).

25 For modern poetry, these phenomena are well described by Bohrer 2015.

26 See McLuhan's 1964, 23 famous sentence "The medium is the message".

sia that extends onto the cosmic;[27] eruptive energy in ecstasy, often associated with fire, pure music and dance, pre-linguistic noise and cry. The chronotope is hybrid, permeable to the extra-dramatic instance of utterance, a shifting zone between inside and outside, ritual and myth. It mediates like the chorus, its hallmark.[28] *Ekstasis* is combined with *enthousiamos*; full of the god and medial noise, one steps out of the usual individuality. The subjective person melts with the community. In Attic theatre, Dionysus manifests himself in performance, rituality and *choreia*, that is, in wild cries, in *theoriai* and *pompai*, accompanied by the shrill melody of the *aulos*, the wild rhythm of *tympana* and ecstatic dance and choral movement, while often specific groups or persons vehemently oppose him in mythical terms.

Mania mediates between the deity of frenzy and his entourage. The maenads set the god in frantic movement in the same way that he, as their choral leader, is responsible for the choral and multimodal performance of his group. This results in the typical fusion and oscillation of perspectives. The god, to some extent, embodies the pure, semiotic voice of the cultic devotees. Epithets like Βάκχος, Βακχεύς, Βακχεῖος, Ἰόβακχος, Εὔιος, Ἴακχος, Ἰήιος, Ἰυγγίης, Σαβάζιος, envision the god as the personification of the inarticulate cries delivered in short and iterated combinations, such as *bakch-*, *eua-*, *eui-*, *iakch-*, *ie-*, *iy-*.[29] The lack of propositional meaning entails an instantiation of "poetic function" according to Roman Jakobson.[30] The utterance in lyric frenzy tends to reference only itself, the performative eruption of energy. Such self-contained moments narrate little, but employ highly ritual forms similar to lyric poetry with a tendency to reference their own chorality and performativity. If any proposition exists, it amounts to myth that frames the ritual.

Already in another publication, I have described how the fusion of word, music and rhythm achieves both an emotional intensity and an insidious loss on the level of signification.[31] Both performers and spectators experience a sense of integrative unity, cohesion and inner meditative centering. The Bacchic chronotope conveys the subjective impression of being included in a greater whole, a feeling of oneness with the cosmos among the performers and partici-

---

27 On holy time, see Pindar *Paean* 6.5 (fr. 52f.5), ἐν ζαθέῳ … χρόνῳ.

28 Gagné and Hopman 2013a, esp. Gagné and Hopman 2013b, esp. 1–28.

29 See Versnel 1970, 16–38, esp. 27–34; on Iacchus, see Graf 1974, 51–66. On the entire argument, see Ford 2011, with a reference to the poetic dimension, 355.

30 Jakobson 1960, esp. 358 [= *Selected Writings* III, 27]. See Tambiah 1985, 165 and Bierl 2001, 287–299, esp. 293 with n. 503, 331–346, esp. 335 with n. 92 (Engl. 2009, 254–265, esp. 259–260 with. n. 503, 296–310, esp. 299 with n. 92). See Bierl 2013b, 36.

31 Bierl 2001, 293–299 (Engl. 2009, 259–264).

pants. Furthermore, I tried to give an account of the effects in cognitive and neurobiological terms. In spillover effects, the flow of stimuli from parts of the central nervous system is diverted to other regions that contain vegetative and sensory centres as well as centres responsible for integrated thought and imagination. The performers (as well as the responsible god) thus become the transmitters and receivers of a message.[32] This paradoxical communication situation is reflected in the reciprocal relationship of χάρις that characterizes hymnic discourse. Ritual thus proceeds on the basis of two mutually complementary and asymmetrical systems. In such excited song and dance, the frontal cortex of the brain, the sensory centres of the cerebral cortex, the left and right hemispheres of the brain and finally hierarchically constructed levels of cortical and subcortical, endocrinal and immunological structures, are stimulated in an energy-creating (ergotropic) or energy-retaining (trophotropic) exchange.[33]

As said, Dionysus displays similarities to Lacan's Other, that is the unconscious, associated to the desire, and this desire of the Other produces language, images, music and dance in the symbolizing process of signification. According to Jacques Lacan, the ego only deceives itself into believing in an individual unity through imaginary means, in opposition to the real. In an intersubjective web, it succumbs to the symbolic, to an alienated Other or Id. It is encoded in linguistic signs on the basis of chains of signifiers by way of the supplementarity in the tropological play of metaphor and metonymy.[34] The human worshipers are separated from God and desire the Other's presence. Close to death, but also abounding in energy, vitality and mystic union, Dionysus constitutes a special form of the Other. Desiring to close the gap between oneself and the deity, this desire, to some extent, becomes the Other's desire, the adoring chorus and spectators engage in symbolic utterance constitutive of a continuous gliding, a "*glissement incessant du signifié sous le signifiant*",[35] which closes the gap between themselves and the Other, creating its imaginary presence.

32 See Tambiah 1985, 145 and 154. On the gods as the ultimate causal impetus behind phenomena experienced in the ritual process, see d'Aquili and Laughlin 1979, 170–171.
33 D'Aquili and Laughlin 1979, 172–180.
34 See Alexiou 2002, 151–171 and Bierl 2007, 255–258.
35 Lacan 1957 (= 1966, 502); Engl. Lacan 2006, 419 ("incessant sliding of the signified under the signifier"); see also Bierl 2006, 85–86.

The Concept of *Chora*

At the same time, the space enclosing the Bacchic *choros* in the Other's epiphany and gliding utterance can be associated with the famous *chora* described in Plato's *Timaeus* (48e–53c). This space proves notoriously elusive, difficult to grasp, a "third species" (τρίτον ... γένος 52a) situated between the ideal forms and their copies (*mimemata*) in the phenomenological world. Somehow it is the unconscious Other, to be understood only "through bastard logic" (λογισμῷ ... νόθῳ 52b) and in a "dreaming state" (ὀνειροπολοῦμεν 52b). It is a receptacle (ὑποδοχή), a nurse (τιθήνη) of all becoming (49a), "a molding-stuff (ἐκμαγεῖον) for everything, being moved and marked by the entering figures" (50c), a wax-material without imprints, a medium, a substrate and catalyzer bringing forms into being and rendering becoming understandable. It possesses only traces (ἴχνη 53b) of the form. It is the space in and from which fire appears as the fiery inflamed or water appears as substance liquefied and drinkable (51b, 52d). It is a "helper" (βοηθός 52c) to make things manifest. God needs to approach this mediating space to effectuate his epiphany (53b).

These quintessential features also describe Dionysus, who suddenly manifests himself in fire and water; born from the maternal womb burned by Zeus' lightning close to the river Ismenus, Dionysus is, in a way, the catalyzer, the medium situated between the worlds, becoming present in his elements. Between man and woman he represents the maternal and creative, working as nurse via vibration and motion. The *chora* as receptacle serves as a winnowing sieve to cleanse the wheat (περὶ τὴν τοῦ σίτου κάθαρσιν), separating it from the chaff, or respectively in a figurative sense, shaking the particles and bringing them into visible formations (52e).[36] Shaking the body means kinesics and dance, thus *choreia*. Dancing belongs to Dionysus who shakes the ground with his feet. Etymology and new studies emphasize the connection between χώρα and χορός.[37] *Choros* also means the dancing floor, and *chora* is again the space enclosing the *choros*, where the group of dancers performs, communing with the divine, and from where its vital energy is processed and made present as *choreia*. Their alleged Indo-European common root **gher-* is the place where this is made possible due to the arrival of a god. According to Julia Kristeva, *chora* is the feminine, erotic and emotive space linked to the body and materiality, and is thus creative. As it has to do with sensation, mood and memory, it is

36 On modern readings of the passage, see Casey 1997 and Sallis 1999.

37 See Chantraine 1968, 12 – he identifies **gher-* as common root, "a place to which something/someone goes or is taken" (Boedeker 1974, 86); see also Boedeker 1974, 85–91; McEwen 1993, esp. 41–78.

socially connected and in touch with the people captured as a body. The depiction of the state in Plato's *Timaeus* is complementary to the purely mental and cold construct of *Republic.* Kristeva argues that magic, carnival, creativity, mystic shamanism and poetry are situated in the *chora.* It is creative as well as a method, a different beginning or form of procedure, an *inventio* in the sensational space. Its signification revolves from the symbolic, logocentric, abstract language back to the purely semiotic, that is language as voice, cry or pre-linguistic emotive language.[38] Other critics emphasize the gliding logic, the creativity through an associational network woven like a traversing *choros* on the dancing floor. Its narrative logic is neither linear nor progressive but it loops both backwards and forwards. It is the place of traces, of emergence, and furthermore a space of memory, of the lively past reenacted for the future.[39] As François de Polignac has shown, *chora* is also the concrete space around the city, the countryside perimeter of about 5 to 8 miles around the *asty* and city centre, the location of many shrines, temples and processions. It represents the space where the holy is situated, moving along bipolar axes to and from the city, being the transforming receptacle where the city's potential energy and vitality are grounded.[40] Just to find the right designation, a name, for this elusive phenomenon is aporetic. *Chora* retracts itself constantly and gives way to new borderline reasoning, "somehow to bastard thinking" (λογισμῷ τινι νόθῳ *Timaeus* 52b), situated at the edge.[41] The place again recalls its constitutive god Dionysus, who produces paradoxical logic of the third kind. Particular confusion lies in his *subjectivity*, his shifting identity as theatrical persona, between "I" and "you".[42] As is well known, his mask is his face – he constantly eludes our grasp with mimetic reenactments.

To summarize, *chora* is a matrix and space in which the idea of the Other is transformed into choral energy, it is a substrate of creative and poetic invention. In Attic theatre, Dionysus becomes particularly present when passing through *chora*, the container of the unconscious, paradox and ambiguous Other. All three, *chora*, *choros* and Dionysus as the decisive deity, mediate between stages, are situated in an in-between zone, making energy erupt. The semiotic overload

---

38 Kristeva 1984, see also the analysis by Rickert 2007, 260–263.

39 McEwen 1993; on a more abstract level, see Derrida 1983; Derrida 1995, esp. 89–127. On Derrida, see Rickert 2007, 263–267.

40 Polignac 1995, esp. 21–88. As Cursaru 2014 underlines, in Greek tragedy χώρα, i.e. space as χῶρος in the specific and restricted sense, designates the "'territorialized' identity" (113) of a polis based on religious and political discourses of authority.

41 See Derrida 1983; Derrida 1995, esp. 89–127.

42 Nagy 2013, 582–585.

creates fusion and flows through excess; oppositions collapse. *Chora* is the receptacle, the space into whose centre the god is led in procession and where he becomes present and epiphanic via choral energy.

The Parodos of Euripides' *Bacchae* and Philodamus' *Delphian Paean*

To develop a truly Dionysiac foil of comparison, I will quickly explore the *parodos* of the *Bacchae* (64–166) as it paradigmatically displays the Bacchic, choral and choraic chronotope. In a recent article,[43] I argued that the epiphanic god's arrival is realized and experienced as choral multimediality in an incoming procession (πομπή). The Bacchic chorus, as a mediator, can cross boundaries, enhancing the quintessential tendencies of fusion in the realm of Dionysus. In the arrival zone, the entrance of the god in song and dance breaks the resistance to the god typically built in the mythic constellation. The retained energy is transformed in the *chora* and released in an all the more violent manner through the god's entourage. In terms of inner-dramatic events, the *parodos* functions as an interface for the further course of the play, where the arrival in the city of Thebes and the expulsion of the opponents into the mountains simultaneously represents the transition to the brutal events on Mt. Cithaeron. After all, the chorus of Lydian *Bacchae,* as a theatrically and aesthetically confusing ensemble, becomes the energetic message, set in a *mise en abyme*, in the rhythmical and ritual performance. Most of all, through choral projection, the movement toward the inside simultaneously and paradoxically becomes one toward the outside, integrating perspectives into the past and future.[44] Thus, the dimensions of time and space, as well as other oppositions, blur in a ritual flow of *choreia* and performativity. Past, present, future and the actual, cultic time of performance fuse. Furthermore, the space and time of myth are reenacted in the ritual of drama. Multiple loops create a sensation of unity and *communitas* in the liminal 'anti-structure'.[45]

In *Bacchae* we indeed have two choruses: the actual chorus of Asian devotees, i.e. the *Bacchae*, and the internal, imaginary offstage chorus, so in fact three choruses, of Theban women, who, having been chased to the mountains,

---

43 See Bierl 2013a; see also Bierl 2011b (in German). See also Kavoulaki 1999.

44 Segal 1982, 78–124, esp. 87 and 245 recognizes that the centripetal force of the inside will be inversed by the centrifugal dynamics of the outside.

45 On *communitas,* see Turner 1974, esp. 274 (definition); on anti-structure (in relation to *communitas*), esp. 45, 46, 50, 272–298.

brutally kill Pentheus, tearing him apart and eating his flesh raw. Semele's three sisters become their virtual choral leaders. The aristocratic women who opposed Dionysus become, by his intervention, cultic followers of the god, that is, wild maenads, perverted ritual bacchants acting out the mythic drama of revenge against the resistance. This second Theban chorus is only imagined offstage in the imaginary space, in the zone beyond the *chora* of the wild mountains created by narration. The audience in the orchestra never sees it, but the chorus on stage dances and sings, just reflecting and at best spurring on the actions of the other chorus.

In short, Dionysus punishes his city through his own ritual, performative and theatrical means. The Lydian worshippers praise Dionysus in εὐοῖ-cries (151), the Euios, the, so to speak, divine embodiment of the ecstatic shout (157). The maenads project themselves onto Mt. Cithaeron in a wildly iterated cry ἴτε βάκχαι, ἴτε βάκχαι (83, 152–153) and simultaneously lead the god "from the Phrygian mountains" (Φρυγίων ἐξ ὀρέων 86) into the city of Thebes, i.e. "into Hellas' broad streets for choral dancing" (Ἑλλάδος εἰς εὐ- / ρυχόρους ἀγυιάς 86–87). In the eyes of the chorus Dionysus represents their chorus leader, just as they envision him as χορηγός and ἔξαρχος (141) of the Theban chorus in the mountains.[46]

In the typical manner of total fusion and reciprocity between performance and space, the *chora* which Dionysus enters is often imagined in a state of frantic dance (αὐτίκα γᾶ πᾶσα χορεύσει *Bacchae* 114).[47] This performative transference to the natural environment underlines the all-encompassing epiphany. We witness a similar totalizing effect, the projection of chorality onto nature, countryside or polis space, in the Delphian paean of Philodamus of Scarphea,[48] lines 19–20: πᾶσα δ' ὑμνοβρύης χόρευ- / ε[ν Δελφῶ]ν ἱερὰ μάκαιρα χώρα ("entirely full of hymns danced the sacred and blessed land of the Delphians"). This happened when the god was born in the past and then he returned to Thebes (lines 5–17). But especially at his first stop on his tour in Delphi "he himself made his starry figure epiphanic" (αὐτὸς δ' ἀστερόεν δέμας / φαίνων lines 21–22), dancing with Delphic maidens on Mt. Parnassus (lines 18–26). In the same way as he had appeared in Delphi in the past with choral celebrations, the group once

46 Dionysus is often notionally envisaged as a virtual divine *choregos* or *exarchos*, see Bierl 2001, 42, 144 n. 101, 145, 147–148 (Engl. 2009, 29, 120 n. 101, 122–124). Numerous passages associate Dionysus with the action of ἄγειν – such as in Eur. *Bacch.* 115, or in the choral projections on Mt. Olympus (*Bacch.* 412–413) and on Mt. Pieria (*Bacch.* 566–570) – both are linked with Euius.

47 See Bierl 2001, 147–148 (Engl. 2009, 123) on Ar. *Thesm.* 995–1000; Kowalzig 2007.

48 Powell 1925, 165–171 and Furley and Bremer 2001 I, 121–128; II, 52–84.

again calls upon itself to receive Dionysus, the Paean and Saviour, in a procession through the streets, accompanied by choruses (ἀλλὰ δέχεσθε Βακχιά- / [σταν] Δι[ό]νυσ[ον, ἐν δ' ἀγυι-] / αἶς ἅμα συγ [χοροῖσ]ι lines 144–146). The aim is to make him present there through sacrifice and *choreia*. From his birthplace Thebes he proceeded to Delphi, the first stop on his tour. He is then envisaged in Eleusis, where he is celebrated as mystic Iacchus, full of light in a *pannychis* (lines 27–36), and later in Thessaly and Pieria before he is summoned to come back to Delphi again. Thus linear movement and procession alternates with a circular loop and with cyclic dance, the dithyramb (line 133–134, 151), through that he should become epiphanic again in Delphi at the new temple, his final destination, so to speak.[49] Moreover, his statue (ἄγαλμα line 137), according to a command by Apollo, is to be drawn into the temple precinct, "attractive, like the bright beams of the rising sun" (lines 136–137) to be later erected watching "over the sacrifice and competition of many dithyrambs" (lines 132–134). The beautiful image, the statue, alternates with his true epiphany in *choreia* (lines 133–134, 144–146). It is as if the *agalma*, exchangeable with its god, could somehow become animate. It is a well-known fact that this hymn, composed around 340 BC, draws heavily on the fifth stasimon of Sophocles' *Antigone* and on the parodos of Euripides' *Bacchae*.

The latter, the ritual entrance song of the chorus of the *Bacchae*, blends in episodes of Dionysus' mythical beginning glossed over by Philodamus (lines 6–10). The aetiological myths of Dionysus's birth – the death of Semele, Zeus' thigh-pregnancy (*Bacchae* 94–98) and Dionysus' second birth (99–104) – justify his special divine authority and the ritual power. At the same time, in reciting the birth myth, the Bacchants underline the processional entrance in a metaphorical and iconic manner. This is, so to speak, the image of the maternal *chora* set in a *mise en abyme*, the female poetic invention put in mythic narration sung by the *choros*. Violence and energy twice erupt from a bodily enclosure, first from Semele's womb, then Zeus' thigh. The city gate, the female womb and the thigh represent the resisting boundaries, which the baby as well as the Bacchic group must rupture. The ecstatic cries that interrupt the syntax represent the semiotic potential of Dionysus Bromius, the Roarer. He represents the pre-linguistic intersubjective desire of unity and thus of the Other. In the *Bacchae* we can see how the phantasma of an integral subject falls in pieces. According to Lacan, the decentered *ex-istence* of man is constituted on language, chains of signifiers produced in the tropological play of metaphor and metonymy. Returning to the linguistic turn based on Ferdinand de Saussure and Roman Jakobson,

---

49 On circular dances, see *Ant.* 59, 124, 133–134.

Lacan argues that human beings close the opened gaps to the Other, to Dionysus, in a gliding signification process, thus overcoming the experience of *sparagmos* through poetry and choral dance in order to create the illusion of an integral self. In the *Bacchae,* Euripides self-referentially unveils this psychoanalytical and anthropological truth and, in the same way, closes the gap in poetic language and religious mysticism. The play exhibits the *sparagmos* of the man who opposes Dionysus, the Other, and the recomposition of his body. Dionysian phantasmagoric unity is conveyed in poetry taking place in the Bacchic-choric chronotope of *chora*, a place of liminality, of transition, the boundary to death. It is not only the zone in which the Other is brought in and acted out in ecstatic choral dance, but also where the opposing forces are mediated, interacting in a violent eruption of energy and media presence.

Sophocles' *Antigone*

Despite the prevalence of political and philosophical interpretations in a Hegelian vein, Sophocles' *Antigone* is a deeply and intrinsically Dionysian tragedy.[50] This must not automatically lead to a Nietzschean reading. Antigone is hardly a Dionysian figure in Kerenyi's sense, who takes the heroine as such because "the cruel touch of death arouses the liveliest life".[51] Yet death plays the crucial role in all interpretational attempts. Without knowing Lacan's famous interpretation in his *Seminar* VII, held in 1959–1960,[52] I emphasized Antigone's association with Dionysus, especially in relation to her one-sided, almost incestuous relationship with her dead brother Polyneices.[53] Antigone's uncompromising attitude toward death links her decisively to Dionysus, who embodies the dichotomy of life and death and has a special relationship with Hades. Lacan sees the special relation through a spatial lens. It is therefore vital to make up leeway and include his thoughts in our discussion of the Bacchic-chor(a)ic chronotope.[54] According to Lacan, Antigone is situated at the boundary between life and death, on the edge of the so-called "second death", a place defined also by symbolic

50 See Kerényi 1935; Bierl 1989; see also Heinrich 2002.

51 Kerényi 1935, 14. Deeply inspired by Nietzsche, Kerényi (1935, 13–14) perceives tragedy in general as a Dionysian experience, in the respect that the tragic nearness of death inspires lust for life. This is the premise for his interpretation (14–19) of *Antigone*.

52 Lacan 1992, 243–287. On secondary literature, see Miller 2007 and Buchan 2012.

53 Bierl 1989.

54 On *Antigone* and *chora*, see already Randhawa 2014; on Dionysus as instance of the semiotic and on the fifth stasimon, see 304–306, 310.

absence. She is driven by a yearning for death. Her liminal position as a suffering, martyr-like hero on the edge of death, fire and *ate* makes her beautiful, rendering her appearance most beautiful, as if she were hit by an arrow of desire that makes her gleam. She emblematizes her yearning for death, a radiant and visible desire (ἐναργὴς ... ἵμερος, Sophocles *Antigone* 795–796).[55] As if sending out a double of herself immune to destruction, she hovers between life and death and in this in-between status she glows with beauty. Like a saint, she administrates the boundary to the desire, becoming its focus.[56] Gliding on the edge of language, between the being and that which has being, in Heidegger's terminology, she succumbs to signification *ex nihilo*, beyond the usual linguistic process, uttering just semiotic cries of lament in tautological manner. This makes her appear a mystified virgin, close to the Other, who must sacrifice her being.[57]

In a quasi-procession Antigone enters the cave where she is immured alive. This is her position at the second death, and through her suicide she finally transcends the boundary to Hades. In this liminal zone, the Bacchic *chora*, particularly in the fourth and fifth stasima, Dionysus will transform into pure energy and fire and subsequently force Antigone to find her destiny, the end of her course.[58] Her death in beauty is cathartic.[59] Like Dionysus and his chronotope, she is positioned in the liminal and synchronic-circular, in a zone of *metabole* and transition. It is a matrix of mediation and catastrophe, of catharsis through *mania* and ecstasy.

The Fifth Stasimon of Sophocles' *Antigone*

In the famous fifth stasimon of Sophocles' *Antigone* (1115–1152), my main example that I will treat in much more detail, the chorus conjures up the Dionysiac landscape through which to send the god once more, processionally, to his homeland Thebes.[60] The plot of *Antigone* reaches its dramatic climax at this

55 See Lacan 1992, esp. 262–269.
56 See Lacan 1992, esp. 257–262.
57 See Lacan 1992, esp. 270–283.
58 Nagy 2013, 593–594 stresses Dionysus' epiphany in fire ("he literally ignites the singing and the dancing as he leaps out, in an elemental burst of flame, from inside the fennel stalk or *narthêx*" [593]) and links it with an Indo-European concept of a god emerging from a reed with lightning, smoke and fire. Lacan 1992, 268–269, 281–282 touches the fourth and fifth stasima.
59 See also Lacan 1992, 243–247.
60 See Jebb 1900, 198–205; Müller 1967, 242–250; Kamerbeek 1978, 186–190; Griffith 1999, 313–322; Brown 1987, 214–217; Furley and Bremer 2001 I, 301–304; II, 272–279; see also, among others, Kerényi 1935, 17; Vicaire 1968, 358–365; Winnington-Ingram 1980, 110–116; Burton 1980,

stage, coming to the decisive turning point, the *peripeteia*. A powerful arc builds until the end of the song, then drops off sharply at the catastrophe's revelation. After a long and obstinate refusal, Creon cedes to Teiresias' warnings. He is finally ready to bury Polyneices' body and to release Antigone from her natural prison. Creon hopes that he can still avert fate after Teiresias prophesizes the king must pay for his injustice towards Antigone with deaths in his own family (1064–1086). Quickly, but too late, he retreats from his wrong deed and orders Antigone freed from her rocky tomb and Polyneices buried (1108–1114). Tragically, Creon's men embark first on the latter order (1196–1204a) so that they arrive too late to save Antigone, who has stepped over the threshold of death. At this point, however, the chorus places all hope of salvation on its god. The Dionysiac excitement of the god's ecstatic worship transfers, ultimately, to the audience. The author directs the audience's emotions in such a way that the imminent upheaval, when it comes, produces maximum tragic effect; casting it all as an illusion intensifies the sudden plunge into the catastrophe – originated in the god's dark, Theban side – for both the chorus and the audience.

The chorus of Theban old men sings the following song:

πολυώνυμε, Καδμείας — στρ. α
νύμφας ἄγαλμα
καὶ Διὸς βαρυβρεμέτα
γένος, κλυτὰν ὃς ἀμφέπεις
Ἰταλίαν, μέδεις δὲ
παγκοίνοις Ἐλευσινίας
Δηοῦς ἐν κόλποις, ὦ Βακχεῦ,
Βακχᾶν ματρόπολιν Θήβαν
ναιετῶν παρ' ὑγρὸν
Ἰσμηνοῦ ῥέεθρον, ἀγρίου τ'
ἐπὶ σπορᾷ δράκοντος.

σὲ δ' ὑπὲρ διλόφου πέτρας — ἀντ. α
στέροψ ὄπωπε
λιγνύς, ἔνθα Κωρύκιαι
στείχουσι Νύμφαι Βακχίδες
Κασταλίας τε νᾶμα.
καί σε Νυσαίων ὀρέων
κισσήρεις ὄχθαι χλωρά τ' ἀ-
κτὰ πολυστάφυλος πέμπει

132–135; Rohdich 1980, 209–214; Segal 1981, 200–206; Oudemans and Lardinois 1987, 151–159; Bierl 1989, 50–54; Bierl 1991, 127–132; Henrichs 1990, 264–269; Henrichs 1994/1995, 77–79; Scullion 1998; Cullyer 2005; Jouanna 2007, 116–132; Kitzinger 2008, 62–70; Ford 2011, 347–348; Macedo 2011; Rodighiero 2012, 152–165; Jiménez San Cristóbal 2013, 276–279.

ἀμβρότων ἐπέων
εὐαζόντων Θηβαΐας
ἐπισκοποῦντ' ἀγυιάς.

τὰν ἐκ πασᾶν τιμᾷς στρ. β
ὑπερτάταν πόλεων
ματρὶ σὺν κεραυνίᾳ
νῦν δ', ὡς βιαίας ἔχεται
πάνδαμος πόλις ἐπὶ νόσου,
μολεῖν καθαρσίῳ ποδὶ Παρνασίαν
ὑπὲρ κλειτὺν ἢ στονόεντα πορθμόν.

ἰὼ πῦρ πνεόντων ἀντ. β
χοράγ' ἄστρων, νυχίων
φθεγμάτων ἐπίσκοπε,
Ζηνὸς γένεθλον, προφάνηθ',
ὦναξ, σαῖς ἅμα περιπόλοις
Θυίασιν, αἵ σε μαινόμεναι πάννυχοι
χορεύουσι τὸν ταμίαν Ἴακχον.

You of many names,
glorious image of the Cadmeian bride,
son of loud-thundering Zeus
you who watch over renowned
Italy and rule
in the folds of Eleusinian
Deo that are open to all, o Baccheus,
dwelling in Thebes, the mother-city of the Bacchae
beside the liquid
stream of Ismenus and over
the seed of the savage dragon.

You the flashing, smoky flame
has seen over the twin peaks
of rock where Corycian Bacchic
nymphs move (in dance),
and the spring of Castalia (has seen you).
And you, the ivied slopes
of Nysean mountains and the green shore
with many grape clusters send,
while immortal words
shout out their euoi-cry, as you are
overseeing the streets of Thebes,

that of all cities you
honour as preeminent, together
with your mother who was struck by lightning.
Now, since the city and its entire people
are held fast under wild sickness,

come with purifying foot across the slope
of Parnassus or the moaning strait.

Io, chorus leader
of stars breathing fire, overseer
of voices in the night,
child, offspring of Zeus, appear in an epiphany,
lord, together with your attendant
Thyiads, who in maddened frenzy the whole night long
set you in dance, the dispenser Iacchus.

The *hymnos kletikos* addressed to Dionysus for epiphany (προφάνηθ' 1149) retains the typical style of an aretalogy, with all the wealth of formulas that go with it.[61] In the end, the song withholds release but opens up the Bacchic chronotope, the *chora* as receptacle and a special, intense moment of holy time, facilitating the god's arrival. Poetry creates a vivid landscape of transition. It is the space from where Bacchus moves from afar to the centre of the here and now of Thebes, becoming present to the Athenian audience as well. As catalyst, the *chora* is able to transform the abstract and symbolic Olympian idea to pure energy responsible for bringing tragedy to its terrible end. It is the time and space where the boundary to fierce fire and death will be transcended. The god of tragedy becomes epiphanic as the elemental and cosmic power at the edge between life and death, while Antigone embarks on her final way, stepping over the boundary of death. The musical and poetic χορός creates a χῶρος that turns out to be a χώρα, a space like a nurse and female womb, where *mimemata* are born from abstract forms. The mimetic chorus invents its own time and space where *logos* and abstract idea transform to fire and pure energy. It is the matrix and receptacle where voice, movement and vital energy outdo Creon's *logos* and purify from it.

The chorus first addresses Dionysus as he "of many names" (πολυώνυμε 1115) in order to attract his attention,[62] as if it were afraid to forget one. Naming is a notorious problem in symbolic language.[63] With the catch-all formula, the chorus first avoids a precise name and the god, as a result, is not reduced to one specific idea but slowly emerges in the poetic *chora*. At the same time, of course, in the hymnic manner clear epitheta and geographic places accumulate

61 See Norden 1913, 143–177.

62 In Aegina he is even called ἐπήκοος, "the listener" (*SEG* XI.4). On Dionysus' names, see Bierl 2013b.

63 This problem haunts also Derrida 1983 and esp. Derrida 1995, 89–127 in regard to *chora* and from a poststructuralist perspective.

and, in relative and participial constructions as well as in you-predications,[64] the god is called upon. Just as *chora* is notoriously difficult to name – it is a so-called a "third term" and follows a "bastard logic" (Plato *Timaeus* 52a–b) –, the chorus appears hesitant to name elusive Dionysus, who defies clear definition. The chorus seems aware of the aporetic status between signifier and the signified. The god, as the Other, can be captured only in traces (53b), through oneiric, unconscious logic (52b). Only some lines later in the hymn does the chorus devise the solution of applying just the names *Bakcheus* and *Iakchos* that lack any proposition or signification and are just the result of the cries of the followers.[65] Possessed by the god, and in frenzy, the singers themselves bear the name *Bakchoi*, just as his mythical female entourage are called Bacchae. Contact with the deity is established by singing a long chain of signifiers through which the chorus tries to bridge the gap of desire for the presence of the Other. In praising Dionysus, they imagine him at rest in possible cult sites from where he might hasten to the aid of his threatened polis, Thebes. As "coming god",[66] who prefers to manifest himself,[67] we see him in a blossoming landscape returning home as in a procession (*pompe*) – in the first antistrophe the places actually send him (πέμπει 1133) home. As a triumphant leader of the *pompe*, Dionysus oversees Thebes' streets (ἐπισκοποῦντ' ἀγυιάς 1136) where he himself is about to arrive. At the same time shouts of divine and inarticulate words, lacking again any proposition, emit their *euai, euoi*-sounds (ἀμβρότων ἐπέων / εὐαζόντων 1134–1135) in jubilation – a bold trope typical of the Dionysian style and chronotope.[68] The acoustic song and speech elements are seen as active instances that perform an action, i. e. they shout *euhoe* or *euie, euie*. In their enigmatic meaninglessness of ecstasy, the words become divine in the same way that the god embodies the cry. Thus he becomes manifest without being actually present, and supervises the procession in the streets that fuse with the Athenian procession of the arriving god as *agalma* (1115). This first specifying noun is consciously kept ambivalent. He is the "pride" and "glory" of Semele, the Cadmeian bride – the mother, as is typical, is addressed first in this female *chora* –, and the "offspring, son"

64 See Norden 1913, 158 and Dorsch 1982, 66–75.

65 Ford 2011, 348 argues that the adjective *polyonymos* could be "hinting at the confusion such confusion can engender, giving the element the sense it has in *polythrooos* or *polyglossos* for overabundant, even confusing speech".

66 See above n. 20.

67 See above n. 17.

68 Usually persons shout *euoi*. Therefore, some textual critics missed the cultic followers as a subject in the genitive absolutus; after Hartung's conjecture ἐπετῶν Pallis tried to emend in ἐπετᾶν); see also Griffith 1999, 320 *ad* 1131–1136.

(1117) of Zeus. In this first apposition, his genealogy is recalled: it is the famous myth of Dionysus' premature birth on the banks of the river Ismenus, when Zeus' lightning torched Semele, who wished to see Zeus in his true form; the premature fetus was then put into Zeus' thigh for the second birth. Fire and water are thus the focus in the first strophic pair. But *agalma* can also mean the statue that is brought into the city through the procession. He is addressed as divine image of the elusive god that comes from Semele and Zeus. Zeus is deeply roaring (βαρυβρεμέτα 1116) because of the thunder that accompanies the lightning, just as his son is a loud Roarer (*Bromios*). Choral song creates first the image of the quintessential divine dancer in arrest, the phantasmagoric *simulacrum*, whereas later on in the song Dionysus will be captured as an activating, frantic performer, the notional leader, the χοραγός (1147), of a concentric ring of choruses reaching even to the cosmic sphere of the stars. This work of wonder (*thauma*) is a statue of great beauty, almost the Lacanian "sublime object", existing for the gaze. Only in others' mind does it become vivid, animated, as if put in motion. In an illuminating contribution on the "metachoral" quality of dancers as artifices, Tim Power compares the choral statuary *agalma* to the Lacanian concept of the *object petit a (= autre).*[69] From the object of monumental beauty desired in gaze emanates a phantasmagoric quality of magic enchantment. The small *autre* (other) stands in Dionysus' case for the big Other. Besides the progressions from the local to the universal and cosmic, from myth to cult and performance, from many to one name, from the afar to the here and now,[70] we can also speak of a culminating movement from the artificial object in arrest towards energetic movement. The statue of the god is brought toward the centre, to Thebes, in a linear procession that culminates in a circular movement in the sky; the fixed image becomes more alive so that the god, interchangeable with it, separates itself from the artifice and acts as quiet spectator of the arrival of his own cultic procession until he changes into the activating agent, dancing and leading others, including nature and objects in choral kinetics.

In a typical relative clause with a you-predication, Dionysus is praised for watching over renowned Italy and reigning over the folds and valleys of the Eleusinian Deo/Demeter (1117b–1121).[71] The female landscape resembles a bosom or

---

69 Power 2011, 89.

70 See Macedo 2011.

71 On the argument in support of the manuscript tradition Ἰταλίαν (1119), on account of its allusion to the mystery cult there, see the scholion; see Henrichs 1990, 267–269 and Burkert 1987, 142 n. 55. In order to limit the horizon and focus on Euboea and Attica, featured prominently in the song, various conjectures were put forward: Οἰχαλίαν by Dawe 1979 in his Teubner edition of *Antigone* (retained in the 1985 revision) and Ἰκαρίαν by Unger in the nineteenth century.

the folding of the womb, both common and open to all. Dionysus clearly assumes the role of Hades, and solid evidence indicates that around 500 BC, in some areas, Dionysus adopted the traits of an underworld deity.[72] Already the hymn hints at his chthonic aspect through the Italian reference (1119) and his representation as the Eleusinian Iacchus/Dionysus (1119b–1121 and 1146–1154), though to be sure only in the purely positive light of the Eleusinian cult. Both the Dionysian mysteries of the Orphic circle in Southern Italy, and later in Magna Graecia, and the belief in Eleusis, whose adherents associated Dionysus as Iacchus with Demeter and Persephone,[73] are distinguished by rites in which the initiates somehow experienced being near Hades.[74] And clearly, Antigone possesses a special affinity towards Hades.[75] In a transferred sense, Dionysus is both led in and leads Antigone to death, which also falls within the invoked realm of Iacchus' responsibility (1152), who in Aristophanes' *Frogs* must lead the Eleusinian initiates to the realm of the blessed.[76] Italy and Eleusis stand for the mystic aspects of salvation that underline the expressed hope.[77]

---

72 A general discussion, esp. of the archaeological evidence as well, is found in Metzger 1944/1945, 314–323 and Casadio 1994. See also Heraclitus' words ὡυτὸς δὲ Ἀίδης καὶ Διόνυσος (DK 22 B 15) that allude to Dionysus' associations with the chthonic underworld, with the realm of death and the hope of an afterlife. These traits are particularly strong in Bacchic-Orphic mystery cults. See esp. the bone tablets from Olbia (463 Bernabé), dated to the fifth century BC, with the opposition ΒΙΟΣ ΘΑΝΑΤΟΣ ΒΙΟΣ ("Life–Death–Life"). On death and rebirth: Pelinna 1–2 = 485–486 Bernabé = 26 a und b Graf and Johnston, line 1: "Now you have died and now you have come into being".

73 On Iacchus, see Versnel 1972, 23–29 and Graf 1974, 43 and 46–69. In the mystery cult of Eleusis, Iacchus was either the son of Demeter or Persephone, or the husband and πάρεδρος of Demeter. The complete identification of Iacchus and Dionysus emerges in the fifth century in Soph. TrGF IV 959; Eur. *Ion* 1074–1086, *Bacch.* 725–726; Philodamus *Paean*, lines 27–36; and schol. Ar. *Ran.* 404.

74 On Italy and the mysteries there, see the lamella from Hipponion (474 Bernabé = 1 Graf and Johnston), which speaks of an underworld procession of μύσται and βάκχοι on the way to the fields of eternal bliss (esp. lines 15–16); see also Burkert 1987, 22 and 142 n. 49. As comparison, the Eleusinian initiates celebrate a feast in Hades in Aristophanes' *Frogs* 440–459; see Graf 1974, 79–94 and Dover 1993, 250–253 *ad* 440–459. On the experience of Hades in Eleusis, see Burkert 1983, 279–280. According to Burkert 1987, 22, the Mysteries of Dionysus, esp. those in Southern Italy, can be seen as an adjunct to Eleusis. This equivalence in the mystery cults is reflected in the already-mentioned reference to Italy (1119); see Burkert 1987, 142 n. 55. On the Italian-Eleusinian mystery aspects of the song, see Henrichs 1990, 264–269.

75 Antigone's affinity with death is expressed as desire. The chorus explains her behaviour thus: θανεῖν ἐρᾷ (220).

76 See Ar. *Ran.* 351–353, 403, 408, 413.

77 The text is highly ambivalent and ironical, shifting between concrete salvation from death and salvation in the mystic sense. The chorus and perhaps the audience hope that Antigone

With the ecstatic outcry, ὦ Βακχεῦ (1121), the god is finally addressed by his name, Bakcheus. In a way, the god emblematizes the energy conveyed in the frantic voice. Βακχᾶν (1122) is directly jointed to Βακχεῦ, the god is identical with his female followers filled with him and his cry. Instead of the relative-clause, the chorus continues now with an attributive participle to complement the predication: "inhabiting the mother-city of the Bacchants" (1122–1123). Again the female side is emphasized. Thebes is his mother-city and not his fatherland, and he dwells "beside the liquid streams of Ismenus ... on the soil of the seeding of the wild dragon" (1123–1125). Next to the fire, Dionysus is associated with the element of water. Moreover, in his motherland, where he should arrive in procession, the soil has not received the usual seed but the teeth of a slain dragon, a mythical allusion to the horrible violence intrinsically linked to his city. While the first strophe envisions the god on a route from the West to his birthplace – or, to some extent, in the eyes of the audience in Athens, the place of the actual performance –, the antistrophe imagines the god now in the cultic centre of Delphi where Apollo is closely associated with Dionysus,[78] situated in the middle of the East-West axis between Southern Italy and Thebes, and on Mt. Nysa. The focus now shifts to seeing (ὄπωπε 1127), what the audience does in the theatre, the place where people watch (derived from θεᾶσθαι). Besides clear acoustic effects, the spectacle of a procession also has a visual base. But here it is not the people that observe the imaginary arrival of the god, but the flashing smoke-flame of his torches. They are lit with his fire and shine through the night on the twin peaks on Mt. Parnassus, where Corycian nymphs – again young girls like Semele – as Bacchic girls (Βακχίδες 1129), inspired by Dionysus, form his chorus with dancing steps. Next to the fire, the water, the spring of Castalia, sees the god (1126–1130). The sublime object of the gaze is the "you" (1126) separated from the *agalma*. Fire and water stand in parallel order to the strophe: the "you" in the first position repeated (1131) in the second period of the first antistrophe (1131–1136), with strong emphasis laid on the personal touch of the god. The strophic pair culminates in a projection of the celebratory arrival of the god: the Bacchic space, also full of the god, sends (πέμπει 1133) and receives him in a procession (πομπή); the streets and the masses are perceived as a sort of second imaginary audience and imaginary chorus in Thebes, while Dionysus is the choral leader of arriving and welcoming masses that fuse at the goal of the trajectory. He is also the one who observes

must not die. At the last moment even Antigone, facing death, seems to wish to step back from her decision. Yet she still desires death, and through the mystic allusions the chorus evokes hopes for a happy afterlife; see also Henrichs 1990, 266–267.

78 See Bierl 1991, 91–94.

and oversees this spectacle of his own homecoming. Thus, despite entering from outside, he has a view over the inner space of the city, the crowded streets.[79] The participle ἐπισκοποῦντ' (1136) supplementing the σε, the "you" that becomes an object receiving an action derived from natural places, is taken up in the address ἐπίσκοπε (1148), the vocative of the noun ἐπίσκοπος ("overseer" 1148) in the second antistrophe.

In an extended warm-up, possible routes of arrival are imagined, opening a huge Greek space of Dionysian cult. As pointed out, one route extends from far to the West, from Southern Italy, renowned for its mystery cults, via Eleusis, the location of Greece's most famous site for mystery initiation, closely linked to the cult of Athens, to Thebes, his dangerous birthplace caught in incest and circularity, the stage of tragic events. During this procession via Eleusis, Attica in the here and now, the Attic nature of the chorus, comes into the fore as well, culminating in the very last word, his Eleusinian appearance as Iacchus (1152). Moreover, the Eleusinian cult is an alternative – all Attic citizens are initiated – that focuses particularly on the Dionysian dimensions of death and afterlife.[80]

The origin of the second possible advent route is the Panhellenic cultic centre of Delphi where Dionysus shares the cult with Apollo, both gods complementing each other.[81] Like in Thebes terrible mythic deaths also took place in Delphi and the Thyiads celebrate wild performative rituals on Mt. Parnassus, comparable to those on Mt. Cithaeron near Thebes.[82] Or Dionysus might arrive from the North or the East, from Mt. Nysa, located in more than a dozen places, but here, probably, Thrace is meant, as in *Iliad* 6.133.[83] Or it is the Nysa of Euboea renowned for its green vegetation[84] – the scholiast to line 1133 associated the green *akta* with the woods in Euboea or still with Mt. Parnassus. Be that as it

79 On the streets as processional way for the arrival of the "coming god" in the city, see Eur. *HF* 783; *Bacch.* 86–87; Philodamus *Paean*, lines 144–146.

80 On Soph. *Ant.* 1146–1152, see Henrichs 1994/1995, 77–78; Bierl 2011a, 323–324; Ford 2011, 345, 347–348.

81 On the interdependence and overlap of both gods in Delphi, see the *Paean* of Philodamus and Cullyer 2005, 6 (with literature).

82 With the naming of Delphi and the θυιάδες (*Ant.* 1151), the choral song is adapted to Athenian cultic ideas, i.e. the Panhellenic worship of Dionysus in Delphi has a specific cultic connection with Athens. Pausanias (10.4.2–3) reports that the Athenians sent their own θίασος of θυιάδες to the mountains above Delphi, where they performed their nightly celebrations together with the Delphic maenads on the heights of Parnassus near the Corycian Cave, the scenery also described in lines 1126–1130; see also Henrichs 1978, 136–137, 152–155.

83 See also Cullyer 2005, 4, 8–18; the scholion *ad* Soph. *Ant.* 1131 associates it with Phocis to keep it consistent with Mt. Parnassus.

84 This is the opinion of most commentators.

may, the movement is now not only horizontal, but vertical, down from the mountains, the place of the *oreibasiai*, toward the city on the plain. A thriving, idyllic landscape sends the god down, sends him home to Thebes, the "coming god", who tends to manifest himself, so that he might still help in the very last moment.

The venues consist of idyllic bays, springs and rivers; the hills of Mt. Nysa are "rich of ivy", the shore is "green" and "rich of vine" (1131–1133). All these attributes project the Dionysian energy of vegetation, which becomes manifest with the brimming growth of vine and ivy, the toxic substances of *mania.*[85] The futility of this hope for salvation and cure in the very last moment, is captured by the adjective "lamenting", specifying the gulf (1145), anticipating imminent catastrophe. In the Bacchic chronotope, nature becomes active, sees, sends and ejects voices. Dionysian ecstasy and *pathos* is somehow transferred to the Greek landscape that in a kind of projection, or "pathetic fallacy", assumes Dionysian traits and agency.[86]

Moreover, many features of the Bacchic *chora* are echoed in the second stasimon.[87] Also in this song that thematizes *hyperbasia* (605), transgression, desire and *ate* (583, 614, 624, 625), "that inescapable complex of delusion, error, crime, and ruin",[88] water, waves (588), shores (591, 592),[89] lamenting landscape connected with the sea (592),[90] light (610) and fire (619)[91] are prominent. "A certain god" (597), one "from the lower deities" (601–602), an anonymous god (624), must be identified as Dionysus, the hidden mastermind in the background (cf. also 278–279). Despite functioning as *Lysios*, he will not come to bring salvation

---

85 On wine and ivy that in cult and myth represent regular cultic attribute of Dionysus, see Blech 1982, 183–201.

86 After Henrichs 1996b, 61 n. 49, I apply this term coined by Copley 1937 from bucolic poetry to this remarkable phenomenon of poetic, pathetic symbiosis with the natural environment in *Antigone.* See also Eur. *HF* 782–784, where "the polished streets (ἀγυιαί) of the seven-gated Thebes" are summoned to "break into dancing (ἀναχορεύσατ'), and fair water of Dirce"; see also Eur. *Bacch.* 114, where in the Dionysiac atmosphere "the whole land will dance at once" (αὐτίκα γᾶ πᾶσα χορεύσει); see in this regard, the *Paean* of Philodamus, lines 19–23 and *Bacch.* 726–727, where "the whole mountain with its beasts participates in the Bacchic dance, and everything was set in rapid motion".

87 Usually the resonances with the parodos are stressed; see e.g. Oudemans and Lardinois 1987, 154–159; Rhodighiero 2012, 162–164; on the resonances with the second stasimon, see Cullyer 2005. On the Dionysian recurrences and network in the entire *Antigone*, see below and Cullyer 2005; Bierl 1989.

88 Griffith 1999, 219.

89 Compare ἀκταί (592) with ἀκτά πολυστάφυλος (1132–1133)

90 Compare στόνῳ βρέμουσιν (592) with στονόεντα πορθμόν (1145).

91 Compare *Ant.* 1126–1127, 1146.

or positive catharsis (cf. 1143); a family just like Oedipus' Theban clan caught in *ate* does not find "solution" (οὐδ' ἔχει λύσιν 598), but a god – it will turn out to be Dionysus – tears it down (597–598), dissolving everything. In the second stasimon, the chorus sings about destruction due to *ate* and transgression. The images of wind,[92] earthquakes (583),[93] shores and waves are just metaphors to emphasize catastrophe in the familial realm. The song displays again a *chora* where Dionysus and the tragic disaster become real. In the fifth stasimon Dionysus is summoned to wander through it and become epiphanic. In passing through the *chora* he takes on shape and will do exactly what the chorus sang in the second stasimon, destroy the family and lead (ἄγει) to *ate* (624). The ἄγειν is just what the chorus leader, the χοραγός, does. He will lead his entourage and bring *ate*, delusion, error, chaos, crime and total ruin. Leading the chorus through the *chora* in a procession, he becomes manifest as a tragic god, in a cyclic dance that includes the forces of the cosmos, he causes death and destruction. So it goes with tragedy: terrible fear and horror.[94] The κάθαρσις is not meant for the dramatic figures but – as in Aristotle's famous treatment in his *Poetics* – for the audience, cleansed from these emotions by watching and empathizing with the terrible events.[95]

If Dionysus takes the route of death and mystery from Italy to Eleusis, he must pass the Corinthian and perhaps even the Saronic gulf – the scholiast mentions even the Sicilian gulf regarding line 1145 – that would both moan and thus eject sound patterns at the prospect of coming events, consequences of the god's arrival in Thebes. But if Euboea is meant in the idyllic description of lines 1131 to 1133, the πορθμός (1145) to be crossed to reach Thebes in Boeotia must be the Euripus Strait, the narrow channel of water separating the green island from the mainland.[96] And if the chorus imagines Mt. Nysa to be in Thrace, it would be one big north-western route through the Aegean sea, over Euboea to Thebes.[97] From Thebes the god had to traverse the wild Mt. Cithaeron to reach the Attic border in Eleutherae. From there Dionysus Eleuthereus arrived annually as a

92 On the motif of wind, see Cullyer 2005.

93 Compare with *Ant.* 153–154, 1273–1274.

94 On the "ironic" or metaphorical epiphany of a violent Dionysus, see Winnington-Ingram 1980, 115; Bierl 1989, 53; Bierl 1991, 130–131; Bierl 2011a, 322–325; Cullyer 2005, 18–20.

95 Macedo 2011, 407, therefore relates the purifying aspect to Dionysus' role as *Lysios* in mystery context.

96 See the scholiast *ad Ant.* 1145; as an alternative he gives the Sicilian sea, attempting to establish coherence with Italy (1119): ἔνιοι τὸν ἀπ' Εὐβοίας εἰς Βοιωτίαν, οἱ δὲ τὴν Σικελικὴν θάλασσαν.

97 For this possibility pleads Cullyer 2005.

statue (*agalma*) to the Athenian theatre in a preliminary procession preceding the dramatic competitions at the City Dionysia. The exact way is unknown but Eleusis was situated on one of the main roads. Devotees, nature and the god, myth and cult, darkness and light, internal and external spaces, Thebes as tragic venue and Athens as the place of the actual performance, they all fuse into a synaesthetic performance of *choreia* and *pompe*. This impression of totality is enhanced by the chaotic accumulation of places, the zig-zag alternation of Thebes, the place of the tragic plot, and other places abroad, focusing on the Apollonian Delphi and ending with the Attic Eleusis.[98] As Rohdich points out, "centrifugal movement" alternates with a "centripetal" one.[99] This feature augments the interplay between the left and right hemispheres of the brain and between ergotropic and trophotropic exchange. The fact that everyday inanimate objects assume their own agency and vitality in the Bacchic chronotope of chorality enhances the impression of fusion and totality. Strangely enough, and against our expectations, it is not the deity animating things through ecstasy. Rather within the *chora* painted in this choral ode, the god stands in the accusative, the affected object, things like the flashing smoke-flame and water act as subjects instead. They see Dionysus. In the same way that shores send him on a procession or words cry *euoi* and render him ecstatic. The Bacchic *chora* and the chorality act as catalysts. In this receptacle, energy is vitalized and transformed into an active force otherwise inherent in the object only as latent and passive potential.

The second strophe emphasizes that Dionysus holds Thebes in the greatest honour of all cities, as it is the inner space from where he watches the arrival of his own *agalma* in a wild and choral procession. And he shares this function with his "lightning-struck mother" Semele (1137–1139). Lines 1115–1117a already anticipated this passage. The Dionysiac *chora* par excellence is female, the fate of his mother again underlined. Semele suffers a death by lightning, yet it is Dionysus who crosses the border to fire and lightning, bringing about the fatal end. Semele's death happened in the mythical past. Since he is related to purifying fire in the past, he should come also now (νῦν δ' 1140). This is the decisive appeal, the *preces* of the cletic song delayed until this very late point – all the previous utterances, the initial *invocatio* and the naming of the possible abodes

98 Thebes 1115–1118a / Italy 1118b-1119a / Eleusis 1119b-1121a / Thebes 1121b-1125 / Delphi 1126–1130 / Nysa 1131–1133 / Thebes 1134–1136 // Thebes 1137–1145 / Delphi 1143, 1150–1151 / Eleusis 1146–1152. Macedo 2011, 407–408 detects a chiastic order: A 1126–1133 (Delphi and Euboea); B 1134–1136 (Thebes); B 1137–1142 (Thebes); A 1143–1145 (Delphi and Euboea).
99 Rohdich 1980, 210.

whence he might arrive, functioning as a surrogate of the *pars epica*, aligned in a long chain of signifiers closing the gap to the Other.

The chorus calls upon Dionysus to have an epiphany, coming and dancing with "cathartic foot" (μολεῖν καθαρσίῳ ποδί 1143). Some critics contend that the plea for purification would not be "tailored specifically to Dionysus" and argue that Apollo as purifier would be chief among those summoned. That Dionysus figures at all in Apollo's presence stems from the fact of Thebes' being his birthplace.[100] Other scholars offer a better solution, pointing out that Dionysian ritual strove for cathartic effect through the music of flutes and ecstatic dancing.[101] If, in his analysis of the genre, Aristotle correctly asserted that tragedy evoked a catharsis in the viewer (*Poetics* 1449b.24–46), then, in a tragedy performed in honour of Dionysus, an invocation of Dionysus as purifying god logically follows. The request is fulfilled, however, differently than the chorus expects. Ultimately, the end of *Antigone* represents purification in its fullest sense: Antigone becomes, as it were, a sacrifice to Dionysus, the source of her divine inspiration, cleansed of her offense against the polis and miraculously uplifted. Creon's "purification", on the other hand, comes with the deaths of his son and wife, payment for his hubris against the god, because, like Agaue in the *Bacchae* (1296, 1374–1376), his insight into the god's great epiphany comes too late.

Thus, the appeal is a most metatragic affair. Just as the entire Theban population desires purification from a violent disease, so the Athenian people in the theatre need from the sickening violence. This expression of νόσος recalls Teiresias' account in lines 1015–1018. It is, however, less the pollution caused by Polyneices' unburied corpse than the inner strife, *stasis*, in the city of Thebes, as analyzed by Haemon earlier.[102] Healing can be accomplished by wild exorcist dances conducted in homeopathic manner, like in the Corybantic cult. It has been shown by Scullion and others that Dionysus is envisaged as a dancer, with emphasis on his foot serving as a *pars pro toto* for choral dance. The ποῦς epitomizes the kinetic movement of the entire body.[103] It can serve also

100 E.g. Förs 1964, 82 and Rohdich 1980, 210, 214, who sees Dionysus just in his positive, almost Apollonian, aspect as polis-god.

101 Vicaire 1968, 363–364 and n. 39 and Scullion 1998. Similarly also, even earlier, Eitrem 1915, 92–93 and Moulinier 1952, 116.

102 See Scullion 1998.

103 See Scullion 1998, 101–104 and Bierl 2001, 83 (Engl. 2009, 64), esp. on the mention of the foot as choral self-reference; e.g. Alcm. fr. 1.48, 78; fr. 3.10, 70; Pind. fr. 52f.18; Aesch. *Eum.* 371, 374; Eur. *HF* 978; *Heracl.* 783; *Tro.* 151, 325, 333–334, 546; *Bacch.* 864, 1230; Ar. *Thesm.* 659, 954,

as an instrument of aggression when someone steps upon one's victim, jumping on the body. The infinitive as imperative is very forceful. But the μολεῖν (1143) also underlines the movement of procession, the coming as arrival. The choral image directly glides into the second antistrophe entirely occupied with dance. The chorus addresses Dionysus-Iacchus as ἰὼ πῦρ πνεόντων / χοράγ' ἄστρων, νυχίων / φθεγμάτων ἐπίσκοπε ("io, chorus leader of fire breathing stars, overseer of nocturnal cries" 1146–1148).[104] The choral god is called to appear as chorus leader of the actual chorus as well as the chorus of the initiates' procession in Eleusis and the maenadic-mystic chorus of Delphi projected simultaneously onto the chorus of stars. He should appear as supervisor of the nocturnal voices, particularly of the Iacchus-shouts that his ecstatic group of Thyiads in Delphi or the initiates in Eleusis emit. The chorus references itself through song, voice and dance, leading to a rapid succession of different choral formations that culminate in the cosmic *chora* where the stars and gods form choruses. Like in those Russian nesting dolls, one chorus builds upon the next in concentric circles: The chorus of the stars the outermost ring, beneath which dance the Delphic Thyiads, then the Eleusinian initiates and, in the innermost circle, so to speak, the Athenian chorus of the actual performance. Their projected leader on all levels is Dionysus.[105] Since the Attic perspective of the audience is always included in the open *chora* of chorality, they call upon Iacchus, the dispenser (τὸν ταμίαν Ἴακχον 1152), who, as politician and official administrator in charge, becomes the leading figure of the violent movement to cleanse the sickness affecting the entire population, the *demos*, in the tragic city of Thebes. The god gives everyone his share. While the song is performed Antigone transgresses the boundary to the netherworld, to Polyneices and to Hades-Dionysus. Both Creon and the voice of the polis hope in vain. Somehow the choreuts sing themselves into a Dionysian delusion. But Dionysus' catharsis in tragedy will have a tragic result. The god is magically drawn in, he does not, contrary to the excessive expectations of the chorus, have an actual epiphany, but will be manifest as the hidden agent bringing destruction to Creon's house and the end of tragedy. As a god – according to the generic laws of tragedy – he will, of course, not appear on stage, but does so in metaphorical and theatrical terms.

The women, in their madness (μαινόμεναι) and their shouting of *iakch*-cries through the night (πάννυχοι 1149), actually dance Iacchus, bringing forth his epi-

---

969; 981–982, 985; *Lys.* 1306, 1309, 1317; *Ran.* 331; *Eccl.* 483; Autocrates fr. 1.1–6 K.-A.; Pratinas *PMG* 708.14; *PMG* 939.6 (Ps. Arion); Timoth. *Pers.*, *PMG* 791.200.

104 Dionysus is often notionally envisaged as a virtual divine *choregos* or *exarchos;* Bierl 2001, 42, 144 n. 101, 145, 147–148 (Engl. 2009, 29, 120 n. 101, 122–124).

105 Henrichs 1994/1995, 78 speaks of a "cascade of choral projection".

phany. They set him in choral movement, by dancing themselves (χορεύουσι 1152). The diction is again unusual. It is not Dionysus who dances, expressed in the middle voice χορεύεσθαι, rather others, his entourage of wild women, who, full of the god, transfer their frantic energy and set the god, the actual source of power, in motion at last, bringing about his arrival in purely cosmic, star-like energy.[106] It is again the *chora* which transforms the god so that he activates his energy. In the last antistrophe we witness the perspective of the Eleusinian procession culminating in the circular dance at the Callichoron Well, the "Well of the Beautiful Dance", also called Parthenion, the "Well of the Maidens", present to the audience in the orchestra,[107] but again fused with the Delphic Thyiads (1151). This *kyklios choros* evoked also starry dances in the sky, etherial *choreia* of mystic light.[108] Night and day intermingle, the song ends in a projected *pannychis*, where Iacchus should appear like fire in the night, to shine, to cure and to kill as well. The transgression to the burning fire already alluded to in lines 1115 and 1139 culminates in the last antistrophe. The destruction might assume a cosmic dimension, while the assault of the Seven with Polyneices' assistance has already been associated with the element of fire (136, 200, 286).

We find a very similar description of the Eleusinian mystic dance in self-referential terms in the parodos of Aristophanes' *Frogs*. In the complementary genre of comedy, Iacchus is again called in a *hymnos kletikos* as the chorus leader. The personification of the ecstatic inarticulate shout of iterated syllables Ἴακχ' ὦ Ἴακχε / Ἴακχ' ὦ Ἴακχε (316–317, cf. 323–325) is invoked as the "deviser of our festal song most sweet (μέλος ἑορτῆς ἥδιστον εὑρών)" (398–399) as well as the "lover of the dance" (402, 413) who is to lead the mystic dancers in a procession – also an allusion of course to the renowned Eleusinian procession – onto the dramatic scene of the mystic underworld meadow and the orchestra in the here and now (Ἴακχε φιλοχορευτά, συμπρόπεμπέ με 402, 413; cf. προβάδην ἔξαγ' 351). Just like in the climax of the fifth stasimon of Sophocles' *Antigone*, Iacchus is "the light-bringing star of the nocturnal rite" (νυκτέρου τελετῆς φωσφόρος ἀστήρ 342); he is summoned to come with burning torches (340) with which to light the meadow in Hades (343) where vegetation sprouts. His epiphany consists again of light and fire. Again we have the same progression from

---

106 On χορεύω in transitive use with a god in the accusative, see also Φοῖβον χορεύων Pind. *Isth.* 1.7; see Henrichs 1996c, 46–47 with n. 59 and Furley and Bremer 2001 II, 278–279 *ad* 1151–1152 (mentioning an obiectum affectum and even effectum). See also Eur. *Ion* 1084–1086 (in the medium); Rodighiero 2012, 154 n. 61; on the meaning of 'to set dancing', see Eur. *HF* 685–686 οὐ καταπαύσομεν / Μούσας αἵ μ' ἐχόρευσαν.

107 See also Eur. *Ion* 1074–1086.

108 See Csapo 2008, esp. 267–272.

linear march to circular dance, from the horizontal to the vertical axis (in the Underworld and in the sky). In a similar concentric chorality he is envisaged as leader of a projected astral chorus guiding "the youth that makes the chorus" to the mystic meadow, "the flowering marshy ground" (ἐπ' ἀνθηρὸν ἕλειον / δάπεδον χοροποιόν, ... , ἥβην 351–352). Again centrifugal movements – the famous procession in the month of Boedromion went from Athens to Eleusis – overlap with centripetal ones. As a matter of fact, Iacchus is again envisaged as coming through the *chora* in a procession, through the famous marshland where the sacred marriage of Dionysus with the Basilinna takes place in the night of the second day of the Anthesteria. In the passage through this receptacle the epiphanic energy of the god materializes again.

Coming back to the fifth stasimon of *Antigone*, the axis of space, the Dionysian *chora* is transferred from the mysteries in Italy and Eleusis to Greece's marvelous landscape, to Eleusis and a mystic dimension of sky-dance. This celestial climax merges Bacchic-Orphic, Pythagorean and Eleusinian perspectives of a fully orchestrated cosmos, a *choreia* of heavenly spheres, stars, gods and men, all united in an eternal, circular movement. The initiates reflect this mystic vision on the gold lamellae. On the leaf of Petelia, dated around 350 BC (476 Bernabé = 2 Graf and Johnston, lines 6–7), an initiate claims: "I am a child of Earth and starry Sky, / but my race is heavenly".[109] On a leaf found in Thurii, a Panhellenic colony rebuilt with Pericles' help in 443 BC, just one year before the performance of *Antigone*, an initiate argues (488 Bernabé = 5 Graf and Johnston, lines 4a–6): "But Moira overcame me and the other immortal gods / and the star-flinger with lightning. / I have flown (ἐξέπταν) out of the heavy, difficult circle (κύκλου ... ἀργαλέοιο) / I have approached the longed-for crowns with swift feet (ποσὶ καρπαλίμοισι ...)". The flying out of a cycle (*kyklos*) is a typical image of choral dance, indicating the moment when the chorus is left in the sky – the liberation came about through typical Bacchic lightning. Moreover, the movement with "swift feet" toward the garland (*stephanos*) is another metaphor for circular movement, obviously now taking place in the Underworld.[110]

---

109 See also the leaf of Thessaly 484 Bernabé = 29 Graf and Johnston, lines 3–4 and the lamella from Pharsalus 477 Bernabé = 25 Graf and Johnston, lines 8–9 (where the second part is changed to "my name is 'Starry' [*Asterios*]"). Regarding the first part, see the leaves from Rethymnon 484a Bernabé = 18 Graf and Johnston, line 3, from Mylopotamos 481 Bernabé = 16 Graf and Johnston, line 3, from Eleutherna 478, 479, 480, 482, 483 Bernabé = 10, 11, 12, 13, 14 Graf and Johnston, line 3 and from Hipponion 474 Bernabé = 1 Graf and Johnston, line 10.

110 See also Csapo 2008, 270–271. For foot as synecdoche for dance, see above n. 103. For flying and gliding, see Henrichs 1994/1995, 106 n. 105; Bierl 2001, 49 and 297 n. 512 (Engl. 2009, 34 and 263 n. 512). A red-figure *astralagos* vase by the Sotades painter (British Museum E 804, 460–

On the axis of time, past and present perspectives merge: since Thebes is his place of birth, where, in the past, tragedy befell Semele, Dionysus honours his city most of all (1137–1139) in the present, and should come "also now" (1040), in anticipation of the future. In terms of space, the Dionysian song proves particularly complex, reaching from the Attic soil of the here and now to a wider Greek cultic space and further to a cosmic space that encompasses the perspective of mystery and death and then even to the city of Thebes of the there and then of the plot reenacted as the here and now through mimesis. The ritual chorus accompanies and enhances the movements of the projected dances of other maenads who drive the god of *mania* mad and convey the energy that leads to death and violence. Ritual festivity collapses into mythic violence; hope and purification into lament and catharsis through *pathos*. Dionysus oscillates between Hades, Thebes, Delphi, Eleusis, Athens and the nocturnal sky. He breathes fire like the wild dragons of Thebes, bringing light and destruction into the nocturnal darkness of the cave, in which Antigone is imprisoned, and into Thebes. Through wild cries and dancing movements he is magically called to appear. He should become epiphanic through his healing foot – the emblem of the arriving god – and dance. Procession and dance are the theatrical modes through which the god manifests himself. The musical performativity reenacted in mimesis overlaps with the real one in the orchestra. He oscillates between all perspectives of time and space, past, present and future, absence and presence, happy festivity and grisly lament, here and there, life and death. The wild performance of the choral song symbolically helps enhance the *peripeteia* and *metabole* from hope to destruction, from life to tragic death. The numerous effects of presence help turn this epiphanic song into a dramatic climax,[111] emblematic, in a way, of the Bacchic-choral chronotope: we have seen that time and space have special blurring effects; everything is hybrid and the song gives the impression of presence, fusion and aesthetic totality.

The Dionysian Chronotope in the Remaining Play

In the end I wish to show that the fifth stasimon is not exceptional in *Antigone*, rather the culmination of a pattern. As I have argued elsewhere, Dionysus is the decisive reference and hidden agent in *Antigone*.[112] In the following I want to ex-

450 BCE), depicts three groups of women from a female chorus who seem to fly. On the garland metaphor, see Caspo 1999/2000, 422.

111 See Beil 2007, 162–168.

112 Bierl 1989; Bierl 1991, 62–67, 127–132. See also Zeitlin 1993, 154–161.

plore in what way Antigone is linked with Dionysus and his *chora*, whereas Creon serves as the opposite. In Thebes, the place of the total Other, as Froma Zeitlin put it,[113] Oedipus' sons, Eteocles and Polyneices, could not share power; they fell into conflict and ultimately slew each other in the famous battle of the Seven Against Thebes. Creon, the new ruler, now tries to take advantage of this situation, emphasizing solidarity by framing it within polis ideology. As his first official act, he issues a decree according every honour to Eteocles, who fought for the polis, but refuses burial for Polyneices, who took up arms against his homeland (192–206).

The Dionysian captures the ambivalent stance toward one's own city as the chorus, in the parodos, exalts the victory over the terrible foe at dawn of the next day. In theological categories it paints a vivid picture of how Thebes, with the assistance of Zeus, the incarnation of righteousness, has beaten back the attackers. In the joy of victory the chorus calls upon its fellow citizens to stage night-long victory celebrations for all the polis deities with Dionysus leading the victory train. With this the members of the chorus affirm this god's power in making us forget previous horrors through the intoxicating ecstasy of dance and music (150–154).[114]

> ἐκ μὲν δὴ πολέμων
> τῶν νῦν θέσθαι λησμοσύναν,
> θεῶν δὲ ναοὺς χοροῖς
> παννυχίοις πάντας ἐπέλ-
> θωμεν, ὁ Θήβας δ' ἐλελί-
> χθων Βάκχιος ἄρχοι.

> After the recent wars create for ourselves forgetfulness,
> and let us approach all the temples
> of the gods with night-long choral performances,
> and let him who shakes up Thebes, Bacchus, lead our dancing!

The god of the play, embodying, in this song, festive joy and political solidarity for the whole city, replicates the Athenian Dionysus of the present performance. Yet in this passage he proves ambivalent towards Thebes, too. The shaking of the ground in night-long choric dance can also be understood to anticipate the total destruction that ensues; the cosmic and elemental star-dance in a *pannychis* as presented in the fifth stasimon. The Dionysian that manifests itself in fire storms

113 Zeitlin 1990 and Zeitlin 1993. On Dionysus in tragedy from Thebes to Athens, see also Bierl 1991, 45–110. On this spatial tension in *Antigone*, see also Zeitlin 1993, 154–161.

114 Consigning sorrow to oblivion constitutes a standard part of Dionysus' positive effects on his human worshippers; Eur. *Bacch.* 188–189 and 282–283.

will ruin and reduce the polis to rubble like an earthquake – in similarly miraculous ways the god becomes epiphanic in the miracle of the palace collapsing in the *Bacchae* (576–603, especially 585–593, and 605–606) –, while the god of tragedy again leads the chorus. Tragedy expresses its own tragic logic in choral and musical terms. Its god is indeed the chorus leader who leads the figures to *ate* (624), and to some extent, to *lysis* (cf. 598). Moreover, the aggressor Polyneices also evinces a Dionysian interpretation. As the chorus describes it, he as "fire-bearer" (πυρφόρος) "breathed upon" (ἐπέπνει) his own city, "in the fury of his mad rush" (μαινομένᾳ ξὺν ὁρμᾷ) like Bacchus (βακχεύων) "with the blast of hateful wind" (135–137).[115] In his ecstatic destructiveness, he becomes the human embodiment of Ares, who in other Greek tragedies also aligns with Dionysus' order-destroying nature.[116] Yet with regard to the local tutelary deity, the chorus stylizes Polyneices simultaneously as an enemy of both Dionysus and the polis. Polyneices' burial drives his sister Antigone into conflict with Creon. The new leader and uncompromising representative of the polis ideology will soon turn out to behave as a kind of raving θεομάχος himself, who like both Pentheus in the *Bacchae* and Lycurgus mentioned in the fourth stasimon of *Antigone* (955–965) adopts Dionysian characteristics even as he struggles with the god. The fire and the breath in Polyneices' nightly attack, as well as the Bacchic *mania* led by Dionysus clearly anticipate the second antistrophe of the fifth stasimon, where Dionysus is called ἰὼ πῦρ πνεόντων / χοράγ' ἄστρων (1146–1147). Dionysus is positioned at the border of fire and terrible light, storm and destruction.

Antigone is linked to her brother Polyneices through an almost incestuous love. At the same time, she is the agent of Dionysus himself. Situated at the very border between life and death, she feels a strong desire to trespass it and die (see θανεῖν ἐρᾷ 220). In Lacanian terms, Antigone is pure desire, the "sublime object", transcending all social categories in her total enjoyment (*jouissance*), going beyond the pleasure principle. Eros and her love of the dead brother, the desire to be united with him through death, serve as leitmotifs in *Antigone* (220, 522–526; *philein* 781–805). The guardian's description about Antigone's first burial with dust deals with the *chora*, the edge of the city, and bears all the signs of a miracle (θαῦμα 254) also typical of Dionysus' epiphanies. The cho-

115 Almost all commentators connect these lines with Capaneus, who at least in the parodos is not explicitly named. As far as I can see, only Segal 1981, 166, 170, 197, 202, Lonnoy 1985, 68 and Zeitlin 1993, 156 connect this Bacchic characterization to Polyneices. I repeat here my argument from Bierl 1989, 47 and Bierl 1991, 63. Capaneus' name can only be derived from the version of the myth handed down from Aeschylus in *Seven against Thebes*.

116 See Eur. *Bacch.* 302–304; see Dodds 1960, 109–110.

rus hints at the right agency, believing the deed was "god-driven" (θεήλατον τ' οὔργον 278–279). It is a δεινόν, something powerful and tremendous (323). In the famous first stasimon (332–375) the chorus takes up this theme of the wondrous man, reflecting on the fact that the only real boundary for men is death – the boundary Antigone desires to transgress. When Antigone, caught in act, is brought in, the chorus leader calls her a τέρας δαιμόνιον ("god-like marvel") (376). The guardian then triumphantly reports Antigone's second burial and how she was caught (407): It was the dangerous hour of midday beneath the burning Mediterranean sun. "Suddenly a whirlwind lifted from the earth a storm of dust" – a natural wonder again – "a trouble in the sky" (417). Dionysus manifests himself in that wonder (*thauma*), in fire, heat and wind. The immediacy, the sudden and surprising act (ἐξαίφνης 417) is typical of his epiphany. The group closed their eyes (μύσαντες), like in a mystery epiphany, before the "divine sickness" (θείαν νόσον 412). Dionysus can either purify or send a plague. In this case it is the little girl Antigone, who "lamented aloud with the sharp cry of a grieving bird" (424–425), fulfilling the basic burial rites again with some dust and libations. Interrogated by Creon, she affirms that she obeyed "the laws which Justice who dwells with the gods (ἡ ξύνοικος τῶν κάτω θεῶν Δίκη) below established among men" (451–452). In her desire for death she says that if she must die before her time, she would count that a gain (461–462). In the chorus' eyes she is "the raw offspring of a wild-raw father" (τὸ γέννημ' ὠμὸν ἐξ ὠμοῦ πατρός 471), Oedipus. The adjective ὠμός hints at the *omophagia*, the custom of eating raw flesh among the god's Bacchants. Additionally, Polyneices's body is polluted by raw-eating dogs (697), who like to tear their victim apart.[117] In Creon's eyes, Antigone is crazy (λυσσῶσα cf. 492; see 633; ἄνους 562; μαίνῃ 765), recalling Dionysus' *lyssa* and *mania*.[118] Contradicting her sister Ismene, Antigone claims her deed was done in accordance with Hades and the ones below (543), the dead. As already established, Hades is identical, in certain respects, with Dionysus;[119] thus Antigone acted with his consent. She seeks

117 On the allusion to a σπαραγμός by wild animals, esp. dogs, on the body of Polyneices, see *Ant.* 1080–1083, esp. σπαράγματ' (1080); see also κυνοσπάρακτον σῶμα Πολυνείκους (1198). On ὠμοφαγία of the corpse, see also: ἐᾶν δ' ἄθαπτον ... δέμας / καὶ πρὸς κυνῶν ἐδεστὸν ... ἰδεῖν (205–206) and κυνῶν βορᾶς (1017).

118 On λύσσα and Dionysus, see Eur. *Bacch.* 851. See also the destructive violence in *Heracles* (*HF* 867–897), where Lyssa, the personification of frenzy, sets Heracles within an ecstatic dance and flute melody, here grotesque and cruel (τάχα σ' ἐγὼ μᾶλλον χορεύσω καὶ καταυλήσω φόβῳ (*HF* 871; cf. 879); see Bierl 1991, 79–89, esp. 85–87, 140–146.

119 Heraclitus (DK 22 B 15) and above n. 72.

death now (555, 559): her status is already in-between, her soul long dead (ἡ δ' ἐμὴ ψυχὴ πάλαι / τέθνηκεν 559–560).

Above we scrutinized the common imagery linking the second stasimon to the fifth. In Haemon's view his father tramples upon the honours of the gods (745), especially those below (749). When Dionysus arrives καθαρσίῳ ποδί (1143), he takes just revenge. With his feet he tramples, so to speak, Creon in turn, destroying his family. The rocky cave imprisoning Antigone, where she is to starve alive, serves as, what Lacan calls, the boundary to the second death. In Creon's words she should learn there that "it is fruitless labor to revere that which is in Hades" (780). This is ironic since Dionysus will strike back. The third stasimon is a song totally dedicated to Eros, to desire, the emotion that drives Antigone. "The one who has experienced you is driven to madness (μέμηνεν)" (790), the Dionysian *mania*. The chorus underlines that "radiant desire swelling from the eyes of the sweet-bedded bride wins" (νικᾷ δ' ἐναργὴς βλεφάρων / ἵμερος εὐλέκτρου / νύμφας 795–797a).[120] Antigone's eyes glow with beauty, with a desire that shines. It is the dangerous Dionysian light that will strike back and conquer his enemy. In her long *kommos* Antigone now embarks on her last journey (897–898) across the boundary to her prison, located at the very brink of real death. Hades-Dionysus leads (ἄγει 811) her to the shore, *akta*, of Acheron (810–812). It is the verb indicating that Dionysus 'leads' the chorus and the individuals to *ate*. Now, in a way, he becomes the hidden leader of a small procession to the *chora*, the Dionysian cave where Antigone will transform her power through suicide. Responding to the chorus' praise she compares herself with Niobe whom "like clinging ivy (κισσὸς ὡς ἀτενὴς)" – the essentially Dionysian plant – "the sprouting stone subdued" (826–827).[121] The chorus criticizes Antigone for the "self-willed desire" that "destroyed" her (αὐτόγνωτος ὤλεσ' ὀργά 875). Bound to Dionysus, the total Other, she views herself now as complete reversal of the usual perspective (813–816, 819, 820, 847, 851–852, 867, 876–877, 879, 881–882). In her so-called calculation (904–915), that many critics, beginning with Goethe, wished to athetize on humanist presumptions, on ethical as well aesthetic predilection of taste, Antigone reaffirms her very special relation to Polyneices (and Dionysus) in a radical stance.[122]

---

120 For Lacan 1992, 268 this expression means Antigone's "desire made visible". See also 281.

121 For Lacan 1992, 268 the myth of Niobe mentioned in *Ant.* 823–833 represents the decisive "image of limit" around which the "the whole play turns". See also Lacan 1992, 281 and Miller 2007, 2.

122 Its authenticity has been defended in recent scholarship; see now Griffith 1999, 277–279 *ad* 904–915. For Lacan 1992, 278–279 this scandalous passage is a proof of Antigone's absolute will

In the fourth stasimon the Dionysian association of the plot comes again to the fore. To console the condemned Antigone, the chorus presents three examples from myth, in which people of significant rank, like the heroine, suffered the terrible fate of being entombed alive. In the first antistrophe (955–965), between the descriptions of Danae (944–954) and Cleopatra (966–987), the chorus mentions the story of Lycurgus, the king of the Edonians, punished because he opposed the introduction of the cult of Dionysus. The chorus therefore speaks an implicit truth for Creon, reporting of Lycurgus that he "tried to check the inspired women and the Bacchic fire, and he angered the Muses who love the flute" (παύεσκε μὲν γὰρ ἐνθέους / γυναῖκας εὔιόν τε πῦρ / φιλαύλους τ' ἠρέθιζε μούσας 963–965)".[123] Logically, we must count Antigone among these women as well. Creon, in a way, follows in Lycurgus' footsteps, opposing the god and his female entourage. Fire and the *euoi*-cries are distinctive features of the god. In the fifth stasimon his star chorus breathes fire (1146–1147). To stop the fire would mean to halt his energy. Bacchants or Muses, just like nymphs, number among his female followers. In many instances they form a dancing chorus accompanied by flute music. This exactly mirrors the chorus of the performance in the Athenian theatre, being a musical institution. We all know that, in tragedy, it is impossible to stop the Dionysian chorus and its music. Antigone is a girl and bride linked to the god. Soon reduced to only womb and body, she is the receptacle for the transformation of Dionysian energy. The parallel myths are situated in the *chora* of the in-between as well. Danae is just a body (δέμας 945), like Semele, that had to endure the light of sky (οὐράνιον φῶς 944) and to exchange it for darkness. The light is poetically associated with the golden rain that impregnated her. In her womb she "administered (ταμιεύεσκε)" (950) to Zeus' semen, just as Iacchus-Dionysus later is called as the *tamias*, the master to give a share of fate. Just as the *chora* of the chorus processes energy, Danae's womb serves as the receptacle whence a new hero, Perseus, is born.

While the fifth stasimon is being performed, Antigone takes her very last journey, committing suicide. Her prison has repeatedly been seen as a bridal chamber (νυμφεῖον 1205) and her imprisonment a marriage to Hades.[124] Creon, the late-comer, notices Haemon's wailing voice, opens the cave and sees a highly erotic scenario: Antigone "hanging by the neck, fastened by a halter of fine linen

---

of transgression that manifests itself in the pure language as signification. See also ibid. 254–256.

123 ἔνθεος refers to the Dionysian ἐνθουσιασμός. On the connection of Dionysus with the Muses, see Aeschylus TrGF 3.60; Eur. *Bacch.* 410–411, 563. See also Solon fr. 26 West and Plato *Laws* 653d.

124 Seaford 1987, esp. 107–108.

threads, while he was embracing her with arms thrown around her waist" (1220–1223). Failing to kill his father, he commits suicide himself, driving his sword in his side. His end metaphorically resembles impregnation with blood, "he clasped the maiden in his faint embrace, and, as he gasped, he shot onto her pale cheek a swift stream of oozing blood" (ἐς δ' ὑγρὸν / ἀγκῶν' ἔτ' ἔμφρων παρθένῳ προσπτύσσεται· / καὶ φυσιῶν ὀξεῖαν ἐκβάλλει ῥοὴν / λευκῇ παρειᾷ φοινίου σταλάγματος (1236b-1239). Sophocles makes the messenger add that by doing so Haemon enacted his marriage rites (τὰ νυμφικὰ / τέλη λαχών) (1241–1242). Hanging underlines her virginity.[125] Nevertheless, she has become a bride impregnated, a receptacle and womb. The hanging girl recalls Erigone, Icarius' daughter.[126] The Athenian welcomed Dionysus as his guest and gave him his daughter in marriage (Ovid *Metamorphoses* 6.125). Dionysus repays the hospitality by introducing wine to Attica. The god instructed his host Icarius on how to grow vine and to make it known. Icarius distributed it further to his shepherds. Completely drunk they killed him, thinking he had poisoned them. His daughter went in search of his father and finally found his body. She hanged herself over his grave. Dionysus sent an epidemic suicide wave among virgins that ended only after the institution of the swing-festival. In the ritual, the myth was transferred to the Aiora, the swinging of young girls and virgins on a swing, celebrated on the last day of the Anthesteria, the Chytroi.[127] In this Sophoclean tableau, Antigone becomes the emblem of a mystic union with Dionysus, with Haemon, the man of blood, as surrogate through whose Dionysiac transformation bitter revenge ensues. According to Burkert, Erigone is to be associated with the Basilinna,[128] the queen of the city given to Dionysus after the "pollutions" of the "defiled day (μιαρὰ ἡμέρα)" of the Choes. The introduction of wine regarded as a bloody sacrifice is expiated by a night celebration of a *hieros gamos*. Located in another place of *chora*, in the Marshes, the "unspeakable", mystic sexual coupling took place in a subterranean *oikos*,[129] recalling Antigone's cave. Only vaguely associated with the Athenian Dionysian, the tableau displays the pattern of a pollution to be cleansed. In tragic logic, blood invokes revenge through further bloodshed. Dionysus arrives through the *chora*, is revitalized through dance and mystic union, which symbolically alluded to sexuality and the triumph of the unspeakable in a mystic beauty. In passing through night, through the sub-

---

125 Loraux 1987, 7–17, 31–32, 38.
126 See also Borgeaud 2005.
127 Burkert 1983, 241–243.
128 Burkert 1983, 243.
129 See Burkert 1983, 234 n. 16.

terranean cave, through Eleusinian mystic dance,[130] with the *chora* as female receptacle and helper, Dionysus is revitalized and readied to strike back.

Conclusion

With the help of modern theory this contribution tried to pin down a new structural interplay of time and space in Greek literature. The Bakhtinian chronotope, Lacanian theory, and Kristeva's *chora*, along with concepts from Jacques Derrida and others, combined with research into *choreia*, Dionysus, rituality, metatheatre, performativity and, last not least, philological and literary close-reading, all aided in determining a specific Bacchic chronotope in archaic and classical choral song culture. It is deeply rooted in chorality and choric culture. Moreover, this space and time configuration involves Plato's famous *chora*, the third way, a receptacle in which and from which products of mimesis, *mimemata*, become vital and take on their vivid energy. Dionysus as the total Other, the unconscious, the god of the middle-ground, of mediation and transformation, serves, in a way, as the emblem of this chor(a)ic constellation. Despite his masculinity, he also has particular links to the female principle. In the poetically achieved Bacchic chronotope, the typical Dionysiac surroundings and atmosphere facilitate his coming into being. From absence, he transforms into pure presence and exuberant energy, expressing himself in epiphany. From *chora* emerges *choros,* his special medium of vitality in performativity. In Bacchic *choreia* we witness little narration, the chorus mostly occupied with its own doings. As in *chora* these songs are highly poetic yet semiotic, not symbolic, connected to the unconscious, and the Other, whose perception takes place in a dream-like manner. Because of the lack of symbolic signification, Dionysus remains enigmatic, notoriously elusive and meeting with resistance. Those who do understand the ecstatic expressions can abandon themselves and merge via worship with the god. The god of presence and epiphany can manifest himself only through his wild sign production. This is true, in particular, for drama based on multimodal performance of words, music and choral dance, where Dionysus is sometimes summoned to have his epiphany through roaring noise, shrill music by *auloi*, violent rhythms and excited dance. We witness a strange whirl of reciprocal interaction between the frantic performance of followers and their god. In a strict reciprocal χάρις-re-

130 The scholiast comments on the "etherial" (αἰθέριος) chorus leader of the stars in line 1146: "according to a certain mystic logos". See κατὰ γάρ τινα μυστικὸν λόγον τῶν ἀστέρων ἐστὶ χορηγός.

lation,[131] his chorus attempts to please and seduce Dionysus through performative behaviour suited to him while he takes pleasure in the chorus' activity. Often he is called on to take over as notional *choregos*, thus driving them even madder. Just as they set him in raging choral motion, so too he does with them.

Furthermore, we encounter a strange tendency to project the totalizing feeling onto other mythical choral groups or even onto the cosmos, the stars, onto the entire environment. Under his influence everything fuses: the entirety of nature is envisaged in frenzied motion, the sky, the stars, the earth and the land. He stands in the middle, the surrounding objects revolving around him in a circular dance. In the end, Dionysus is nothing more than the underlying substance, the abstraction of ecstatic, inarticulate signs lacking proposition, signs that crystallize to strange names and *epikleseis* of the god responsible for that extraordinary experience. As a hypostasized expression of ecstatic performance, Dionysus is present for the insider, the initiates – thus his association with mysteries –, whereas for the outsider it is purely insane behaviour absent any aesthetic meaning.

The Bacchic chronotope is permeable, hybrid, fluid and shifting. Moreover, his *chora* is a space of arrival in procession where the god transgresses boundaries in sudden epiphany. He manifests himself in multimodal ways in performance, and his energetic vitality, oscillating between gay festivity and dangerous violence, becomes perceptible in elemental and cosmic power. He erupts as fire, water, air or earth, lightning and thunder, earthquakes, storm and sprouting vegetation. He is inebriating wine or ivy and wild dance. Moreover, his chor(a)ic chronotope is located in the Other, in the realm of death, in the marshes, on the shores and in the sea: he must pass through the *chora* in order to become suddenly present in epiphany.

After I identified briefly the major features of the Bacchic chronotope in the parodos of *Bacchae* that I had addressed elsewhere, I focused on the fifth stasimon of *Antigone*, an exemplary tragic song in this regard. Of course, the concept can be applied to many choral songs featuring Dionysus. The concept developed above can open an entire research agenda, but time and place forbids me to pursue the Bacchic chronotope in other instances.

Having given a detailed analysis in a close reading of the fifth stasimon of *Antigone*, I concluded in giving an account of the web of Dionysiac references and Bacchic patterns that constitute this play. We have seen that *Antigone* is a most Dionysiac play. The fifth stasimon is thus only the final stage on a circuitous path leading to the catastrophic end. Antigone is connected with the god,

---

131 Bierl 2001, esp. 140–150 (Engl. 2009, 116–125).

she acts in his *chora* of mystic wonder and is interconnected with death. She desires the Other and finally transgresses the boundary to death. Dionysus is, from the very beginning, or latest from the parodos, present in his *chora*. He continually draws closer to his opponent Creon, changing from potentiality into energetic and elemental power. In this regard, it is logical that the chorus calls for him to appear (προφάνηθ' 1149). His force should be cathartic. As a matter of fact, the chorus is caught in the vain hope that the city-god will reverse the grim situation of the diseased polis. But at the same time, the chorus, as authoritative voice, says typically the right thing, too. It envisages the god arriving in a *pompe* that he welcomes himself as the notional chorus leader.

The Bacchic chronotope is a whirl of concentric choruses extending even to the cosmic level. In this regard, the chronotope is highly metatragic and a powerful *mise en abyme*, since it references back to the choral performance executed in honour of Dionysus and displayed in the orchestra of the Athenian theatre of Dionysus. The god thus becomes epiphanic in the orchestra, in the here and now, but also somehow in the mimetically produced world of the mythic past. Present in dance and performativity, he remains absent in presence. After passing through the *chora* he is transformed into an energetic force, to fire, that will take revenge on Creon. All things considered, the Bacchic chronotope, the special configuration of space and time, is a vital element to understand texts performed in a Dionysian context and occasion.

Bibliography

Alexiou, M. 2002. *After Antiquity. Greek Language, Myth, and Metaphor.* Ithaca and London.

Bakhtin, M.M. 1981. *The Dialogic Imagination.* Translated by C. Emerson and M. Holquist. Austin.

Beil, U.J. 2007. Bedecken, Beflecken, Beschwören. Präsenzeffekte in der Antigone des Sophokles. In Kiening 2007b, 147–178.

Bernabé, A. 2005. *Poetae epici Graeci. Testimonia et fragmenta. Pars II. Fasc. 2: Orphicorum et Orphicis similium testimonia et fragmenta.* Munich and Leipzig (= Bernabé).

Bierl, A. 1989. Was hat die Tragödie mit Dionysos zu tun? Rolle und Funktion des Gottes am Beispiel der Antigone des Sophokles. *WJA* 15: 43–58.

Bierl, A.F.H. 1991. *Dionysos und die griechische Tragödie. Politische und 'metatheatralische' Aspekte im Text.* Tübingen.

Bierl, A. 2001. *Der Chor in der Alten Komödie. Ritual und Performativität* (unter besonderer Berücksichtigung von Aristophanes' *Thesmophorlazusen* und der Phallosliedcr fr. 851 *PMG*). Munich and Leipzig 2001 (see the revised second English edition Bierl 2009).

Bierl, A. 2006. Räume im Anderen und der griechische Liebesroman des Xenophon von Ephesos. Träume? In *Mensch und Raum von der Antike bis zur Gegenwart*, ed. A. Loprieno (Colloquium Rauricum 9), 71–103. Munich and Leipzig (see the revised second English edition Bierl 2014).

Bierl, A. 2007. Mysterien der Liebe und die Initiation Jugendlicher. Literatur und Religion im griechischen Roman. In *Literatur und Religion, II. Wege zu einer mythisch-rituellen Poetik bei den Griechen*, eds. A. Bierl, R. Lämmle and K. Wesselmann, 239–334. Berlin and New York.

Bierl, A. 2009. *Ritual and Performativity. The Chorus in Old Comedy.* Translated by Alex Hollmann. Washington, DC and Cambridge, MA.

Bierl, A. 2011a. Dionysos auf der Bühne: Gattungsspezifische Aspekte des Theatergottes in Tragödie, Satyrspiel und Komödie. In Schlesier 2011, 315–341.

Bierl, A. 2011b. Prozessionen auf der griechischen Bühne: Performativität des einziehenden Chors als Manifestation des Dionysos in der Parodos der Euripideischen *Bakchen*. In *Medialität der Prozession. Performanz ritueller Bewegung in Texten und Bildern der Vormoderne. Médialité de la procession. Performance du mouvement rituel en textes et en images à l'époque pré-moderne*, eds. K. Gvozdeva and H.R. Velten, 35–61. Heidelberg.

Bierl, A. 2012. Dionysos' Epiphany in Performance. The God of Ecstatic Cry, Noise, Song, Music, and Choral Dance. In *Dionysus: Myth, Cult, Ritual. Electra* 2: 1–12 online publication (published April 2013) http://electra.lis.upatras.gr/index.php/electra/index

Bierl, A. 2013a. Maenadism as Self-Referential Chorality in Euripides' *Bacchae*. In Gagné and Hopman 2013a, 211–226.

Bierl, A. 2013b. Der vielnamige Dionysos. Wesen und Funktion des Gottes im Spiegel seiner Beinamen. In *Dionysos. Rausch und Ekstase*, ed. M. Philipp (Catalogue of the Exhibition, Bucerius Kunst Forum Hamburg and Staatliche Kunstsammlungen Dresden 2013/2014), 30–39. Munich.

Bierl, A. 2014. Space in Xenophon of Ephesus: Love, Dream, and Dissemination. Publication in electronic form on the CHS-website http://chs.harvard.edu/CHS/article/display/5637

Blech, M. 1982. *Studien zum Kranz bei den Griechen* (RGVV 38). Berlin.

Boedeker, D.D. 1974. *Aphrodite's Entry into Greek Epic* (Mnemosyne, Suppl. 31). Leiden.

Bohrer, K.-H. 2015. *Das Erscheinen des Dionysos. Antike Mythologie und moderne Metapher.* Berlin.

Borgeaud, P. 2005. D'Antigone à Erigone. In *Antigone et le devoir de sépulture*, ed. S. Muriel, 44–58. Geneva.

Brown, A. 1987. *Sophocles. Antigone.* Edited with translation and notes. Warminster.

Buchan, M. 2012. Sophocles with Lacan. In *A Companion to Sophocles*, ed. K. Ormand, 492–504. Malden, MA and Oxford.

Burkert, W. 1983. *Homo Necans. The Anthropology of Ancient Greek Sacrificial Ritual and Myth.* Translated by P. Bing. Berkeley, Los Angeles and London.

Burkert, W. 1987. *Ancient Mystery Cults.* Cambridge, MA.

Burton, R.W.B. 1980. *The Chorus in Sophocles' Tragedies.* Oxford.

Casadio, G. 1994. Dioniso italiota. Un dio greco in Italia meridionale. *AION (filol.)* 16: 79–104.

Casey, E. 1997. *The Fate of Place. A Philosophical Journey.* Berkeley and Los Angeles.

Chantraine, P. 1968. *La formation des noms en grec ancien.* Revised edition. Paris.

Copley, F.O. 1937. The Pathetic Fallacy in Early Greek Poetry. *AJPh* 58: 194–209.

Csapo, E. 1999/2000. Later Euripidean Music. *ICS* 24–25: 399–426.

Csapo, E. 2008. Star Choruses: Eleusis, Orphism, and New Musical Imagery and Dance. In *Performance, Iconography, Reception: Studies in Honour of Oliver Taplin*, eds. M. Revermann and P. Wilson, 262–290. Oxford.

Cullyer, H. 2005. A Wind That Blows from Thrace: Dionysus in the Fifth Stasimon of Sophocles' *Antigone*. *CW* 99.1: 3–20.

Cursaru, G. 2014. Χώρα au cœur des enjeux politico-religieux et de la rhétorique patriotique dans les tragédies grecques. Étude de cas: *Œdipe à Colone* de Sophocle. *Mythos* n.s. 8: 113–136.

D'Aquili, E.G. and C.D. Laughlin Jr. 1979. The Neurobiology of Myth and Ritual. In *The Spectrum of Ritual. A Biogenetic Structural Analysis*, eds. E.G. d'Aquili, C.D. Laughlin Jr. and J. McManus, 152–182. New York.

Dawe, R.D. (ed.) 1979. *Sophocles: Tragoediae*, II. Leipzig.

Derrida, J. 1983. The Principle of Reason: The University in the Eyes of Its Pupils. *Diacritics* 13.3: 2–20.

Derrida, J. 1995. *On the Name*. Edited by T. Dutoit, translated by D. Wood, J.P. Leavey Jr. and I. McLeod. Stanford.

Dodds, E.R. 1960. *Euripides. Bacchae*. Edited with introduction and commentary. Second edition. Oxford.

Dorsch, K.-D. 1982. *Götterhymnen in den Chorliedern der griechischen Tragiker. Form, Inhalt und Funktion*. Dissertation Münster.

Dover, K.J. 1993. *Aristophanes. Frogs*. Edited with introduction and commentary. Oxford.

Eitrem, S. 1915. *Opferritus und Voropfer der Griechen und Römer*. Christiania.

Ford, A.L. 2011. Dionysos' Many Names in Aristophanes' *Frogs*. In Schlesier 2011, 343–355.

Förs, H. 1964. *Dionysos und die Stärke der Schwachen im Werk des Euripides*. Dissertation Tübingen University. Bamberg.

Frank, M. 1982. *Der kommende Gott: Vorlesungen über die neue Mythologie*. Part 1. Frankfurt a.M.

Furley, W.D. and J.M. Bremer 2001. *Greek Hymns*, I: *The Texts in Translation*; II: *Greek Texts and Commentary* (Studien und Texte zu Antike und Christentum 9–10). Tübingen.

Gagné, R. and M.G. Hopman (eds.) 2013a. *Choral Mediations in Greek Tragedy*. Cambridge.

Gagné, R. and M.G. Hopman 2013b. Introduction: The Chorus in the Middle. In Gagné and Hopman 2013a, 1–34.

Gernet, L. 1953. Dionysos et la religion dionysiaque: éléments hérités et traits originaux. *REG* 66: 377–395.

Gödde, S. 2011. 'Fremde Nähe'. Zur mythologischen Differenz des Dionysos. In Schlesier 2011, 85–104.

Graf, F. 1974. *Eleusis und die orphische Dichtung Athens in vorhellenistischer Zeit* (RGVV 33). Berlin.

Graf, F. and S.I. Johnston 2007. An Edition and Translation. In *Ritual Texts for the Afterlife. Orpheus and the Bacchic Gold Tablets*, eds. F. Graf and S.I. Johnston, 1–49. London and New York (= Graf and Johnston).

Griffith, M. (ed.) 1999. *Sophocles. Antigone*. Cambridge.

Haß, U. 2014. Die zwei Körper des Theaters. Protagonist und Chor. In Tatari 2014, 139–159.

Haß, U. and M. Tatari 2014. Eine andere Geschichte des Theaters. In Tatari 2014, 77–90.

Heinrich, K. 2002. Der Staub und das Denken. Zur Faszination der Sophokleischen *Antigone* nach dem Krieg. In *Sophokles. Antigone*, ed. G. Greve, 25–58. Tübingen.

Henrichs, A. 1978. Greek Maenadism from Olympias to Messalina. *HSPh* 82: 121–160.
Henrichs, A. 1982. Changing Dionysiac Identities. In *Jewish and Christian Self-Definition, III. Self-Definition in the Graeco-Roman World*, eds. B.F. Meyer and E.P. Sanders, 137–160 and 213–236. Philadelphia.
Henrichs, A. 1984. Loss of Self, Suffering, Violence: The Modern View of Dionysus from Nietzsche to Girard. *HSPh* 88: 205–240.
Henrichs, A. 1990. Between Country and City: Cultic Dimensions of Dionysus in Athens and Attica. In *Cabinet of the Muses. Essays on Classical and Comparative Literature in Honor of Thomas G. Rosenmeyer*, eds. M. Griffith and D.J. Mastronarde, 257–277. Atlanta.
Henrichs, A. 1994/1995. 'Why Should I Dance?': Choral Self-Referentiality in Greek Tragedy. *Arion* Third Series 3.1: 56–111.
Henrichs, A. 1996a. Dionysus. *OCD*[3]: 479–482.
Henrichs, A. 1996b. Dancing in Athens, Dancing on Delos: Some Patterns of Choral Projection in Euripides. *Philologus* 140: 48–62.
Henrichs, A. 1996c. *"Warum soll ich denn tanzen?". Dionysisches im Chor der griechischen Tragödie* (Lectio Teubneriana 4). Stuttgart and Leipzig.
Henrichs, A. 2008. Dionysische Imaginationswelten: Wein, Tanz, Erotik. In *Dionysos. Verwandlung und Ekstase*, eds. R. Schlesier and A. Schwarzmeier, 19–27. Berlin and Regensburg.
Henrichs, A. 2011. Göttliche Präsenz als Differenz: Dionysos als epiphanischer Gott. In Schlesier 2011, 105–116.
Jakobson, R. 1960. (Closing Statement:) Linguistics and Poetics. In *Style in Language*, ed. T.A. Sebeok, 350–377. Cambridge, MA (reprint in *Roman Jakobson. Selected Writings*, III, ed. S. Rudy. The Hague, Paris and New York 1981, 18–51).
Jebb, R.C. 1900. *Sophocles. The Plays and Fragments*. With critical notes, commentary, and translation in English prose, III: *The Antigone*. Cambridge.
Jiménez San Cristóbal, A.I. 2013. The Sophoclean Dionysos. In *Redefining Dionysos*, eds. A. Bernabé, M. Herrero de Jáuregui, A.I. Jiménez San Cristóbal and R. Martín Hernández, 272–300. Berlin/Boston.
Jouanna, J. 2007. L'hymne chez Sophocle. In *L'hymne antique et son public*, ed. Y. Lehmann, 109–132. Turnhout.
Kamerbeek, J.C. 1978. *The Plays of Sophocles. Commentaries, III: The Antigone*. Leiden.
Kavoulaki, A. 1996. *ΠΟΜΠΑΙ: Processions in Athenian Tragedy*. Dissertation Oxford.
Kavoulaki, A. 1999. Processional Performance and the Democratic Performance. In *Performance Culture and Athenian Democracy*, eds. S. Goldhill and R. Osborne, 293–320. Cambridge.
Kerényi, K. 1935. *Dionysos und das Tragische in der Antigone* (Frankfurter Studien 13). Frankfurt a.M.
Kiening, C. 2007a. Mediale Gegenwärtigkeit. Paradigmen – Semantiken – Effekte. In Kiening 2007b, 9–70.
Kiening, C. (ed.) 2007b. *Mediale Gegenwärtigkeit*. Zurich.
Kitzinger, M.R. 2008. *The Choruses of Sophokles' Antigone and Philoktetes. A Dance of Words*. Leiden and Boston.
Kowalzig, B. 2007. 'And Now All the World Shall Dance!' (Eur. *Bacch.* 114). Dionysus' Choroi between Drama and Ritual. In *The Origins of Theater in Ancient Greece and Beyond. From Ritual to Drama*, eds. E. Csapo and M.C. Miller, 221–251. Cambridge.

Kramer, K. and J. Dünne 2009. Einleitung. Theatralität und Räumlichkeit. In *Theatralität und Räumlichkeit. Raumordnungen und Raumpraktiken im theatralen Mediendispositiv*, eds. J. Dünne, S. Friedrich and K. Kramer, 15–32. Würzburg.
Kristeva, J. 1984. *Revolution in Poetic Language*. New York.
Lacan, J. 1957. The Instance of the Letter in the Unconscious, or Reason Since Freud. In Lacan 2006, 412–441 (originally published in *La Psychoanalyse* 3, 1957: 47–81; reprint of the French original in Lacan 1966, 493–528).
Lacan, J. 1966. *Écrits*. Paris.
Lacan, J. 1992. *The Ethics of Psychoanalysis. The Seminar of Jacques Lacan*. Ed. J.-A. Miller. Book VII, translated with notes by D. Porter. London (orig. Paris 1986).
Lacan, J. 2006. *Écrits. The First Complete Edition in English*. Translated by B. Fink. New York and London 2006.
Lonnoy, M.-G. 1985. Arès et Dionysos dans la tragédie grecque: Le rapprochement des contraires. *REG* 98: 65–71.
Loraux, N. 1987. *Tragic Ways of Killing a Woman*. Translated by A. Forster. Cambridge.
Macedo, J.M. 2011. In between Poetry and Ritual: The Hymn to Dionysus in Sophocles' *Antigone* (1115–54). *CQ* 61: 402–411.
McEwen, I.K. 1993. *Socrates' Ancestor. An Essay on Architectural Beginnings*. Cambridge, MA.
McLuhan, M. 1964. *Understanding Media. The Extensions of Man*. New York (second edition 1966; reissued Cambridge, MA 1994).
Metzger, H. 1944/1945. Dionysos chthonien d'après les monuments figurés de la période classique. *BCH* 69/69: 296–339.
Miller, P.A. 2007. Lacan's Antigone: The Sublime Object and the Ethics of Interpretation. *Phoenix* 61.1/2: 1–14.
Moulinier, L. 1952. *Le pur et l'impur dans la pensée des Grecs d'Homère à Aristote*. Paris.
Müller, G. 1967. *Sophokles. Antigone*. Erläutert und mit einer Einleitung versehen. Heidelberg.
Murrey, L. 2015. *Hölderlin's Dionysiac Poetry. The Terrifying-Exciting Mysteries*. Cham.
Nagy, G. 2013. *The Ancient Greek Hero in 24 Hours*. Cambridge, MA and London (http://nrs.harvard.edu/urn3:hul.ebook:CHS_NagyG.The_Ancient_Greek_Hero_in_24_Hours.2013).
Nancy, J.-L. 2014. Theater als Kunst des Bezugs. In Tatari 2014, 91–108.
Norden, E. 1913. *Agnostos Theos. Untersuchungen zur Formengeschichte religiöser Rede*. Leipzig.
Otto, W.F. 1933. *Dionysos. Mythos und Kultus*. Frankfurt a. M. (seventh edition 2011).
Oudemans, T.C.W. and A.P.M.H. Lardinois 1987. *Tragic Ambiguity: Anthropology, Philosophy and Sophocles' Antigone*. Leiden.
Polignac, F. de 1995. *Cults, Territory, and the Origins of the Greek City-State*. Translated by J. Lloyd, foreword by C. Mossé. Chicago and London (orig. Paris 1984).
Powell, J.U. 1925. *Collectanea Alexandrina*. Oxford (reprint Chicago 1981).
Power, T. 2011. Cyberchorus: Pindar's Κηληδόνες and the Aura of the Artificial. In *Archaic and Classical Choral Song. Performance, Politics and Dissemination*, eds. L. Athanassaki and E. Bowie (Trends in Classics, Suppl. 10), 67–113. Berlin and New York.
Randhawa, B.P. 2014. The Unblest Room: Kristeva's Chora in Sophocles's *Antigone*. *Philosophy and Literature* 38: 293–313.
Rickert, T. 2007. Toward the *Chôra*: Kristeva, Derrida, and Ulmer on Emplaced Invention. *Philosophy and Rhetoric* 40: 251–273.
Rodighiero, A. 2012. *Generi lirico-corali nella produzione drammatica di Sofocle*. Tübingen.

Rohdich, H. 1980. *Antigone. Beitrag zu einer Theorie des Sophokleischen Helden.* Heidelberg.
Sallis, J. 1999. *Chorology: On Beginning in Plato's* Timaeus. Bloomington.
Schlesier, R. 1997. Dionysos I. Religion. *DNP* 3: 651–662.
Schlesier, R. (ed.) 2011. *A Different God? Dionysos and Ancient Polytheism.* Berlin and Boston.
Scullion, S. 1998. Dionysos and Katharsis in *Antigone. ClAnt* 17: 96–122.
Seaford, R. 1987. The Tragic Wedding. *JHS* 107: 106–130.
Seaford, R. 2012. *Cosmology and the Polis. The Social Construction of Space and Time in the Tragedies of Aeschylus.* Cambridge.
Segal, C. 1981. *Tragedy and Civilization. An Interpretation of Sophocles.* Cambridge, MA and London.
Segal, C. 1982. *Dionysiac Poetics and Euripides' Bacchae.* Princeton (expanded second edition with a new afterword by the author, 1997).
Segal, C. 1983. Greek Myth as Semiotic and Structural System and the Problem of Tragedy. *Arethusa* 16(1–2): 173–198.
Segal, C. 1986. *Interpreting Greek Tragedy: Myth, Poetry, Text.* Ithaca, NY.
Sourvinou-Inwood, C. 1989. Assumptions and the Creating of Meaning: Reading Sophocles' *Antigone. JHS* 109: 134–148.
Tambiah, S.J. 1985. *Culture, Thought, and Social Action. An Anthropological Perspective.* Cambridge, MA and London.
Tatari, M. (ed.) 2014. *Orte des Unermesslichen. Theater nach der Geschichtsteleologie.* Zurich and Berlin.
Turner, V. 1974. *Dramas, Fields, and Metaphors. Symbolic Action in Human Society.* Ithaca and London.
Vernant, J.-P. 1965. *Mythe et pensée chez les Grecs. Études de psychologie historique.* Paris; cited after third new edition, revised and enlarged, 1985.
Vernant, J.-P. 1981. Le dieu de la fiction tragique. In Vernant and Vidal-Naquet 1986, 17–24.
Vernant, J.-P. (with F. Frontisi-Ducroux) 1983. Figures du masque en Grèce ancienne. In Vernant and Vidal-Naquet 1986, 25–43 (originally published in *Journal de Psychologie* 1/2, 1983: 53–69).
Vernant, J.-P. 1985. Le Dionysos masqué des *Bacchantes* d'Euripide. In Vernant and Vidal-Naquet 1986, 237–270 (originally published in *L'Homme* 93, 1985: 31–58).
Vernant, J.-P. 1986. Conclusion. In *L'association dionysiaque dans les sociétés anciennes* (Collection de l'École Française de Rome 89), 291–303. Rome.
Vernant, J.-P. and P. Vidal-Naquet (eds.) 1986. *Mythe et tragédie en Grèce ancienne*, II. Paris.
Versnel, H.S. 1970. *Triumphus. An Inquiry into the Origin, Development and Meaning of the Roman Triumph.* Leiden.
Versnel, H.S. 1972. ΙΑΚΧΟΣ: Some Remarks Suggested by an Unpublished Lekythos in the Villa Giulia. *ΤΑΛΑΝΤΑ* 4: 23–38.
Vicaire, P. 1968. Place et figure de Dionysos dans la tragédie de Sophocle. *REG* 81: 351–373.
Wiles, D. 2003. *A Short History of Western Performance Space.* Cambridge.
Winnington-Ingram, R.P. 1980. *Sophocles: An Interpretation.* Cambridge.
Zeitlin, F.I. 1990. Playing the Other: Theater, Theatricality, and the Feminine in Greek Drama. In *Nothing to Do with Dionysos? Athenian Drama in Its Social Context*, eds. J.J. Winkler and F.I. Zeitlin, 63–96. Princeton.

Zeitlin, F.I. 1993. Staging Dionysus between Thebes and Athens. In *Masks of Dionysus*, eds. T.H. Carpenter and C.A. Faraone, 147–182. Ithaca and London.

Elena Iakovou

The Re-enactment of the Past in the Present and the Transformation of Space in Sophocles' *Oedipus Tyrannus*

Introduction

Sophocles' *Oedipus Tyrannus* sheds light on two interesting narrative perspectives of the Oedipus myth: the re-enactment of the past in the present and the spatial transformation of the story.[1] In *Oedipus Tyrannus* the hero needs to bring together all the pieces of his past to discover his true identity.[2] As De Jong remarks, protagonists in tragedies (like *Oedipus Tyrannus*) function as secondary embedded and intra-dramatic narrators, because they recount events from the past.[3] The chronological order of *Oedipus Tyrannus* is as follows:[4] (i) The god Apollo gives the prophecy to Oedipus' father Laius that if he has a son, that son will kill him.[5] (ii) Laius has a baby with his wife Jocasta, decides to expose his baby son and gives it to his Theban servant. The latter hands it over to a Corinthian herdsman, who brings it to the Corinthian royal couple. (iii) From a drunkard Oedipus discovers that he is not the real son of his parents; as a result, he goes to Delphi and receives an oracle that he will commit patricide and incest. (iv) At the crossroads Oedipus unknowingly kills his father Laius. (v) He saves the city of Thebes by solving the riddle of the Sphinx. (vi) He wins the widow queen of Thebes, Jocasta (his mother). In this play, however, this narrative

1 I extend my warmest thanks to Professor Heinz-Günther Nesselrath and Professor Patrick Finglass for their helpful criticisms and assistance. A special debt of gratitude is owed to the organizers, Menelaos Christopoulos and Athina Papachrysostomou of the conference 'Time and Space in Greek Myth and Religion' (July 2015, Patras, Greece) for giving me the opportunity to present my paper at the conference.
2 See Kraus 1994, 294–299.
3 See De Jong 2004, 255.
4 For a schematic overview of the action of play see also Flashar 1976, 356–357.
5 On the role of Apollo in Oedipus' life see Kovacs 2009a, 359–368; also 359: " ... in the world of the play all of Oedipus' actions are perfectly free, but that in the case of the parricide and incest Apollo created a situation where Oedipus, a free agent acting on the information available to him, unwittingly carried out Apollo's designs".

DOI 10.1515/9783110535150-008

is presented in "smaller and larger retrospective narratives, which are recounted by different characters and in anachronical order".[6]

The whole process that Oedipus undertakes to discover his identity is initiated by his quest to find the murderer of Laius. This inquiry ("Who is Laius' killer?") results in two further crucial questions: "Is Oedipus the murderer?" and "Am I the murderer?"[7] This narratological progression reflects Oedipus' journey to self-discovery and his gradual downfall since the answers to these questions will not only reveal the murderer of Laius, but also the polluter who caused the plague. De Jong rightly argues: "the direction of the retrospective narratives is regressive, moving from a more recent past to the remote past: Oedipus begins his search with the question of the murder of Laius but ends with his birth and the even more vital point of his identity."[8] Every step that Oedipus takes, supposedly forward, is in fact a step back in time. Sophocles' use of the dramatic time in his plays highlights his interest in the theme of mortals struggling with the difficulties of their lives in which the past is re-enacted in the present.[9]

Temporal Re-enactments / Turning Back in Time

Knowledge plays a significant role in *Oedipus Tyrannus* and is the result of separate but connected events of the past which are re-enacted in the present.[10] When Oedipus brings together the pieces of his life's puzzle that connect events from his infancy (and childhood) and early manhood with the discovery of his present life and circumstances, he will be revealed as a person who has a double

---

6 See De Jong 2007, 277 (also 278).

7 This tripartite formula is suggested by Cameron 1968, 36–58.

8 De Jong 2007, 278.

9 De Jong 2007, 292. Further examples of Sophoclean tragedies in which important information about the past is revealed in the present: *Aj.* 18–33 (Odysseus talks about Ajax's madness and slaughter of the cattle); 36–65 (Athena confirms Ajax's actions); Tecmessa also recounts how Ajax killed the animals with a sword (214–220, 233–244, 284–327); 748–783 (the messenger reports Calchas' words about Athena's rage against Ajax). *Trach.* 555–581 (Deianira recounts how she gained access to the poison, i.e. from Nessus), 1159–1171 (Hercules reveals the prophecies about his death). *OC* 324–454 (Oedipus and his daughter Ismene talk about the quarrel between Eteocles and Polynices); 1291–1325 (in front of his father and his sister Antigone Polynices points out the reasons for fighting against his brother).

10 See Segal 1998², 147–148; also 159 "he fuses, or confuses, the diachronic and the synchronic axes. By deepening the temporal perspective through the motif of discovering and remembering a long-forgotten past, he also calls attention to the representational power of drama, by which a single action unfolding before us on the stage can contain symbolically the meaning of an entire lifetime".

contradictory role: he is the king of Thebes, who vanquished the Sphinx and saved the city from the plague, but he is also the absolute outcast, a polluter and transgressor who broke social and religious rules.[11]

Oedipus Tyrannus begins with Oedipus' eagerness to discover Laius' murder (lines 108–109). But the path he takes in his search for the murderer of the former king of Thebes and ex-husband of his wife will lead him to the past of his own life and his true origins. This path supplies Oedipus with the answers that he seeks, though they are catastrophic for his present. Segal clarifies that the road to his past "proves to be not single but manifold, just as Oedipus himself proves to be not one but many."[12] His eager investigations to find *a* murderer will result in tracking *the* murderer who is no-one else but he. By remembering the incident of his killing of an old man at the crossroads (lines 800–813) Oedipus contributes unknowingly to the discovery of the true killer. His "successful" search leads, however, to the revelation of Oedipus' other identities: he is the exposed child of the royal couple of Thebes and the husband / son of his mother Jocasta.

The temporal re-enactment of the past in Oedipus' present life, as now the king of Thebes and husband of Jocasta, is already expressed in the very first line of the play. Here Oedipus as an attentive father addresses his fellow citizens, who came to him to ask for his assistance against the plague, and combines in one sentence the old and new generation of Cadmus' descendants: Ὦ τέκνα, Κάδμου τοῦ πάλαι νέα τροφή (line 1).[13] The dramatic re-enactment of the past in the present is also evoked a few verses later, where the priest and leader of the suppliants points out the range of various ages.[14] The leader of the suppliants is an elderly priest (15–19). Between the 'young' and 'old' generation of Cadmus Oedipus appears in the middle since, by committing incest with his mother, he embodies two generations, that of Laius and that of his sons.[15]

Moreover, the seer Tiresias functions as a mediator of the past and present (and especially of the future) already in the first episode (216–462).[16] On the

11 See Segal 1998², 148.
12 Segal 2007, 218.
13 "O children, of Cadmus old the newest breed". Translations of the Greek passages of *OT* are slightly adapted from Lloyd-Jones' edition (1998²).
14 See also Segal 2007, 219.
15 See Segal 2007, 219. See Seaford 2007, 23–28, esp. 25–28, who explains the confusion of generations in *OT* due to the tension between Laius and Oedipus and the desire of the latter to accumulate his wealth; money and power seem to increase his autonomy and self-sufficiency both as an individual and a king.
16 On the role of Tiresias in the Theban saga see Torres 2014, 339, 344–347. See also Ugolini 1995, 117–142 and 186–201.

one hand, he brings on stage the past by revealing Apollo's prophecies that Oedipus is the murderer of Laius (350–353, 362, 367–368); on the other hand, he combines the past with the future since the revelations about Oedipus' past are followed by prophecies of his future and the consequences he will suffer after the revelations of his true origins (413–414, 427–428, 452–460).[17] Tiresias' question to Oedipus: ἆρ' οἶσθ' ἀφ' ὧν εἶ; ("do you know from whom you are?"; line 415) reminds of us the reason why the latter went to consult Apollo.[18]

Among the most important of the past actions re-enacted in the present of Oedipus are his visit to Delphi and his encounter with Laius at the crossroads (787–827); and his encounter with the drunken man at the feast at Corinth who accused him of being a bastard (779–786). To this one can add the important initiative taken by the Corinthian and Theban shepherds to save Oedipus when he was just a baby (1156–1181). These past events have their counterparts in the present, as they appear in the course of the play.

To begin with, his quarrel with a powerful man, as his father Laius was, is now re-enacted in his argument with his maternal uncle Creon (512–677). The thought that he might only be the putative son of Polybus and Merope drove him to seek out Apollo's oracle for the truth (785–786). However, the god did not reveal to him the truth about his parentage; instead he sent him away (788–789). The terrifying prophecies about his future crimes compelled Oedipus to abandon his putative birthplace in order to avoid their fulfillment (789–793). But his very first attempt to flee Corinth actually led him to the committing of his first crime, parricide, when the driver of Laius and the king himself tried to drive him forcibly off the road (800–805). Oedipus' anger at Laius' provocation brings

---

17 Lines 413–414: σὺ καὶ δέδορκας κοὐ βλέπεις ἵν' εἶ κακοῦ, / οὐδ' ἔνθα ναίεις, οὐδ' ὅτων οἰκεῖς μέτα ("though you have sight, you do not see what a state of misery you are in, nor where you live, nor with whom you share your home"); 427–428: ... σοῦ γὰρ οὐκ ἔστιν βροτῶν / κάκιον ὅστις ἐκτριβήσεταί ποτε ("for there is none among mortals that will ever be rooted out more miserably than you"); 452–460: ξένος λόγῳ μέτοικος· εἶτα δ' ἐγγενὴς / φανήσεται Θηβαῖος, οὐδ' ἡσθήσεται / τῇ ξυμφορᾷ· τυφλὸς γὰρ ἐκ δεδορκότος / καὶ πτωχὸς ἀντὶ πλουσίου ξένην ἔπι / σκήπτρῳ προδεικνὺς γαῖαν ἐμπορεύσεται. / φανήσεται δὲ παισὶ τοῖς αὑτοῦ ξυνὼν / ἀδελφὸς αὑτὸς καὶ πατήρ, κἀξ ἧς ἔφυ / γυναικὸς υἱὸς καὶ πόσις, καὶ τοῦ πατρὸς / ὁμοσπόρος τε καὶ φονεύς ... ("he is thought to be a stranger who has migrated here, but later he shall be revealed to be a native Theban, and he shall not enjoy his fortune; for he shall travel to a foreign land blind instead of seeing, poor instead of rich, feeling his way with his stick. And he shall be revealed as being to his children with whom he lives both a brother and a father, and to his mother both a son and a husband, and to his father a sharer in his wife and a killer ...").

18 Liu 2010, 63–64. Gregory 1995, 146, remarks: "Oedipus never forgot the original question which drove him to Delphi; it was not heedlessness, but the assumption that all danger was limited to Corinth that led him unwittingly to fulfill the Delphic prophecy".

back to memory his similar behaviour towards Creon, whom he assails with the insults of being the murderer of Laius and the robber of his kingship (534–535, 538–542). Both scenes depict Oedipus on the one hand as the perpetrator who is dominated by increasing anger and rage and on the other hand his "victims", Creon and Tiresias, as the defendants who try to dismiss the vehement and indignant accusations of Oedipus.[19]

Another temporal re-enactment of the past in the present is the scene with the drunken man of Oedipus' youth (779–780) which evokes the present confrontation between Oedipus and Tiresias, who earlier in the plot (already in the first episode) reveals to him the truth about his origins, albeit in riddling language (350–379, 408–462). The insults of the drunken man forced Oedipus to set off on the path for his self-discovery. The drunken man delivered the information about Oedipus' putative parentage in the past and Tiresias not only confirms it, but provides him with insights into his future life, thus revealing that Oedipus committed parricide, regicide and incest. In the first case, his persistent inquisitiveness to seek the truth (he asks for oracular advice) sets in motion the fulfilment of his terrible destiny. In the second case, Oedipus reacts angrily to Tiresias, and his wrath culminates in his conflict with Creon.

Equally, the two shepherds and Oedipus are reunited in a crucial moment for the plot, upon his re-discovery of his identity (923–1046; 1119–1181).[20] But although last time they saved the newborn Oedipus, this time they bring the king to his destruction. The result of this re-encounter is that Oedipus will realize that he possesses two identities: a Corinthian (as it is expressed by the Corinthian shepherd) and a Theban one (as it is expressed by the Theban shepherd).

The confrontation between Oedipus, the Corinthian messenger/shepherd and the Theban shepherd/servant can also be associated with the symbolic meaning of the tripartite structure of the crossroads.[21] The three roads to (or from) Daulis, Delphi and Thebes signify the past, present, and future that collide with each other (733–734). Oedipus confronts unknowingly his past, when he encounters his real father. The Corinthian messenger/shepherd recounts that it was on Cithaeron, where the Theban shepherd disobeyed Laius' order to expose the baby and handed over the newborn Oedipus to his fellow-shepherd from Corinth, who then passed him on to the childless Corinthian royal couple. Thebes

19 Cf. lines 337–338, 343–344, 350–353, 356, 362–363, 366–367, 369, 372–373, 376–377, 408–428, 435–436, 447–462, 571, 574–574, 583–615, 673–675. On Creon's behaviour in the confrontation with Oedipus see also Kyriakou 2011, 458–464.

20 See also Rehm 2012a, 334–335.

21 See Rehm 2012a, 334–335.

and Corinth are thus represented by the two shepherds, whereas Thebes and Cithaeron illustrate Oedipus' fate *par excellence*.

Another significant element of temporal re-enactment are the six separate pilgrimages (real or planned) to the Delphic oracle that function as time indicators in the play, as Rehm explains.[22] All six oracular consultations are mentioned in the course of play as further examples of the association of the past with the present. Without forgetting the drunkard's information, Oedipus leaves Corinth to consult the god Apollo (774–789). At Delphi it was prophesized that Oedipus would kill his father and marry his mother, hence Oedipus chose not to return home to Corinth (791–797). The second report on the pilgrimage to Delphi occurred, when Oedipus was attacked by Laius at the crossroads. The latter was also travelling to consult the Delphic oracle (114–115) and this was Laius' second effort to ask for oracular advice; the first occurred when he and his wife Jocasta received the prophecy that their son was destined to murder his father (711–714).[23]

A further pilgrimage in *Oedipus Tyrannus* takes place in the prologue where Oedipus informs the suppliants that he has sent his brother-in-law Creon to Delphi to receive information about the plague in Thebes (69–77). In the last scene, Creon is cautious and reluctant concerning the action that needs to be taken in view of the harrowing and abominable revelations, and so he wants to send a mission to Delphi to make sure that the punishment for Laius' murderer still holds (1438–1445, 1518–1519).[24] Oedipus, however, who remains authoritatively assertive, tries to convince Creon to comply with the recent requests of Apollo (i.e. the polluting criminal should be expelled), because he wants to put an end to his misery through the fulfillment of Apollo's most recent oracle.[25]

Supplication as a motif, too, plays a significant role in *Oedipus Tyrannus* as part of the various temporal re-enactments:[26] The Chorus of Thebans gather around the altar in front of the palace to implore king Oedipus for help and protection (1–5, 40–57). Oedipus promises to save (once again) the city from the plague, as soon as Creon provides him the required information from the Delphic oracle (11–13, 59–64, 68–72, 76–77). The Chorus then enter onstage praying to

22 On this topic see Rehm 2012b, 420–421.

23 In those verses Jocasta does not mention if Laius was also told that his son would marry his mother.

24 Creon now assumes the regency and is thus in charge of family and city. Cf. the Chorus' announcement of the last entry of Creon that he is left to protect the land (1416–1418). See also Kyriakou 2011, 462 with n. 46.

25 See lines 1436–1437, 1440–1441, 1449–1454, 1518–1519.

26 See also Rehm 2012b, 422–423.

various gods (151–215), and describe the Theban wives and mothers who have come to supplicate to get rid of the plague (182–185). But Oedipus also plays the role of the suppliant since near the end of the play pleas for Creon's mercy (1432–1434, 1508–1510).[27] Oedipus asks not only to be exiled to Mount Cithaeron, but also requests a chance to embrace his daughters, born of incest, and implores Creon to look after them in the future. Creon holds no grudges toward his former accuser (1422–1423), and sensitively brings Oedipus' daughters along to comfort their father in his misery (1473–1477).[28] These scenes of supplication depict the tragic destiny that surrounds Oedipus, his transformation from a powerful king and benefactor for Thebes into a pitiful beggar and suppliant.[29] The constant flashbacks in time and space make it difficult for Oedipus to disentangle the riddles of his past, although he had already solved the riddle of the Sphinx.

Spatial Re-enactments / Revisiting Old Places

(Thebes / Corinth / Exile = Cithaeron)

The drama focuses throughout on significant places of Oedipus' life, which represent the social transformation of the hero from an adopted son and illegitimate

27 Lines 1432–1434: πρὸς θεῶν, ἐπείπερ ἐλπίδος μ' ἀπέσπασας, / ἄριστος ἐλθὼν πρὸς κάκιστον ἄνδρ' ἐμέ, / πιθοῦ τί μοι· πρὸς σοῦ γάρ, οὐδ' ἐμοῦ, φράσω ("for the gods, since beyond all expectation you have come in all your goodness to me, the worst man, grant me a favour: I will speak for your own good, not mine"); lines 1508–1510: ἀλλ' οἴκτισόν σφας, ὧδε τηλικάσδ' ὁρῶν / πάντων ἐρήμους, πλὴν ὅσον τὸ σὸν μέρος. / ξύννευσον, ὦ γενναῖε, σῇ ψαύσας χερί ("but take pity on them, seeing them deprived of everything at their age, except so far as you provide. Nod your assent, noble man, and touch them with your hand").

28 It is striking that Oedipus does not undertake any arrangements for his sons' future, only for his daughters'. His sons do not even appear onstage; Oedipus only mentions that Creon should not take care of them, because they are men and able to support themselves (1459–1461). See Kyriakou 2011, 467–468.

29 The motif of supplication frames the other Sophoclean play *OC:* at the beginning of the play Oedipus successfully supplicates the Erinyes for asylum since his death will be beneficial to those who have provided him with protection, but simultaneously harmful to those who sent him into exile (38–110). Oedipus also asks Theseus for his safety and burial on Athenian soil (551–649), whereas he himself rejects the supplications of his son Polynices and casts a curse upon him (and his other son Eteocles) instead (1156–1180, 1254–1396). On a brief summary of the play and its aspect as a suppliant drama see Hesk 2012, 167–173 and 179–181. See Markantonatos 2007, 123–140.

heir into a legitimate king: in the remote past Oedipus decides to give up the security of his home in Corinth and becomes a traveller beyond the boundaries of his putative country. He comes to Delphi and from there goes to Thebes, where he becomes the glorious hero who vanquished the Sphinx and found his true origins. This is a journey that takes him from youth to maturity, from singlehood to fatherhood, in sum, from anonymity to fame.

This social transformation, however, begins already in Corinth, when he will be exposed as a baby on Mount Cithaeron, and will become the child of the royal couple of Corinth. As the adopted child of Polybus and Merope, Oedipus crosses social and spatial boundaries as his identity shifts and goes to another place (before he enters Thebes), where he will solve the riddle of the Sphinx. During this spatial journey he arrives outside Delphi, the crucial point of Oedipus' life, since there he encounters his past and the man who exposed him, his father. At the end of this journey, he will be transformed into a blind man with his civic and social identity revealed, but this newly discovered double identity is now connected with a third identity that he is the taint of an incestuous man, of an absolute outcast among people.

These spatio-temporal, back-and-forth shifts challenge Oedipus' main skill: his clairvoyance and his ability to see things that are distant in time (and space). Although he can solve the Sphinx's enigma on the temporal development of human being, he is unable to unfold the storyline of his own identity. The past is twofold for Oedipus: the immediate past (i. e. his quest to find the murderer of Laius), and the earliest past (i.e. the discovery of his own origins and parentage). It is no longer of importance to him to shed light on his glorious actions of the past (saviour of Thebes due to his victory over the Sphinx), but to gain knowledge of his parents' actions.[30]

Once the truth about Oedipus is revealed, he is presented as wishing to stop time and space: time should have stopped at his birth; his place should have been on Cithaeron. This is how he repeatedly requests to receive now his penalty for his crime, exile or death.[31] His fervent wish is to return to Cithaeron, the place which symbolizes *par excellence* the abandonment, but still it is far away for his putative and real birth place, Thebes and Corinth (1449–1454).[32]

30 See Kyriakou 2011, 444.

31 See lines 1340–1341, 1409–1415, 1436–1437, 1451–1454, 1518. According to Apollo's oracle (96–101) Laius' murderer should suffer the same punishments, exile or death. Tiresias had already prophesized a terrible exile for the polluter (454–456).

32 "But as for me, do not require the city of my father to have me living in it, but let me live in the mountains, where there is that mountain of my own that is called Cithaeron, which my moth-

These words also express Oedipus' wish to have died on Cithaeron, instead of being saved and suffering these calamities. More importantly, he remarks that no disease or any kind of physical difficulty will destroy him; he will overcome all the terrible sufferings, as he survived as an exposed and neglected infant (1455–1457). Oedipus is able to predict his terrible sufferings based on the knowledge that he gained from his own past.[33]

The play, however, comes to an end not on Cithaeron but in Thebes. His brother-in-law and the new ruler of Thebes Creon surprisingly forbids him to leave the city, until Apollo is consulted once more (1429–1445). However, Oedipus' stay in the palace is explicitly not meant to be a punishment or an insult: οὐχ' ὡς γελαστής, Οἰδίπους, ἐλήλυθα, / οὐδ' ὡς ὀνειδιῶν τι τῶν πάρος κακῶν (1422–1423).[34] Creon claims that he would have banished Oedipus from Thebes, were it not necessary – in his point of view – to defer to Apollo's judgment and advice (1438–1443, 1518–1519). The brother-in-law now becomes the ruler of the city, therefore he inevitably needs to operate with caution and vigilance, contrary to his predecessor's audacity and self-confidence.[35]

Until Creon receives the delegation from Delphi and Apollo's divine mandate, Oedipus is forced to stay in the palace. Creon compels his servants to take the polluter indoors: ἀλλ' ὡς τάχιστ' ἐς οἶκον ἐσκομίζετε (1429).[36] Creon thus brings the former exposed child, now the polluted adult man, back to his birthplace and true origins: i.e. the palace. Oedipus will now reenter his house, his real birthplace not as a secure and an honoured member of the city and family and as a relieved person who discovered the truth about his life, but as a person who will always carry the stigma of a polluted outcast.[37]

Hence, Oedipus' journey takes a particular turn. His life has come full circle: it began and ended at the same place in this play, i.e. the palace. It is the place of pollution and incestuous marriage; this place, nonetheless, turns out to have a

er and father fixed as my appointed tomb for me when I was still alive, so that I may die from them who tried to kill me".

33 See also Kyriakou 2011, 444–445.

34 "I have not come as someone who ridicules you, Oedipus, nor in order to reproach you because of any of those past evils".

35 Creon's superiority is divulged also by the fact that he grants Oedipus' wish to let him feel for the last time his beloved daughters / sisters (1473–1477).

36 "Take him into the house as quickly as possible".

37 Doubts about the authenticity of the ending of the sophoclean tragedy *OT* (1424–1523) were raised by Boivin de Villeneuve 1729, 372–384. See also Dawe 2006², 192–193; Kovacs (2009b), 53–70. See Finglass 2009, 42–62 who rightly argues for authenticity of the above section and shows that the lines 1524–1530 are not genuine. See also Sommerstein 2011, 85–93.

more intimate signification: Jocasta's womb.[38] Oedipus becomes a king because he marries the queen Jocasta, and not by hereditary right, and at the same time he is the child of his wife (and father of his siblings), as the Chorus express it: ἰὼ κλεινὸν Οἰδίπου κάρα, / ᾧ μέγας λιμὴν / αὐτὸς ἤρκεσεν / παιδὶ καὶ πατρὶ / θαλαμηπόλῳ πεσεῖν (1207–1210).[39] He was exposed as a baby (not even three days after his birth, see lines 717–718), and then returned back to that space to become the husband of his own mother.

The palace is not only the site of primal pollution and incestuous marriage, not only the place where Oedipus was conceived and where he begot offspring with his own mother – at the same time it becomes a kind of tomb. In its private chambers, away from the spectators' eyes, Jocasta commits suicide and Oedipus gouges his eyes out as a penalty for his crimes; his self-blinding is effected via the brooches of his mother and wife, who gave him the light of life. These pins of his mother now deprive him of the light of the sun. At the same time, though, the death of his mother and his own blindness give him a special kind of light, of an insight and a new sense of 'seeing', since he can finally 'see' the truth of his birthplace and identity. The removal of the pins suggests the act of undressing the queen in her marriage chamber as she lies there (1241–1243, 1266–1267). This gesture, then, is a horrible re-enactment of the first night of their union as a man and wife.[40]

Before the play's closure the messenger's speech depicts Jocasta's suicide; the messenger relying on the elements of sound and memory reconstructs the scene in the very interior of Jocasta's bridal chamber: ὅπως γὰρ ὀργῇ χρωμένη παρῆλθ' ἔσω / θυρῶνος, ἵετ' εὐθὺ πρὸς τὰ νυμφικὰ / λέχη, κόμην σπῶσ' ἀμφιδεξίοις ἀκμαῖς (1241–1243).[41] Nonetheless, the messenger's memory (1239–1240)[42] is less painful than that of Jocasta who recalls her remote past with Laius: πύλας δ', ὅπως εἰσῆλθ', ἐπιρράξασ' ἔσω, / καλεῖ τὸν ἤδη Λάϊον πάλαι νεκρόν, / μνήμην παλαιῶν σπερμάτων ἔχουσ', ὑφ' ὧν / θάνοι μὲν αὐτός, τὴν δὲ τίκτουσαν λίποι / τοῖς οἷσιν αὐτοῦ δύστεκνον παιδουργίαν (1244–1248).[43] Jocasta recalls the

38 See Taplin 2015, 6.

39 "Alas famous Oedipus, whom the same wide harbour served as child and as father, on your bridal bed".

40 See Segal 1998^2, 127.

41 "When in her passion she passed through the door, she rushed straight towards her marriage bed, tearing her hair with the fingers of both hands". See Segal 1998^2, 155–156.

42 See also Segal 1998^2, 155–156.

43 "And when she entered, she dashed the doors together behind her, called on Laius, now long a corpse, remembering of the sowing (seeds, i.e. love-making) of long ago, which had brought him death, leaving her to bring forth a progeny accursed by one that was his own". On Jocasta's memory of her past see Kyriakou 2011, 445–446.

night when she conceived Oedipus with Laius. She closes the doors of her bridal chamber and calls on her first husband Laius and their first union. He is now dead, but he left her behind as the mother and wife of his own child, who then begot his own children with her. As Segal remarks, this scene demonstrates a symbolic re-enactment of Jocasta's second (incestuous) marriage with the son who shall replace his father.[44] The details of that night with Laius parallel the ugliness of the present scene with Oedipus (the revelation of the truth that Oedipus is her biological son).

A further spatial re-enactment in the play occurs when Oedipus bursts into the palace to find his mother and then he 'strikes' his eyes when he sees her laying upon the ground: βοῶν γὰρ εἰσέπαισεν Οἰδίπους ..., v. 1252; ἄρας ἔπαισεν ἄρθρα τῶν αὑτοῦ κύκλων (v. 1270).[45] His breaking through the closed doors of Jocasta's chamber symbolically re-enacts the incest, since the 'double doors' recall the 'double bedding' and 'double field' that describe the incest just before (1249–1250: γοᾶτο δ' εὐνάς, ἔνθα δύστηνος διπλῆ / ἐξ ἀνδρὸς ἄνδρα καὶ τέκν' ἐκ τέκνων τέκοι, and 1257: κίχοι διπλῆν ἄρουραν οὔτε καὶ τέκνων).[46] The word for the 'sockets' of his eyes ἄρθρα (1270) in his self-blinding is the word used also for the 'joints' of his ankles in both Jocasta's and the Corinthian messenger's accounts of the exposure (line 718 and line 1032).[47] The pierced feet and eyes are bound together and form a sort of ring composition, since his tragic destiny began with pierced ankles and completed with the pierced eyes.

There is a suggestive parallel to this scene in another Sophoclean play, *Women of Trachis.* Deianeira, having discovered that the gift from Nessus is in fact poison that is killing Hercules, goes inside the house and bids farewell to her marriage bed (912–922). Afterwards, she removes the golden pin that holds her *peplos*, revealing her upper body and stabs herself with a sword (923–926).[48] Her suicide resembles her wedding night with Hercules.[49] Deianei-

---

44 Segal 1998², 156.

45 Line 1252: "For Oedipus burst in crying out loud ..."; line 1270: "and lifting them [sc. the pins] struck his own eye-balls". See Segal 1998², 156.

46 Lines 1249–1250: "She wept over the bed where in double misery she had brought forth a husband by her husband and children by her child"; line 1257: "But the field that had yielded two harvests, himself and his children". See Rehm 2002, 221.

47 See Rehm 2002, 221. Already by Hurlbut 1903, 141–143.

48 Cf. Euripides' Jocasta, who uses also the sword to commit suicide (*Phoen.* 1455–1459). Suicide by sword is committed also by men: Sophocles, *Aj.* 821–865 (Ajax); *Ant.* 1231–1243 (Haimon). On the theme of tragic female figure committing suicide see Foley 2001, 280–283.

49 Rehm 1994, 77.

ra's unusually masculine form of death (women in Greek tragedy, like Jocasta[50] or Phaedra, generally commit suicide by hanging) continues the instability of male and female roles that runs throughout the play;[51] but it also marks the motif of death-in-love, of a marriage to death that usually takes place in the *oikos*.[52] Both scenes make use of the sexual symbolism involved in loosening a woman's robe in her interior chambers, near the conjugal bed.

Nonetheless, the symbolic power of the palace is still to be contested. We saw above its association with 'womb', 'tomb', as a 'place of humiliation'; however, at the end of the play the palace becomes the 'hiding place,' 'refuge,' or, most importantly, 'family asylum', because Creon in the final scene of the play sends Oedipus into the inner chambers.[53] It is the place that procures Oedipus' protection – at least for the time being – from civil indignity. Moreover, this temporary shelter offers him, the incestuous and parricidal man, the opportunity to attempt his reintegration within the family framework.[54]

The Theban herdsman's life, as Segal points out, is characterized by a spatial shift in Oedipus' life, "from house to mountain, from a figure at the centre of the palace life to a figure at the margins of the city (756), in the mountains".[55] The shepherd's life is connected with the shift of space and time, and parallels Oedipus' life. The herdsman might be related to Oedipus' tragic life concerning his emotions of pity and fear, however Sophocles presents him as the opposite of Oedipus.[56] The shepherd usually chooses to avoid the difficult situations and this is the reason why he runs away. For example, in the remote past he escaped

50 In Seneca's *Oedipus* Jocasta also chooses a masculine way of death: she takes the sword and stabs herself in the incestuous belly (1032–1039).

51 On the moral autonomy given to women in Greek tragedy and the fatal consequences see Foley 2001, 112–116.

52 For example, in Soph. *Ant.* Haimon decides to follow his fiancé Antigone into death; Hades is both their marital and funeral place (804–805, 810–816, 891–894, 1204–1207, 1219–1225, 1231–1243). With his death Haimon forges a marriage link with Antigone; his mother Eurydice, on the other hand, stabs herself with her own hand as a means of assimilation of the blood bond with her son Haimon, dissolving the conjugal union with Creon (1315–1316: παίσασ' ὑφ' ἧπαρ αὐτόχειρ αὑτήν, ὅπως / παιδὸς τόδ' ᾔσθετ' ὀξυκώκυτον πάθος – "with her own hand she struck herself beneath the liver, so that she experienced the suffering of her son, loudly to be lamented"). See Rehm 1994, 62–71. In *Trach.* Deianira experiences a similar situation, in which her suicide in the marital chamber underscores the connection between marriage and death. Rehm 1994, 76–83.

53 Thus Revermann 2003, 797.

54 See also Rehm 2002, 221–222 and 228–229.

55 See Segal 2007, 222.

56 See Segal 2007, 222.

when Oedipus attacked his king Laius at the crossroads (118 – 119; 756),[57] but in the present, he undertakes the exact same evasion, when Oedipus now forces him to reveal the truth (1129 – 1131, 1146 – 1159, 1165, 1169). Oedipus had also a similar reaction to the shepherd when the former received in the past the terrible oracle at Delphi that he would commit parricide and incest (787– 797). Oedipus and Jocasta are obliged to face their abnormal and incestuous past that is re-enacted in the present.

Spatio-Temporal Poetics

The above exploration of the spatio-temporal turns and ironies of the play leads to the following consideration: that the human condition of Oedipus can be associated with the Sphinx and the solution of its riddle. The crossroads, too, carry a significant meaning for Oedipus' fate. Jocasta recounts Laius' death at the hands of robbers (716, 729 – 730), but this new piece of crucial information for Oedipus' tragic life prompts her husband's memory to recall the fatal encounter with an elderly arrogant traveller near the junction of those three roads (801–813).

The problem of the exact number of the robbers who killed Laius highlights the dichotomy of 'one' and 'many', since the former immediately proves Oedipus' guilt and the latter his innocence (842– 847): λῃστὰς ἔφασκες αὐτὸν ἄνδρας ἐννέπειν / ὥς νιν κατακτείνειαν. εἰ μὲν οὖν ἔτι / λέξει τὸν αὐτὸν ἀριθμόν, οὐκ ἐγὼ 'κτανον· / οὐ γὰρ γένοιτ' ἂν εἷς γε τοῖς πολλοῖς ἴσος· / εἰδ' ἄνδρ' ἕν' οἰόζωνον αὐδήσει, σαφῶς / τοῦτ' ἐστὶν ἤδη τοὔργον εἰς ἐμὲ ῥέπον.[58] This ambiguity of the number (singular or plural) evokes a mathematical equation that is essential to solve the riddle(s) of Oedipus' life.[59] The murderer of Laius appears to be, though, both 'one' and 'many', because Sophocles uses the word λῃστής both in singular and plural (122, 124).[60] The herdsman who witnessed the murder of

---

57 Lines 118 – 119: θνῄσκουσι γάρ, πλὴν εἷς τις ὃς φόβῳ φυγὼν / ὧν εἶδε πλὴν ἓν οὐδὲν εἶχ' εἰδὼς φράσαι ("they were all killed, except one, who ran away in terror and could tell nothing of what he saw for certain, except one thing"); line 756: οἰκεύς τις, ὅσπερ ἵκετ' ἐκσωθεὶς μόνος ("a servant, who was the only one to come back safe"). See Segal 1998², 153 – 154.

58 "You said that he told you that robbers had killed him; so if he still gives the same number, I was not the killer, for one is not the same as many. But if he speaks unmistakably of one solitary man, then at once undoubtedly the balance tilts towards me".

59 See Knox 1957, 151 with n. 141 and 154 with n. 148; also Dawe 2006², *ad* 845; Karakantza 2011, 157 – 158; Liapis 2012, 90. For a thorough survey on the disparity of the singular and plural number of the robber(s) see Goodhart 1978, 57 – 61 and 67 – 70. Also Ormand 1999, 131 – 138.

60 See Dawe 2006², 7.

Laius had claimed that there were several murderers (122–123; see lines 106–107). Oedipus, though, knew that he had been alone when killing an unknown man at the crossroads. The herdsman is not the only witness of the 'many' robbers; Tiresias had also been a kind of witness – at least he knows for certain who had murdered Laius (362). So the number of witnesses becomes more than one and the 'many' murderers are reduced to 'one'. Sophocles constantly and consequently presents this bipolar distinction and aims to clarify that for his hero 'one' entails literally 'many' identities.[61]

When Oedipus recounts the story of his putative identity and past to his wife Jocasta, he reinforces the distinction between 'one' and 'many' (771–813). Due to the remarks of the drunken man, Oedipus is confronted immediately with the problem of his double identity (779–786). Is he really the son of the royal couple of Corinth? His 'one' identity is now divided into 'many' identities. Thus, the hero is not certain if he can be 'one' or 'many'. In the course of the play, his past and present, though, supply him with many identities, since his investigation to discover his self brings him to the fact that he is 'one' person who embodies "a series of *dédoublements* of roles ... : husband *and* son to his mother, father *and* brother to his children, and so on".[62] On the symbolic literary fabrication of so many double identities and roles for Oedipus the whole Sophoclean tragedy is built. The multiplication of the 'one' to 'many' becomes blatant in the process of "renaturalizing Oedipus, of proving and reproving Oedipus' identity as Laius' son and as Laius' murderer, even as the *Oedipus* itself can suggest that the number of parents of Oedipus is multiple and, most notoriously, that the number of murderers of Laius is also multiple".[63]

In the last scene, Oedipus as a blind man must remain in the palace and expects someone to escort him. This scene recalls the solution of the riddle of the Sphinx: The identity of the creature that starts walking on four feet, then on two, then on three is 'Man' according to Oedipus' correct answer. Tiresias had already prophesized that Oedipus would turn out to be a blind beggar seeking his way with a stick (v. 456). While walking with the help of the servants of Creon (v.

61 See Segal 1981, 214–216 with n. 21 and Segal 2001², 91. Also Rokem 2010, 45–46 on the philosophical connotations of the opposition between 'one' and 'many'.

62 Liapis 2012, 90.

63 King 2012, 405. It has been rightly stated that Sophocles portrays a narrative sequence of events in the discourse of the play and leads the actions of Oedipus, who yields to the cause of the events, to their actual meaning and significance, by obtaining a narrative coherence (Culler 2001², 192–196). The interpretation of the significance of the events contributes and determines the revelation of the truth, therefore "meaning is not the effect of a prior event but its cause". Culler 2001², 194.

1429), Oedipus in the final scene "recalls the three-legged creature of the riddle".[64]

Until the end of the play Oedipus is struggling to discover his real identity, and part of this quest is that he constantly remembers his past actions, so, only when he puts together all the pieces, he is able to solve the puzzle of his life. He explicitly recalls once again the crucial places of his past in a chronological order: Cithaeron, Corinth, the crossroads (1391–1402). All these places are the personified addressees of Oedipus: firstly, he accuses Cithaeron of not providing him a shelter when it was needed (1391–1393); it would have been better, if he had died immediately at this place where he was not only be abandoned by his parents, but also by Cithaeron itself. Secondly, he rails bitterly against adopted father Polybus and his temporary home Corinth since they 'nurtured' him, as he says, as 'something handsome with evils festering underneath' (1394–1397).[65] Finally, Oedipus compares the crossroads to a gigantic anthropomorphic beast that drinks his and Laius' blood and witnesses Oedipus' horrible crimes (1398–1403).

Sophocles uses in *Oedipus Tyrannus* the physical confrontation between mother and son to dramatize the revelation of the incest. The Chorus describe Jocasta to the messenger as the 'wife and mother of that one's children', and the word order in Greek implies, 'this is his wife and mother'.[66] The fact that both Jocasta and Oedipus do not know yet their true relationship enhances the dramatic irony that characterizes the whole play. Nonetheless, Oedipus has already suspected in the past that he might not be the legitimate son of Polybus (line 780: ... πλαστὸς ὡς εἴην πατρί, "with not being my father's child").

It is striking that only the identity of his father raises doubts in Oedipus. Hence, now, on the path towards his self-discovery, Oedipus craves information from his mother about his true parentage. Instead, Jocasta advises Oedipus to ignore the oracle, by underestimating its veracity and adducing as proof the prophecy that was given to her former husband Laius in the remote past: his son would kill him, therefore Laius bound the ankles of his newborn child and exposed him on the mountains (717–719; see lines 851–858). Jocasta adds to this piece of information (i.e. that she had already given birth to a child prior to the arrival of Oedipus at Thebes) further details that allude to the true

---

64 Revermann 2003, 799.

65 Ὦ Πόλυβε καὶ Κόρινθε καὶ τὰ πάτρια / λόγῳ παλαιὰ δώμαθ', οἷον ἆρά με / κάλλος κακῶν ὕπουλον ἐξεθρέψατε ("O Polybus and Corinth and what was called the ancient home of my fathers, what a beautiful thing you brought me up underneath the evil").

66 Line 928: γυνὴ δὲ μήτηρ θ' ἥδε τῶν κείνου τέκνων. See Goldhill 2009, 22–23 and McClure 2012, 376. See schol. ad Soph. *OT* 928 that underlines the juxtaposition of γυνή and μήτηρ for Jocasta.

relationship between her and Oedipus.[67] For example, Jocasta mentions that Oedipus bears a resemblance to her deceased former husband Laius: μορφῆς δὲ τῆς σῆς οὐκ ἀπεστάτει πολύ (743).[68]

Immediately, after receiving this crucial information Oedipus begins to suspect that he might actually be the murderer of Laius, and Jocasta again expresses her qualms about the validity of oracles, supporting her point of view that "the killing of Laius shall be never proven as it was predicted" (852–853) and "the poor child never killed him [sc. his father], but the latter himself perished before the son" (855–856: καίτοι νιν οὐ κεῖνός γ' ὁ δύστηνός ποτε / κατέκταν', ἀλλ' αὐτὸς πάροιθεν ὤλετο). Despite Jocasta's attempts to restrain Oedipus' investigations for his true identity the latter finds out from the Corinthian herdsman that he is not kin of Polybus and Merope (1016–1020). Jocasta, however, apprehends the truth much earlier than her son, and wishes him to never find out who he really is (line 1068): εἴθε μή ποτε γνοίης ὃς εἶ ("may you never find out who you are").

Indeed, from his next interlocutor, the Theban herdsman, Oedipus discovers the identity of his father: the Theban king Laius exposed his infant.[69] Once more, Jocasta has the superior knowledge of this event, as the herdsman utters: "she inside could best tell you how it was".[70] Whereas Jocasta had earlier pronounced that Laius mutilated the ankles of his own child and then exposed it on the mountains,[71] the Theban shepherd divulges now that the queen herself abandoned the infant.[72] Indeed, this new piece of information, that Jocasta, not Laius, exposed the baby, accelerates the process of the anagnorisis since Oedipus will finally discover the truth about his origins: "I who am revealed as cursed in my birth, cursed in my marriage, cursed in my killing".[73]

The significant places for Oedipus, like Cithaeron, Corinth, and Thebes embody his mother in some extent: Cithaeron symbolizes the maternal abandonment, Corinth functions as the site of his adoptive mother, and Thebes illustrates

67 See McClure 2012, 377.

68 "And his appearance was not far from yours".

69 Line 1171: κείνου γέ τοι δὴ παῖς ἐκλῄζεθ' ... ("it was said to be his child").

70 Lines 1171–1172: ... ἡ δ' ἔσω / κάλλιστ' ἂν εἴποι σὴ γυνὴ τάδ' ὡς ἔχει.

71 Lines 718–719: ... καί νιν ἄρθρα κεῖνος [sc. Λάιος] ἐνζεύξας ποδοῖν / ἔρριψεν ἄλλων χερσὶν εἰς ἄβατον ὄρος ("when Laius fastened his ankles and had him cast out by the hands of others upon the trackless mountain").

72 Line 1173: (Οι.) ἦ γὰρ δίδωσιν ἥδε [sc. Ἰοκάστη] σοι; (Θε.) μάλιστ', ἄναξ (Oed. "Was it she who gave it to you?" Sheph. "Yes, my Lord").

73 Lines 1184–1185: ὅστις πέφασμαι φύς τ' ἀφ' ὧν οὐ χρῆν, ξὺν οἷς τ' / οὐ χρῆν ὁμιλῶν, οὕς τέ μ' οὐκ ἔδει κτανών. See McClure 2012, 377.

the maternal roots.[74] Especially, Thebes becomes the place violated by Oedipus; in acknowledging this fact he uses agricultural language in the play's final lines, as he talks to his children/siblings: "What misery is absent? Your father killed his father; he had issue of his mother, from whom he himself had sprung, and begot you from the source of his own being".[75] Thus, Oedipus acknowledges that his incestuous marriage has embroiled him in another more terrible situation, namely in a horrified augmentation of identities: Oedipus is not only a son of Laius and Jocasta, but also a husband of his own mother; his children are not simply his progeny but his siblings too; his mother is both wife and a mother to him.[76]

Final Remarks

Oedipus Tyrannus is a play about time and space, about narratological retrospection of events and analeptic scenes. Oedipus' revisiting of the spatio-temporal events of his life, major actions, major places, major decisions, leads to one realization. The crucial time is that of birth, between the light of knowledge and the darkness of ignorance, and the crucial place is the womb (and later the tomb). However, the end of the play does not focus on either past times or past places but on a kind of future. Oedipus now enters the third stage of the human condition, old age, but with old age and experience he also gains a new view of life: in the last scene of the play, Oedipus is characteristically presented as a blind man who needs help to find his way indoors.

I have discussed how Sophocles' *Oedipus Tyrannus* manipulates the dimensions of time and space in order to show the frailty of the human condition. In *Oedipus Tyrannus* the present action which is connected with the reconstruction

---

74 See McClure 2012, 378. Also Karakantza 2011, 154–155, who rightly states that Cithaeron embodies not only the site of Oedipus' exposure, but also the source of this re-birth and discovery of his true identity: The Chorus acknowledges indeed Cithaeron as the "new" mother and provider to Oedipus (1092: καὶ τροφὸν καὶ ματέρ' αὔξειν – "and shall exalt you as the nurse and mother'"), whereas the hero identifies himself as the child of Tyche (1080–1082: ἐγὼ δ' ἐμαυτὸν παῖδα τῆς Τύχης νέμων / τῆς εὖ διδούσης οὐκ ἀτιμασθήσομαι. / τῆς γὰρ πέφυκα μητρός ... "but I regard myself as child of the event that brought good fortune, and shall not be dishonoured"). See Karakantza 2011, 162–163.

75 Lines 1496–1499: τί γὰρ κακῶν ἄπεστι; τὸν πατέρα πατὴρ / ὑμῶν ἔπεφνε· τὴν τεκοῦσαν ἤροσεν, / ὅθεν περ αὐτὸς ἐσπάρη, κἀκ τῶν ἴσων / ἐκτήσαθ' ὑμᾶς, ὧν περ αὐτὸς ἐξέφυ ("what misery is absent? Your father killed his father. He ploughed his mother, from whom he himself was sowed, and begot you on equal terms, from where he himself sprung").

76 See also McClure 2012, 379.

of past events can be seen as a re-enactment of the Sphinx's enigma about the human condition, exemplified by Oedipus, who leaves the play as a three-legged creature. However, the end of the drama does not prevent the spectators from seeing the sequel of the story. Indeed, what they see in the exodus might be a humiliated blinded Oedipus stripped of the glories of his past; but by travelling in space and time Oedipus has gained in wisdom and clairvoyance, but also pays a price, i.e. his self-blinding and expulsion from Thebes. To some extent, Oedipus is reborn through these spatio-temporal re-enactments as a man with new abilities as he is presented in *Oedipus at Colonus*, in which he is the "master story-teller" (by contrast with *Oedipus Tyrannus*, in which the horrible events of his life are presented to him by various narrative characters).[77]

Bibliography

Boivin de Villeneuve, J. 1729. *Discours sur la tragédie de Sophocle, intitulée Οἰδίπουν Τύραννον. Oedipe Roy, Memoires de litterature tirés des registres de l'Academie Royale des Inscriptions et Belles Lettres. Depuis l'année M.DCCXVIII. jusques & compris l'année M.DCCXXV.* Tome sixième, Paris.

Cameron, A. 1968. *The Identity of Oedipus the King: Five Essays on the Oedipus Tyrannus.* New York and London.

Culler, J. 2001^2. *The Pursuits of Signs. Semiotics, Literature, Deconstruction.* Ithaca and New York.

Dawe, R.D. 2006^2. *Sophocles: Oedipus Rex.* Cambridge.

De Jong, I.J.F. 2004. Sophocles. In *Narrators, Narratees, and Narratives in Ancient Greek Literature. Studies in Ancient Greek Narrative*, eds. I.J.F. De Jong, R. Nünlist and A. Bowie, vol. 1, 255–268. Leiden and Boston.

De Jong, I.J.F. 2007. Sophocles. In *Time in Ancient Greek Literature: Studies in Ancient Greek Narrative*, eds. I.J.F. De Jong and R. Nünlist, vol. 2, 275–292. Leiden and Boston.

Diggle, J. (ed.) 1994. *Euripidis Fabulae: Helena, Phoenissae, Orestes, Bacchae, Iphigenia Aulidensis, Rhesus*, vol. 3. Oxford.

Finglass, P.J. 2009. The Ending of Sophocles' *Oedipus Rex. Philologus* 153.1: 42–62.

Flashar, H. 1976. Die Handlungsstruktur des König Ödipus. *Poetica 8:* 355–359.

Foley, H.P. 2001. *Female Acts in Greek Tragedy.* Princeton and Oxford.

Goldhill, S. 2009. Undoing in Sophoclean Drama: Lusis and the Analysis of Irony. *TAPhA* 139: 21–52.

Goodhart, S. 1978. Ληστάς ἔφασκε: Oedipus and Laius' Many Murders. *Diacritics 8:* 55–71.

Gregory, J. 1995. The Encounter at the Crossroads in the Sophocles' Oedipus Tyrannus. *JHS* 115: 141–146.

Hesk, J. 2012. Oedipus at Colonus. In *Brill's Companion to Sophocles*, ed. A. Markantonatos, 167–189. Leiden and Boston.

77 Markantonatos 2002, 32.

Hurlbut, S.A. 1903. An Inverted Nemesis. On Sophocles, *Oedipus Rex*, 1270. CR 17: 141–143.
Hutchinson, G.O. (ed.) 1985. *Aeschylus. Septem contra Thebas (with Introduction and Commentary).* Oxford.
Karakantza, E. 2011. In Quest of the Father in the Narratives of Origin and Movement in *Oedipus Tyrannus. Mètis* 9: 149–164.
King, B.M. 2012. Masculinity and Freedom in Sophocles. In *A Companion to Sophocles*, ed. K. Ormand, 395–407. Oxford.
Knox, B.M.W. 1957. *Oedipus at Thebes.* New Haven.
Kovacs, D. 2009a. The Role of Apollo in *Oedipus Tyrannus*. In *The Play of Texts and Fragments. Essays in Honour of Martin Cropp*, eds. J.R.C. Cousland and J.R. Hume, 357–368. Leiden and Boston.
Kovacs, D. 2009b. Do We Have the End of Sophocles' Oedipus Tyrannus? *JHS* 129: 53–70.
Kraus, M. 1994. Erzählzeit und erzählte Zeit im König Ödipus des Sophokles. In *Orchestra: Drama, Mythos, Bühne*, eds. A. Bierl and P. von Möllendorf (in collaboration with S. Vogt), 289–299. Stuttgart and Leipzig.
Kyriakou, P. 2011. *The Past in Aeschylus and Sophocles.* Berlin and Boston.
Liapis, V. 2012. Oedipus Tyrannus. In *A Companion to Sophocles*, ed. K. Ormand, 84–97. Oxford.
Liu, C. 2010. *The Motif of Fate in Homeric Epics and Oedipus Tyrannus.* California.
Lloyd-Jones, H. and N.G. Wilson (eds.) 1992^2. *Sophoclis fabulae.* Oxford.
Lloyd-Jones, H. (ed.) 1998^2. *Sophocles. Ajax, Electra, Oedipus Tyrannus (with Translation).* Cambridge, MA and London.
Markantonatos, A. 2002. *Tragic Narrative: A Narratological Study of Sophocles' Oedipus at Colonus.* Berlin.
Markantonatos, A. 2007. *Oedipus at Colonus. Sophocles, Athens, and the World.* Berlin and New York.
McClure, L. 2012. Staging Mothers in Sophocles' Electra and Oedipus the King. In *A Companion to Sophocles*, ed. K. Ormand, 367–380, Oxford.
Ormand, K. 1999. *Exchange and the Maiden: Marriage in Sophoclean Tragedy.* Austin.
Rehm, R. 1994. *Marriage to Death. The Conflation of Wedding and Funeral Rituals in Greek Tragedy*, Princeton, NJ.
Rehm, R. 2002. *The Play of Space. Spatial Transformation in Greek Tragedy.* Princeton and Oxford.
Rehm, R. 2012a. Sophocles. In *Space in Ancient Greek Literature. Studies in Ancient Greek Narrative*, ed. I.J.F. de Jong, 325–340. Leiden and Boston.
Rehm, R. 2012b. Ritual in Sophocles. In *Brill's Companion to Sophocles*, ed. A. Markantonatos, 411–27. Leiden and Boston.
Revermann, K. 2003. Spatio-temporal Dynamics in Sophocles' Oedipus the King. *University of Toronto Quarterly* 72.4: 789–800.
Rokem, F. 2010. *Philosophers and Thespians: Thinking Performance.* California.
Seaford, R. 2007. Money and the Confusion of Generations. In *Generationenkonflikte auf der Bühne. Perspektiven im antiken und mittelalterlichen Drama*, ed. T. Baier, 23–28, Tübingen.
Segal, C. 1981. *Tragedy and Civilization: An Interpretation of Sophocles.* Cambridge, MA and London.

Segal, C. 1998². *Sophocles' Tragic World. Divinity, Nature, Society.* Cambridge, MA and London.
Segal, C. 2001². *Oedipus Tyrannus: Tragic Heroism and the Limits of Knowledge.* New York.
Segal, C. 2007. Life's Tragic Shape: Plot, Design, and Destiny. In *Bloom's Modern Critical Interpretations. Sophocles' Oedipus Rex. Updated Edition*, ed. H. Bloom, 205–224, New York.
Sommerstein, A.H. 2011. Once More the End of Sophocles' Oedipus Tyrannus. *JHS* 131: 85–93.
Taplin, O. 2015. *Sophocles Oedipus the King and Other Tragedies.* Oxford.
Torres, J. 2014. The Theban Seer. *Trends in Classics* 6.2: 339–356.
Ugolini, G. 1995. *Untersuchungen zur Figur des Sehers.* Tübingen.
West, M.L. (ed.) 1998². *Aeschyli tragoediae: cum incerti poetae Prometheo.* Stuttgart.
Zwierlein, O. (ed.) 1986. *L. Annaei Senecae Tragoediae incertorum auctorum Hercules [Oetaeus] Octavia.* Oxford.

Athina Papachrysostomou

Time- and Space-Travelling in Greek Middle Comedy

Greek Middle Comedy – just like the entire ancient Greek literature – is quintessentially permeated and substantively underpinned by the fundamental as much as multifaceted notions of *time* and *space*. These two notions have been extensively studied – on an individual basis – in two collective volumes; *Time in Ancient Greek Literature* (2007) edited by De Jong and Nünlist, and *Space in Ancient Greek Literature* (2012) edited by De Jong. The comic genre is represented, in both volumes, by a chapter on Aristophanes by A.M. Bowie. The latter's contributions are instructive and carefully structured; yet, both the need and the challenge to extend this chrono-topic analysis to the rest of the comic genre (especially the fragmentarily surviving material) remain largely unanswered.

The present chapter aspires to take up the challenge and respond to this need (at least in part), and thereby add a pebble to the (impressive, yet potentially puzzling) mosaic that Greek Comedy is. The chapter focuses on the much debated era of Greek Middle Comedy[1] and studies a certain aspect of the dramaturgical function of *time* and *space* (as literary parameters) during this period. In particular, the present analysis seeks to demonstrate how the comic playwrights of the fourth century BC can – almost capriciously – choose to annul, temporarily suspend or drastically transcend the boundaries of both *time* and *space*. To this end, five comic fragments (by three different playwrights) are being close read below. These fragments were designedly selected, for they archetypically feature Middle Comedy's whimsical transcendence (and even defiance) of reality's temporal and spatial frameworks; as a result (certainly, not an accidental one) of this transcendence, the comic characters are portrayed as *travelling* in *time* and in *space*, within a practically unified / singular (i.e. *time-less* and *space-less*) universe.

Before we proceed, a vital *caveat* should be outlined, given the exclusively fragmentary status of Middle Comedy: the study of fragments can be an exciting but also a highly speculative occupation, which – by nature – is destined to lead

1 Middle Comedy is often treated as a controversial period regarding its existence, its appellation (*Middle*), as well as its (distinctive/singular) content; for a detailed discussion and further bibliography see Papachrysostomou 2008, 10–23; Papachrysostomou 2011. Chronologically, Middle Comedy is conventionally defined at one end by the production of Aristophanes' *Plutus* (388 BC) and at the other end by Menander's first stage appearance (*Orge*, 324/323 or 321 BC).

DOI 10.1515/9783110535150-009

to tentative and / or ambiguous conclusions. The fragment's isolation from its original context, the excerpter's *ad hoc* reasons for citation, the manuscripts' errors of transmission are some of the factors that can inadvertently mislead the researcher, and conduce to confusion or, worse, to inaccurate postulations.[2] Yet, fragmentary literature constitutes for many a fascinating and challenging field. Provided that scholars dealing with fragments (a) are willing to take certain risks and "think outside the box", (b) are aware of the fact that they may prove themselves erroneous, in case more papyri are unearthed, and (c) do not need to persuade the entire scholarly community about the veracity of their findings, but simply base their deductions on some reasonable evidence, I for one strongly believe that a scholar's duty is to scrutinise all fragmentary material and subsequently pursue all implications, allusions, and other information contained therein to the furthest possible extent. Of course, when working with fragments, one always needs to bear in mind that certainty is impossible and that all conclusions remain open to potential challenge, review and even radical overturn.[3]

To begin with, it can be argued that transcending, ignoring and otherwise violating the laws of (literary) nature regarding *time* and *space* constitute manifestations of the pattern of "anti-realism" that palpably permeates the comic genre. Ruffell has recently explored the multiplicity, diversity and versatility of Old Comedy's fictional and utopian worlds.[4] Expectedly (i.e. within the framework of overarching continuity that is inherent to the comic genre), the elements of fiction and anti-realism are also present in Middle Comedy. Yet, during the latter period an additional, distinctive feature conspicuously comes to the foreground and combines nicely with the fictitious / anti-realistic background; this feature is myth.[5] Its dramaturgical uses are revealed to be multifarious and comically exploitable in many, unpredictable, ways. Most importantly, myth becomes the vehicle that facilitates the *travelling* of comic heroes through time and space; simultaneously, it provides a most apposite backdrop for the time-less and space-less universe(s) created thereby. The playwrights of Middle Comedy con-

---

2 The case-study carried out by Dover 2000 demonstrates how imperfect and distorted our understanding of Aristophanes' *Frogs* would be, if no manuscript of the entire play had survived but, instead, we had only the indirect, fragmentary, tradition to rely on.

3 See further Most 1985, 36–41; Arnott 2000; Kraus 2002.

4 Ruffell 2011, 1–53, 157–213, 314–360; Ruffell 2014. Simultaneously, as Ruffell repeatedly points out, these unreal worlds function as vehicles for specific political messages and / or social manifestos; and this is how the comic playwrights commonly use them. See also the relevant analyses by Bowie 2007; Bowie 2012; Lowe 2006; Bakola 2010, 230–296.

5 For an in-depth analysis of the essence of Greek myth and its key features, see Bremmer 1987.

stantly engage in an intense experimentation and dialogue with myth (and its multiple versions); and, in so doing, they prove inventive and innovative. Contrary to tragedy's serious treatment of myth, the parody of mythological tradition becomes one of Comedy's quintessential features and a defining attribute of particularly the fourth century BC. In several cases the traditional myth is distorted and is given a comic version that is full of twists and unexpected turns.[6] There is no way for us to know the details of the absurd twist that every playwright gave the myth in each case, but Aristotle feeds our imagination by offering an example of the extent that myth burlesque could generally take in Comedy (*Po.* 1453^{a}37–39): οἳ ἂν ἔχθιστοι ὦσιν ἐν τῷ μύθῳ, οἷον Ὀρέστης καὶ Αἴγισθος, φίλοι γενόμενοι ἐπὶ τελευτῆς ἐξέρχονται, καὶ ἀποθνῄσκει οὐδεὶς ὑπ' οὐδενός ("those who are the worst enemies in myth, like Orestes and Aegisthus, leave the stage at the end having become friends and no one is killed by anyone").

My specific interest here is to highlight how mythological parody often entails the incongruous transfer of mythical figures into the everyday life of fourth century Athens. The result is preposterously anachronistic, as *mythical time* and *space* (notions that are already vague enough to begin with) get intertwined and fused with contemporary Athenian reality. In these cases, the comic stage accommodates a fictitious, as much as absurd, time-less and space-less world. This world features elements that are either similar to or heavily reminiscent of contemporary Athenian reality, while, at the same time, this world is being irrationally intruded by one or more mythical figures. The contemporary and the mythical sphere infiltrate into one another and function inseparably, dynamically and idiosyncratically within Comedy's indeterminate and indeterminable time and space. Contemporary people can be discerned surreptitiously lurking behind mythic characters, and current socio-political events can be detected beneath mythic episodes. Mythical figures are pulled out of the heroic world and are straightforwardly plunged into the everyday life of fourth century Athens; they are also given a comic twist, so that they behave and look like ordinary Athenians. In most cases, although the title suggests mythical content and plot, the actual play probably had a contemporary setting (characters, place, time), just like *Protesilaos* by Anaxandrides (*PCG* 2,259–264; see Millis 2015, 194–237), *Galateia* by Alexis (*PCG* 2,44–46; see Arnott 1996, 139–149) and *Neoptolemos* by Theophilus (*PCG* 7,703–704; see Papachrysostomou 2008, 263–268), to mention but a few examples.

6 See Webster 1970², 16–19, 82–85; Nesselrath 1990, 188–241; Nesselrath 1995; Bowie 2010, 143–159; Konstantakos 2014.

Although this trend of myth travesty becomes prominent during the period of Middle Comedy, its origins can be traced back to Aristophanes and other poets of Old Comedy. Time- and space-travelling of mythical figures is believed to have been practised – to an unknown extent – by Cratinus in *Ploutoi* (429 BC; see *PCG* 4,204–213; in the particularly conspicuous fr. 171 the Titans visit fifth century Athens[7]), Eupolis in *Demoi* (412 BC; see *PCG* 5,342–376; the play brings Solon, Aristides, Miltiades, and Pericles together on stage in a time-less political ensemble[8]), Plato in *Phaon*[9] (391 BC; *PCG* 7,508–717), and Aristophanes in *Kokalos*[10] (387 BC; see *PCG* 3.2,201–207). Middle Comedy inherits the trend of myth burlesque and propels it to extremes – as this period normally does with the majority of trends and motifs that Old Comedy had only peripherally tackled.[11]

The following fragments of Middle Comedy offer a key insight into this elaborate approach of mythical tradition by the comic playwrights. First in order is the play *Ἀθάμας* (*Athamas*) by Amphis.[12] The play-title strongly suggests a mythological theme; i.e. the title-figure being the homonymous mythical king of Boeotia.[13] However, the play's single surviving fragment (fr. 1) is wholly non-mythical

7 For an overview of the usage of myth in Cratinus in general, see Bakola 2010, 180–229.

8 On the play's mythical and political parameters, see Storey 2003, 111–174, 391–394; Martin 2015.

9 Most intriguing and most typical of this myth-and-reality amalgam is fr. 189, where a character (in all probability, the legendary boatman Phaon) is portrayed as consulting the newly-published cookbook (ὀψαρτυσία) by Philoxenus, a poet of the fourth century BC. See further Pirrotta 2009, 338–376.

10 Although the title-figure directly plunges the play into the mythical realm (Kokalos was the mythical king of the city of Kamikos in Sicily; see D.S. 4.76–79, Hdt. 7.169, Ov. *Met.* 8.261–262, and Sophocles' fragmentarily surviving play *Kamikoi*), at least two surviving fragments from this play reveal pieces of tangible, contemporary, Athenian reality: fr. 364 features a sympotic context with old women going into raptures about black wine from the island of Thasos; and fr. 370 features the term κορινθιάζομαι ("to behave licentiously, like a Corinthian hetaira"), which relates directly with the socio-economic phenomenon of hetairai; see below, n. 15.

11 The same is true for the comic 'sub-trends', on which see Papachrysostomou 2012/2013. See also Papachrysostomou 2008, 18–23.

12 Amphis (*PAA* 126100) was active during the first half of the fourth century BC and, possibly, during the third quarter of that century too; see Papachrysostomou 2016, 11, 15.

13 See further Papachrysostomou 2016, 20–29. On myth details see Apollod. 1.9.1–2, 3.4.3; D.S. 4.47; Paus. 1.44.7–10, 9.34.7; Tz. *ad Lyc.* 22; Ov. *Met.* 4.464–542. See also Gantz 1993, 176–180, 183–184.

in its essence, since it touches on two issues of social (and legal) interest in ancient Athens: wives[14] and hetairai.[15]

Amphis fr. 1 (*Ἀθάμας*):
εἶτ' οὐ γυναικός ἐστιν εὐνοϊκώτερον
γαμετῆς ἑταίρα; πολύ γε καὶ μάλ' εἰκότως.
ἡ μὲν νόμῳ γὰρ καταφρονοῦσ' ἔνδον μένει,
ἡ δ' οἶδεν ὅτι ἢ τοῖς τρόποις ὠνητέος
ἄνθρωπός ἐστιν ἢ πρὸς ἄλλον ἀπιτέον

And so, is not a hetaira more well-disposed
than a wedded wife? Very much so and reasonably enough.
For a wife, even if she is disdainful, gets to stay in the house by custom,
while a hetaira knows that she must either buy
a man with her manners or go off to another one

Son of Aeolus and ruler of Boeotia, Athamas was known for his three wives (Nephele, Themisto, and Ino, who gave him eight children in total) and for his tragic fate. The vicious conspiracies of Ino and Themisto eventually left him childless and wifeless; cf. Sophocles fr. 4: ὡς ὢν ἄπαις τε κἀγύναιξ κἀνέστιος ("how being childless, wifeless, and homeless").[16] Athamas' story inherently possesses the potential for comic adaptation;[17] i.e. it is conceivable that, within a comic context, Athamas was pictured being fed up with his unluckiness with wives and, consequently, seeking refuge and consolation in the loving arms of hetairai. The above fragment is a quasi-philosophising soliloquy[18] that argues

14 The bibliography on women's status in Athens is practically inexhaustible; see e.g. Thompson 1972; Dover 1974, 95–98, 209–213; Pomeroy 1975, 57–148; Pomeroy 1995; Reinsberg 1989; Schaps 1979; Gould 1980; Keuls 1985, 98–128; Winkler 1990; Cohen 1991, 70–170; Fantham 1994; Hunter 1994, 9–42; Arthur-Katz 1995.

15 On hetairai see – from among the plethora of relevant studies – Hauschild 1933; Keuls 1985, 153–186, 267–273; Konstan 1987; Nesselrath 1990, 318–324; Davidson 1993; Davidson 1997, 73–77; Kurke 1997; Faraone and McClure 2006 (all collected essays are of interest); Henry 1985; Henry 2006; Auhagen 2009; Glazebrook and Henry 2011.

16 Tragedy dealt repeatedly with Athamas' story; Aeschylus (*TrGF* 3,123–125), Sophocles (*TrGF* 4,99–102) and Astydamas (*TrGF* 1,200) produced homonymous tragedies, Xenocles composed a homonymous satyr play (*TrGF* 1,153), and Euripides wrote a play entitled *Ἰνώ* (*TrGF* 5.1,442–455), after one of Athamas' wives. The Roman tragic poets Ennius and Accius also wrote plays entitled *Athamas*.

17 Antiphanes also wrote a comic play entitled *Ἀθάμας* (*PCG* 2,318); yet, the one surviving fragment contains no contemporary allusions.

18 For the comic trend of 'fake-philosophising', i.e. the tendency of comic characters towards generalisation in argument, which borders with the exercise of fake philosophy, see Papachrysostomou 2012/2013, 167–173.

a paradox, i.e. that hetairai are more loving than wives (despite and because of the fact that their affection is for hire); this comic paradox makes perfect sense against the established mythical tradition regarding the ill-starred Boeotian king. It is reasonable to believe that the play narrated how (the mythical figure of) Athamas, utterly ruined by his three legally wedded wives, resorted to (fourth century Athenian) hetairai as an alternative; plus, Athamas himself may well have been the speaker of this fragment.[19] The comic context allows the unhindered communication of myth and reality, reconciles any (real-life) incompatibilities, and favours the interaction between the two spheres. For the dramatic purposes of Amphis' play, the mythical figure of Athamas travels in time and in space, from the mythical Boeotia to contemporary Athens. What is additionally striking is that the comic poet chooses to present Athamas in a distinctively Athenian way and his case of bad luck with wives from a typically Athenian point of view; i.e. the mythical figure of Athamas is redesigned and drawn comparable to a wealthy Athenian bourgeois, as he finds comfort in the loving and affectionate hetairai.

Another remarkable figure is Odysseus, who was fairly popular among the poets of Middle Comedy. Odysseus is the title-figure of two plays by Alexis (*Ὀδυσσεὺς ἀπονιπτόμενος: Odysseus washing off clean*, *PCG* 2,110 – 111; and *Ὀδυσσεὺς ὑφαίνων: Odysseus weaving*, *PCG* 2,111 – 112) and of one play by Anaxandrides (*Ὀδυσσεύς*; *PCG* 2,253 – 255). A surviving fragment of Alexis (fr. 159, from *Ὀδυσσεὺς ὑφαίνων*) features a contemporary Athenian context: an attack against fishmongers and their scheming ways (a most typical theme of fourth- and third-century comedy[20]); see Arnott 1996, 465 – 470. In Anaxandrides' version of *Odysseus* the Homeric hero visits the city of Athens; fr. 35 is a long address to Athenians, probably delivered by Odysseus himself, in what appears to be "an amalgam of legend and reality", as Millis puts it (2015, 155). Regarding the hero's treatment by Amphis (*Ὀδυσσεύς*; *PCG* 2,225), there is again good reason to believe that the play consisted of myth travesty and anachronistic transfer of the plot to the comic poet's contemporary era. The one surviving fragment from Amphis' play (fr. 27) features certain stereotypical preparations for a symposion. Keeping his traditional identity as a shipwrecked sailor, Odysseus is probably the expected distinct guest about to be hosted by an extravagant upper-class symposiarch; the latter is identified in the text as the "master" (δεσπότης), instructing a number of slaves to spruce up the symposion-room.

---

19 Webster (1970², 83) argues that Middle Comedy plays featuring mythic figures in their titles had these figures delivering a prologue speech. This might have been the case in Amphis' *Athamas* as well.

20 See Papachrysostomou 2016, 194 – 196.

Amphis fr. 27 (*Ὀδυσσεύς*):
(A.) ἐρίοισι τοὺς τοίχους κύκλῳ Μιλησίοις,
ἔπειτ᾽ ἀλείφειν τῷ Μεγαλλείῳ μύρῳ,
καὶ τὴν βασιλικὴν θυμιᾶτε μίνδακα.
(B.) ἀκήκοας σύ, δέσποτ᾽, ἤδη πώποτε
τὸ θυμίαμα τοῦτο;

(A.) (dress) the walls all around with Milesian wool,
then anoint (the guests) with Megalleian unguent,
and burn the royal mindax-incense.
(B.) My master, have you ever heard before of
this kind of incense?

The play-title (*Ὀδυσσεύς*) explicitly tells in favour of a mythical / heroic background for the play. But the play's surviving fragment introduces a substantial element of contemporary reality; decorating a room with various garments (coverlets, carpets, rugs, etc.) and anointing guests with unguent unambiguously reveal a symposion context. In addition, the mention of two specific items, i.e. wool from the city of Miletus and Megalleian perfume, makes the setting blatantly and undeniably contemporary; the dramatic illusion of Odysseus' heroic age is abruptly terminated and the play is instantly bridged with the here-and-now, i.e. the fourth century BC and beyond, when these expensive items (Milesian wool and Megalleian perfume, both unknown to Homer) circulated widely in the Greek world and were considered most luxurious. In particular, Milesian wool was of high quality and enjoyed a great reputation, particularly for its softness (Ael. *NA* 17.34).[21] The Megalleian perfume, first mentioned by the Old Comedy playwrights Strattis (fr. 34) and Aristophanes (fr. 549), bears his inventor's name (Megallus from Sicily; *PAA* 636610) and is reported to have been a strongly scented and a highly luxurious unguent.[22] The third item, the royal mindax incense (of Persian origin), is a hapax term (its description as *royal* along with its quality as *hapax* constitute adequate evidence to turn this incense into a *status symbol*). Whatever the actual relationships may have been in the play and regardless of whether Odysseus or Alcinous or even the Cyclops was the expected eminent guest,[23] the setting is unquestionably contemporary; and all heroic / mythic characters appearing on stage have ludicrously transcended and travelled across the boundaries of time and space.

---

21 Suffice to say that the clothes of the Sybarites were said to be made out of Milesian wool (Ath. 12.519b); see Gow 1952^{2}, ii.300 – 301.

22 On its manufacture see Theophr. *Od.* 29 – 30, Dsc. 1.58.3 and Plin. *HN* 13.13.

23 For further analysis, see Papachrysostomou 2016, 170 – 176.

The next example comes from the Middle Comedy poet Aristophon[24] and his play *Πειρίθους* (*Peirithous*); here the mythical background of Centauromachy blends with the reality of fourth century Athens, as the following fragment reveals.

Aristophon fr. 7 (*Πειρίθους*):
(A.) καὶ μὴν διέφθαρταί γε τοὖψον παντελῶς·
κλεῖδες μὲν ὀπταὶ δύο παρεσκευασμέναι
(B.) αἷς τὰς θύρας κλείουσι; (A.) θύννειοι μὲν οὖν
(B.) σεμνὸν τὸ βρῶμα. (A.) καὶ τρίτη Λακωνική

(A.) And besides, the dish is utterly spoiled;
two roast keys all prepared.
(B.) Those with which they lock the doors? (A.) Tunny-keys, of course!
(B.) A noble dish. (A.) And a third, Laconian key

Tragic plays entitled *Peirithous* had already been produced in the fifth century BC by Achaeus (*TrGF* I, 20 F36) and Critias (*TrGF* I, 43 F1–14), featuring the ill-fated wedding ceremony of the Thessalian hero and Hippodameia.[25] But Comedy's treatment of the story of Centauromachy[26] was bound to entail burlesque and blatant anachronisms. Although the mythical element of the plot is established already in the title, the play's one surviving fragment suggests a contemporary banquet context. It is vital to remember that food is part and parcel of the comic discourse in general; during especially the period of Middle Comedy, food is much celebrated, unrestrainedly pursued, and extravagantly consumed. The one foodstuff that comes to the foreground is fish,[27] and it cannot be a coincidence that the discussion in this fragment focuses on a couple of roast shoulder-bones of tuna (with a pun on the double meaning of the term κλείς: "door-key" and "collar/shoulder-bone"[28]).

---

24 Aristophon (*PAA* 176015) was active in the mid-fourth century BC. His first victory occurred between 358 and 350 BC (*IRDF* 2325E.46). See Papachrysostomou 2008, 101–149.

25 On myth details see D.S. 4.70, 4.63 and Apollod. 2.5.12; for a different version see Plu. *Thes*. 31.4, 35.1–2 and Paus. 1.17.4.

26 Centauromachy features in various artistic illustrations: on the Parthenon's south metopes, a mural in Theseion, the west pediment of Zeus' temple in Olympia, and numerous vases. See *LIMC* VIII Suppl. s.v. *Kentauroi et Kentaurides*, 382, 384, 404; Boardman 1989, figs. 50, 185, 186; Gantz 1996, i.277–282.

27 See Davidson 1995; Wilkins 2000, 293–304; Papachrysostomou 2016, 195–196.

28 *Laconian keys* are first attested in Ar. *Thesm*. 421–423 and their peculiarity rests with the fact that a door locked with such a key could open only from the outside, not from the inside; i.e. the person(s) inside the house, or any other building, were in effect locked in. See Barton 1972; Whitehead 1990.

Apart from fish, the other telltale feature that lands the fragment in the very midst of fourth-century Athens is the identity of speaker A; his παρὰ προσδοκίαν sorrow about the spoiling of fish (possibly as a result of Centauromachy) and his exclusive focus on food (rather than on the killed Centaurs and Lapiths, as one would normally expect) strongly argue in favour of him being a professional cook (perhaps hired by Peirithous to look after the wedding feast). The character of (professional) cook (μάγειρος) emerges as a prominent stereotypical figure during the period of Middle Comedy.[29] The present character's frustration about the spoiling of food is best interpreted if indeed *he* was the one who prepared the dish. His overwhelming preoccupation with food (rather than with anything else) is rendered understandable in a context where the preceding battle has affected negatively the "right timing" (καιρός) for serving and eating the dish.[30] Against the mythical setting of Centauromachy, it is food (one of Middle Comedy's defining attributes) that overwhelmingly takes centre stage. Within the mythical context the presence of the cook figure constitutes an anachronism in itself, and – through this figure – the world of fourth century Athens is made perceptible and almost tangible.

Once again, we find ourselves situated mid-way between myth and reality. The title-figure of Peirithous is pulled out of the mythical context and, through a journey in time and space, makes a stage appearance, probably along with Theseus, within a typically Athenian milieu, being portrayed as a bon-viveur and nonchalant bourgeois, with extravagant spending habits, such as costly predilection for fish. The consumption of fish in particular was widely considered a socio-economic status statement in itself, with far-reaching implications, as Davidson has repeatedly demonstrated (see Davidson 1993, 1995, and 1997 passim).

The last, twofold, example comes from the play *Οἰνοπίων* (*Oinopion*) by the playwright Philetaerus.[31] According to mythical tradition, Oinopion was the son of Dionysus (or Theseus) and Ariadne, and ruler of the island of Chios, where he was believed to have introduced the cultivation of vines.[32] Despite the mythical element introduced already by the title-figure, the context entailed by the play's

29 On the cook figure see Dohm 1964, 67–275; Nesselrath 1990, 297–309; Dalby 1995, 121–124; Arnott 1996, 116; Wilkins 2000, 87, 387–410; Dobrov 2002; Papachrysostomou 2016, 115.

30 The right timing and knowing how to handle time (i.e. when to serve the courses and when to remove them) appears to be quintessential for cooks in Middle Comedy; e.g. Dionysius fr. 2, Alexis fr. 153.7–13 (see Arnott 1996, 450), Sosipater fr. 1.48–56.

31 Philetaerus (*PAA* 924630) was one of Aristophanes' sons. He won his first victory between 372 and 366 BC (*IRDF* 2325E.38). See Papachrysostomou 2008, 221–247.

32 See Theopompus 115 F 276 *FGrH*, Plu. *Thes.* 20.2, D.S. 5.79.1.

two surviving fragments is unequivocally contemporary. The speaker in fr. 13 sounds like a guru giving a lesson on pleasure and inviting people to indulge as much as possible during lifetime:

Philetaerus fr. 13 (*Οἰνοπίων*):
θνητῶν δ' ὅσοι
ζῶσιν κακῶς ἔχοντες ἄφθονον βίον,
ἐγὼ μὲν αὐτοὺς ἀθλίους εἶναι λέγω·
οὐ γὰρ θανὼν δήπουθ' ἂν ἔγχελυν φάγοις,
οὐδ' ἐν νεκροῖσι πέττεται γαμήλιος

All those mortals
who live miserably, although they have plentiful means of living,
I for one consider to be wretched;
for once you die, you can hardly eat eels,
nor is a bride-cake cooked among the dead

Again, since Oinopion features in the title, we can confidently establish that this mythical character was central to the play's plot and had a speaking part. He could even be the one delivering the fragment above, given that his very name, *Oinopion*, is a speaking one, i.e. "the one who drinks wine"[33] (wine obviously being amongst the pleasures the speaker is seeking). Contemporary reality and mythical tradition intertwine once again. Keeping his mythical identity, Oinopion travels in time and in space, as he is being transferred by Philetaerus into a contemporary context, where he is portrayed behaving and speaking like a fourth century, manifestly wealthy, Athenian who is fond of eels and indulges in all sorts of pleasures. Both the incitement to pursue ephemeral pleasures and the reference to eels (ἔγχελυς) and wedding cakes (γαμήλιος; *sc.* πλακοῦς) constitute concrete and distinctive features of Athenian reality, which mingles inextricably with mythical tradition. It is intriguing that among all gastronomical pleasures the speaker chooses the eel, *the* luxury food *par excellence*. Dicaeopolis in Aristophanes' *Acharnians* 885–894 goes into raptures over an eel, priced at three drachmas (line 962),[34] and some fifty years later twelve drachmas are said to be due just to catch a sniff of one (Antiphanes fr. 145).[35]

33 Alternatively, it may derive from οἴνοψ, "wine-coloured" (so Welcker 1824, 549).
34 At approximately the same time the daily wage of an unskilled labourer would hardly exceed the amount of two drachmas. For an exhaustive analysis of wages in Athens, see Loomis 1998.
35 See Davidson 1997, 8–10, 186–187; Olson and Sens 2000, 48–53 (comm. on Archestratus fr. 10).

Fr. 14 (from the same play by Philetaerus) is even more interweaved with current Athenian affairs, since the speaker mentions by name Στρατόνικος (Stratonicus), a fourth century BC musician.

> Philetaerus fr. 14 (*Οἰνοπίων*):
> ὁ μάγειρος οὗτος Πατανίων προσελθέτω
> πλείους Στρατονίκου τοὺς μαθητάς μοι δοκεῖ
> ἕξειν Πατανίων
>
> Let this cook Patanion come forward
> I think that Patanion will have more disciples
> than Stratonicus

The particular cook (Patanion) is a fictitious character;[36] as mentioned above regarding Aristophon's *Peirithous*, the cook figure constitutes an anachronistic presence within a play with a mythical title. But what is most striking is the ὀνομαστί reference to the historic figure of Stratonicus, which straightforwardly links the fragment and its context to contemporary Athenian reality. Stratonicus (Stephanis 2310) was a music teacher and innovator who lived and flourished in Athens from approximately 410 to 360 BC.[37]

The above five fragments exemplify Comedy's witty fusion of myth and reality, of a fuzzy and indeterminable "there and then" with a very concrete "here and now". Through a discontinuous version of dramatic illusion (exclusively found in Comedy), the comic genre manages to bridge the mythical and hardly definable past with the contemporary and almost tangible present. The result is not only preposterously hilarious (as expected, vis-à-vis tragedy's serious treatment of the same stories), but also in most cases this weird amalgam is thought-provoking and can be socio-economically instructive, for it yields further and intriguing implications regarding the contemporary reality that refers to. Costly fish (tuna in *Peirithous* and eels in *Oinopion*), banquets (similar to the one being set up in *Odysseus*), and pursuits of hetairai (like the one Athamas longs for) are easily recognisable features of the bourgeois world of fourth century Athens, wherein numerous mythical characters are plunged by the comic poets. Comedy's attempt to link a play that is immersed in myth (already through its title) to reality can – indirectly – prove highly valuable for reasons of socio-economic analysis of the era. For, in order to achieve the above (i.e. associate

---

36 So named after πατάνιον, a kind of dish (a hapax proper name and also an appropriate one for a cook); see Poll. 10.107 and Hsch. s.v. πατάνια.

37 So Maas in *RE* IV.A1 s.v. Stratonikos nr. 2. See Callisthenes 124 F 5 *FGrH*, Macho fr. 11, Ath. 8.347f–352d.

myth with reality), the comic playwrights, naturally (and unconsciously), seek to imprint in their plays traits and features that are not simply immediately recognisable as contemporary, but, in addition, are the most representative ones of current reality. In other words, Middle Comedy reflects and projects as "contemporary" the most conspicuous and predominant trends (socio-economic, cultural and other) of the respective era. Functioning as a mirror of contemporary reality, Middle Comedy selectively focuses on what quintessentially constitutes the defining attributes of the contemporary socio-economic ambience in Athens. The task to further follow Comedy's clues and hints in the direction of mapping the complex socio-economic grid of fourth century Athens is a challenging one that falls outside this chapter's scope but still one that I plan to pursue in detail in the not so distant future.

Bibliography

Arnott, W.G. 1996. *Alexis: The Fragments. A Commentary.* Cambridge.

Arnott, W.G. 2000. On Editing Comic Fragments from Literary and Lexicographical Sources. In *The Rivals of Aristophanes: Studies in Athenian Old Comedy*, eds. F.D. Harvey and J.M. Wilkins, 1–13. London.

Arthur-Katz, M. 1995. Ideology and "the Status of Women" in Ancient Greece. In *Women in Antiquity: New Assessments*, eds. R. Hawley and B. Levick, 21–43. London and New York.

Auhagen, U. 2009. *Die Hetäre in der griechischen und römischen Komödie.* Munich.

Bakola, E. 2010. *Cratinus and The Art of Comedy.* Oxford.

Barton, I.M. 1972. Tranio's Laconian Key. *G&R* 19: 25–31.

Boardman, J. 1981/2009. *Lexicon Iconographicum Mythologiae Classicae*, 8 vols. (and 2 vols. suppl.). Zürich. (abbr. as *LIMC*)

Boardman, J. 1989. *Athenian Red Figure Vases. The Classical Period.* London.

Bowie, A.M. 2007. Aristophanes. In De Jong and Nünlist 2007, 305–317.

Bowie, A.M. 2010. Myth and Ritual in Comedy. In *Brill's Companion to the Study of Greek Comedy*, ed. G.W. Dobrov, 143–176. Leiden and Boston, MA.

Bowie, A.M. 2012. Aristophanes. In De Jong 2012, 359–373.

Bremmer, J. 1987. What is a Greek Myth? In *Interpretations of Greek Mythology*, ed. J. Bremmer, 1–9. London.

Cohen, D. 1991. *Law, Sexuality, and Society: The Enforcement of Morals in Classical Athens.* Cambridge.

Dalby, A. 1995. *Siren Feasts: A History of Food and Gastronomy in Greece.* London.

Davidson, J.N. 1993. Fish, Sex and Revolution in Athens. *CQ* 43: 53–66.

Davidson, J.N. 1995. *Opsophagia:* Revolutionary Eating at Athens. In *Food in Antiquity*, eds. J. Wilkins, D. Harvey and M. Dobson, 204–213. Exeter.

Davidson, J.N. 1997. *Courtesans and Fishcakes.* London.

De Jong, I.J.F. 2012. *Space in Ancient Greek Literature: Studies in Ancient Greek Narrative.* Leiden and Boston, MA.

De Jong, I.J.F. and R. Nünlist. 2007. *Time in Ancient Greek Literature*. Leiden and Boston, MA.
Dobrov, G.W. 2002. Μάγειρος Ποιητής: Language and Character in Antiphanes. In *The Language of Greek Comedy*, ed. A. Willi, 169–190. Oxford and New York.
Dohm, H. 1964. *Mageiros: die Rolle des Kochs in der griechisch-römischen Komödie*. Munich.
Dover, K.J. 1974. *Greek Popular Morality in the Time of Plato and Aristotle*. Oxford.
Dover, K.J. 2000. Foreword: Frogments. In *The Rivals of Aristophanes: Studies in Athenian Old Comedy*, eds. F.D. Harvey and J.M. Wilkins, xvii–xix. London.
Fantham, E. 1994. Women in Classical Athens: Heroines and Housewives. In *Women in the Classical World: Image and Text*, eds. E. Fantham, H.P. Foley, N.B. Kampen, S.B. Pomeroy and H.A. Shapiro, 68–127. New York.
Faraone, C. and L. McClure. 2006. *Prostitutes and Courtesans in the Ancient World*. Madison, WI.
Gantz, T. 1996 (originally published in a one-volume edition, 1993). *Early Greek Myth: A Guide to Literary and Artistic Sources*, 2 vols. Baltimore and London.
Glazebrook, A.M.J. and M. Henry. 2011. *Greek Prostitutes in the Ancient Mediterranean, 800 BCE – 200 CE*. Madison, WI.
Gould, J. 1980. Law, Custom and Myth: Aspects of the Social Position of Women in Classical Athens. *JHS* 100: 38–59.
Gow, A.S.F. 1952^2. *Theocritus. Edited with a Translation and Commentary*, 2 vols. Cambridge.
Hauschild, H. 1933. *Die Gestalt der Hetäre in der griechischen Komödie* (Ph.D. Diss.). Leipzig.
Henry, M. 1985. *Menander's Courtesans and the Greek Comic Tradition*. Frankfurt am Main.
Henry, M. 2006. *Neaera: Writing a Prostitute's Life*. London.
Hunter, V.J. 1994. *Policing Athens: Social Control in the Attic Lawsuits, 420–320 B.C.* Princeton.
Kassel, R. and C. Austin. 1983– . *Poetae Comici Graeci*. Berlin and New York. 8 vols. (abbr. as *PCG*)
Keuls, E.C. 1985. *The Reign of the Phallus: Sexual Politics in Ancient Athens*. New York.
Konstan, D. 1987. Between Courtesan and Wife: Menander's *Perikeiromene*. *Phoenix* 41: 122–139.
Konstantakos, I.M. 2014. Comedy in the Fourth Century I: Mythological Burlesques. In *The Oxford Handbook of Greek and Roman Comedy*, eds. M. Fontaine and A.C. Scafuro, 160–180. Oxford and New York.
Kraus, C.S. 2002. Introduction: Reading Commentaries/Commentaries as Reading. In *The Classical Commentary: Histories, Practices, Theory*, eds. R.K. Gibson and C.S. Kraus, 1–27. Leiden.
Kurke, L. 1997. Inventing the 'Hetaira': Sex, Politics, and Discursive Conflict in Archaic Greece. *ClAnt* 16.1: 106–150.
Loomis, W.T. 1998. *Wages, Welfare Costs, and Inflation in Classical Athens*. Ann Arbor.
Lowe, N. 2006. Aristophanic Spacecraft. In *Playing around Aristophanes: Essays in Celebration of the Completion of the Edition of the Comedies of Aristophanes by Alan Sommerstein*, eds. L. Kozak and J. Rich, 48–64. Oxford.
Martin, R. 2015. Solon in Comedy. *Trends in Classics* 7.1: 66–84.
Millis, B.W. 2015. *Anaxandrides: Introduction, Translation, Commentary* (Fragmenta Comica 17). Heidelberg.
Millis, B.W. and S.D. Olson. 2012. *Inscriptional Records for the Dramatic Festivals in Athens: IG II^2 2318–2325 and Related Texts*. Leiden. (abbr. as *IRDF*)

Most, G. 1985. *The Measures of Praise*. Göttingen.
Nesselrath, H.-G. 1990. *Die attische Mittlere Komödie*. Berlin and New York.
Nesselrath, H.-G. 1995. Myth, Parody, and Comic Plots: The Birth of Gods and Middle Comedy. In *Beyond Aristophanes: Transition and Diversity in Greek Comedy*, ed. G.W. Dobrov, 1–28. Atlanta.
Olson, S.D. and A. Sens. 2000. *Archestratos of Gela: Greek Culture and Cuisine in the Fourth Century BCE: Text, Translation, and Commentary*. Oxford and New York.
Papachrysostomou, A. 2008. *Six Comic Poets. A Commentary on Selected Fragments of Middle Comedy*. Tübingen.
Papachrysostomou, A. 2011. Καλειδοσκόπιο στη Μέση Κωμωδία: Η νομοτέλεια της αλλαγής στο δράμα. In *Αττική Κωμωδία: Πρόσωπα και Προσεγγίσεις*, eds. Th. Pappas and A. Markantonatos, 90–102. Athens.
Papachrysostomou, A. 2012/2013. Continuity and Change in the Comic Genre or How It All Ended Up With Menander: The Case of Sub-Trends. *Ordia Prima* 11–12: 165–189.
Papachrysostomou, A. 2016. *Amphis: Introduction, Translation, Commentary* (Fragmenta Comica 20). Heidelberg.
Pirrotta, S. 2009. *Plato Comicus: die fragmentarischen Komödien: ein Kommentar*. Berlin.
Pomeroy, S.B. 1975. *Goddesses, Whores, Wives, and Slaves: Women in Classical Antiquity*. New York.
Pomeroy, S.B. 1995. Women's Identity and the Family in the Classical *Polis*. In *Women in Antiquity: New Assessments*, eds. R. Hawley and B. Levick, 111–121. London and New York.
Reinsberg, C. 1989. *Ehe, Hetärentum und Knabenliebe im antiken Griechenland*. Munich.
Ruffell, I.A. 2011. *Politics and Anti-Realism in Athenian Old Comedy: The Art of the Impossible*. Oxford and New York.
Ruffell, I.A. 2014. Utopianism. In *The Cambridge Companion to Greek Comedy*, ed. M. Revermann, 206–221. Cambridge and New York.
Schaps, D.M. 1979. *Economic Rights of Women in Ancient Greece*. Edinburgh.
Snell, B., R. Kannicht and S. Radt. 1971–2004. *Tragicorum Graecorum Fragmenta*, 5 vols. Göttingen. (abbr. as *TrGF*)
Stephanis, I.E. 1988. *Διονυσιακοὶ Τεχνῖται*. Herakleion. (abbr. as Stephanis)
Storey, I.C. 2003. *Eupolis: Poet of Old Comedy*. Oxford and New York.
Thompson, W.E. 1972. Athenian Marriage Patterns: Remarriage. *CSCA* 5: 211–225.
Traill, J. 1994–. *Persons of Ancient Athens*, 21 vols. Toronto. (abbr. as *PAA*)
Webster, T.B.L. 1970^2. *Studies in Later Greek Comedy*. Manchester.
Welcker, F.G. 1824. *Die Aeschylische Trilogie Prometheus und die Kabirenweihe zu Lemnos*. Darmstadt.
Whitehead, D. 1990. The Lakonian Key. *CQ* 40: 267–268.
Wilkins, J. 2000. *The Boastful Chef: The Discourse of Food in Ancient Greek Comedy*. Oxford.
Winkler, J.J. 1990. Laying Down the Law: The Oversight of Men's Sexual Behavior in Classical Athens. In *Before Sexuality: The Construction of Erotic Experience in the Ancient Greek World*, eds. D.M. Halperin, J.J. Winkler and F.I. Zeitlin, 171–210. Princeton.

Part III: **Empirical and Imaginary Chronotopes**

Marion Meyer

The *martyria* of the Strife for Attica – *martyria* of Changes in Cult and Myth. Space and Time in the West Pediment of the Parthenon

My contribution discusses issues of space and time in visual communication, taking the west pediment of the Parthenon[1] as an example. Images seem to be the convenient medium for relations of space, not those of time. There is, however, a danger of misunderstanding either of them – in ancient and in modern times.

Architectural sculpture of the classical period was expected to present narrative images[2] – images which refer to tales that the beholders had to know in order to understand what they saw. So the questions to be asked are: What does the image show, and what does it tell?

Thanks to an anonymous draftsman in the entourage of the Marquis de Nointel there is a drawing of the pediment made in 1674, thirteen years before the Parthenon was severely damaged during the Venetian siege (fig. 1a – b),[3] and thanks to Pausanias we know what was represented: an *eris* (a quarrel) of Poseidon with Athena about the land.[4]

The earliest literary reference to this *eris* is Herodotus who says that on the Acropolis, in the temple of Erechtheus, there were an olive tree and a *thalassa* (salt water), and the Athenians said that Poseidon and Athena set them up as *martyria* when they quarreled about the land.[5] Herodotus thus links the *eris* to

1 Brommer 1963, 30 – 61, 115 – 121, 126 – 140, 158 – 170 fig. 15 pl. 63 – 131, 152; Berger 1976, 122 – 128 pl. 29; Berger 1977, 124 – 134 fig. 3 – 5 pl. 35 – 36 fold-out III (reconstruction); Palagia 1993, 7 – 17, 40 – 59, 61 fig. 3 – 5, 7a; 22, 71 – 86, 90 – 96, 98 – 120; Palagia 2005, 225 – 234, 242 – 253 fig. 77, 80, 89 – 90; Valavanis 2013, 116 – 117, 148 – 149 fig. 170, 199 – 200.

2 The Parthenon frieze – representing the Athenians practising cult – is the conspicuous exception. Neils 2001; Neils 2005, 198 – 223; Schneider 2010, 259 – 279; Fehr 2011. *Contra:* Connelly 2014. For narrative images see Giuliani 2003.

3 Brommer 1963, 115 – 116 pl. 64 – 65.1; Palagia 1993, 41 fig. 3 – 4; Palagia 2005, 226 – 228 fig. 77a-b. For the false attribution to Jacques Carrey see de Rycke 2007, 721 – 753.

4 Paus. 1.24.5: τὰ δὲ ὄπισθεν ἡ Ποσειδῶνος πρὸς Ἀθηνᾶν ἔστιν ἔρις ὑπὲρ τῆς γῆς ("in the rear side there is Poseidon's strife with Athena about the land").

5 Hdt. 8.55: ἔστι ἐν τῇ ἀκροπόλει ταύτῃ Ἐρεχθέος τοῦ γηγενέος λεγομένου εἶναι νηός, ἐν τῷ ἐλαίη τε καὶ θάλασσα ἔνι, τὰ λόγος παρὰ Ἀθηναίων Ποσειδέωνά τε καὶ Ἀθηναίην ἐρίσαντες περὶ τῆς χώρης μαρτύρια θέσθαι ("on the acropolis there is a temple of the so-called earthborn

DOI 10.1515/9783110535150-010

Figure 1a: West pediment of the Parthenon (drawing of 1674)

a specific site and to physical marks visible on this site. Later sources locate the olive tree in the precinct of Pandrosus west of the Erechtheion and the salt water in the west cella of the Erechtheion.[6] Speaking about the divine protagonists both Herodotus and Pausanias mention Poseidon first and thus suggest that the *eris* was the god's interest or initiative.

The west pediment of the Parthenon is the earliest visual representation of this *eris*. The composition focuses on the gods in the very center. They can easily be identified even in the present state of preservation: Athena wears the *aegis*, and Poseidon is accompanied by marine figures underneath his charioteer (in the drawing of 1674, see fig. 1b) or underneath his horse (Acropolis Museum, see below). The protagonists are shown in a chiastic pose and in abruptly chang-

Erechtheus, with the olive and the *thalassa* inside, about which the Athenians say that Poseidon and Athenaia set them up as testimonies when they quarreled about the land").

6 The νηός of Erechtheus mentioned by Herodotus (see n. 5) must have been the sacred space later occupied by the west cella of the Erechtheion (which housed the *thalassa*, according to Paus. 1.26.5) and the adjacent Pandroseion (where the olive tree stood, see Philochorus *FGrH* 328 F 67; Apollod. 3.14.1). For the situation before the Erechtheion was built see the reconstruction by A. Papanikolaou in Greco 2010, 132, 137 fig. 57 (M.C. Monaco).

Figure 1b: West pediment of the Parthenon (drawing of 1674)
© Paris, Bibliothèque nationale de France

ing motions, conveying an ambiguous encounter: They are approaching each other (their legs overlapping) and shying away (their bodies bending in opposite directions), and they are looking backwards (turning their heads towards each other). Poseidon is marked as the aggressor by intruding into Athena's space: He transgresses the (invisible) borderline which cuts the pediment into two halves, and he steps into the section of the goddess.[7]

Both divinities are flanked by chariots which serve two functions: By their rapid movement towards the center they lead the spectators' view in that direction. As the horses are being stopped and restrained the chariots point to a sudden shift of movement, too. Thus, the chariots echo and underline the clash of the two protagonists. This is a purely visual strategy. The designers of the pediment did not intend to tell that both gods had jumped off their chariots in full speed and managed to pass the horses.[8]

---

7 Berger 1976, 124–126 pl. 29; Berger 1977, 127 fold-out III: Poseidon's right foot exceeds the borderline by ca. 0.45 m.

8 For an interpretation of the *eris* as a competition for speed see below. Poseidon's charioteer is his companion Amphitrite; on Athena's side one would expect Nike, see Palagia 2005, 232

E. Simon reconstructed Zeus' thunderbolt emerging between the opponents' heads and separating both figures before the Pella hydria was found which does show a huge golden thunderbolt between Athena and Poseidon.[9] The vase was painted (in Athens) about a generation after the Parthenon was finished. It presents, of course, another interpretation of the episode and cannot serve as testimony for the composition of the earlier image. I wonder whether a thunderbolt was actually shown in the pediment or whether the protagonists retreat as a reaction to Zeus' invisible interference. The rearing horses, at any rate, enhance the impression of a sudden reaction.

The striding male and female figure in the background, partly overlapped by the horses, are convincingly interpreted as Hermes and Iris, the messenger gods who link the immortal and the mortal spheres.[10] Both corners are filled with more than a dozen mortals who are seated, crouching or reclining. The individual identities of these figures, most of them females and children, have been disputed, but there is a consensus that they represent heroes and heroines of Attica, evoking the very beginning of life and cult in this land.[11] They convey what the *eris* is all about: The gods are interested in the people of Attica.

There are only poor traces of the *thalassa* and none at all of the olive tree. The *ketos* depicted in the drawing of 1674 (fig. 1b) is lost.[12] Fragments of a sea snake are probably part of a triton which served as support for one of Poseidon's horses.[13] Remains of a tree trunk with a snake apparently do not belong to the original setting.[14]

Although the *martyria* cannot be satisfactorily restored it is understood that they were not the objects of the gods' action or attention. The positions of the

fig. 80. Berger 1996, 355, however, pointed out that there would not be sufficient space for wings; he suggested Zeuxippe, mother of Boutes.

9 Simon 1980, 239–255 fig. 1. – Hydria Pella, Arch. Mus. 80514 (ca. 400 BCE): BAPD 17333; Drougou 2000, 147–216 pl. 30–39 color pl. I-IV; Palagia 2005, 244 pl. 6; Tiverios 2005, 299–319 fig. 1–10; Tiverios 2009, 159–163 fig. 1–2; Marx 2011, 33–36 pl. 5.1–3; Jubier-Galinier 2012, 278–289 fig. 3–5; Junker and Strohwald 2012, 71–76 fig. 57a-b; Neils 2013, 595–613 fig. 1–9; Isler-Kerényi 2015, 180–182 fig. 96.

10 Palagia 2005, 232 fig. 80.

11 Palagia 1993, 61; Harrison 2000, 279–285; Pollitt 2000, 224–226; Fehr 2004, 145–146; Kreuzer 2005, 194–200 fig. 1–2; Palagia 2005, 232 fig. 80, 244–253.

12 The head of a *ketos* attributed by Yalouris 1984, 281–283 pl. 28–29 does not belong, see Palagia 2005, 250 n. 128.

13 Brommer 1963, 49–50 pl. 118,2; Palagia 1993, 40, 42, 47–49 fig. 103. The function can be deduced from the support for Athena's horse (see fig. 1a) which was identified with the torso Akr. 879, see Palagia 1993, 45, 47–48 fig. 5, 22, 90–91.

14 Brommer 1963, 41, 96–97, 164 no. 4 pl. 102, 152; Hurwit 2004, 130–131 fig. 92. The original tree is supposed to have been made of bronze: Simon 1980, 250–252; Palagia 2005, 246.

right arms of both figures and their gaze preclude that. Unlike on both later Attic vases[15] the olive tree cannot possibly have been placed between the gods (whose legs overlap) nor anywhere near Athena's right arm. It will have stood in the background, near the center but behind Athena and Poseidon (and would eventually have left enough room for a thunderbolt appearing above the goddess' left shoulder).[16] The marine figures (triton, *ketos*) were far away from Poseidon's trident. The *martyria* are not being set up, they are there, and they have the function of visualizing the powers and specific capacities of both gods. They characterize Athena as the goddess of civilization and cultivation and Poseidon as the god of natural forces (like the sea). Furthermore, by recalling the actual olive tree and salt water seen by Herodotus (and others) some thirty meters north of the Parthenon, they locate the scene on the Acropolis and give permanent proof of the gods' interest and commitment for this site.

To sum up: The pediment shows the climax of the *eris* (and thus foreshadows Athena's victory): The spatial relation of the protagonists is narrative and indicative of their opposition and separation. The chariots are attributive and serve to enforce both the movements towards and away from each other. The *martyria* are attributive, too. They visualize the respective powers and capacities of both deities. The figures in the corners describe the setting, evoke the mythical past of Attica and represent the worshippers.

It remains to be asked what the image actually 'tells' – which narrative(s) it evokes. Herodotus refers to the *eris* as if it were well-known. It will turn out that it was indeed well-known, but not necessarily linked to one specific tale.

Herodotus and Pausanias suggest that Poseidon started the *eris*, and in the pediment the god is marked as the aggressor. There is, in fact, a tradition of Poseidon's claim for a *chora*. Seven such cases are attested. The evidence is late – except for Attica – , but consistent. Poseidon challenges various divinities in various places – Apollo in Delphi, Helios in Corinth, Hera in Argos, Zeus on Aegina, Athena in Athens, Dionysos on Naxos, Athena in Troizen. The point is that he always loses – except in Troizen where the conflict ends in a tie.[17] Poseidon's claim

---

15 The hydriai in Pella (see n. 9) and in St. Petersburg are the only additional images of the *eris*. – St. Petersburg, Hermitage, P 1872.130 (ca. 350 BCE): BAPD 6988; Tiverios 2005, 301–302, 307, 312, 316, 319 fig. 11; Marx 2011, 33–36 pl. 6.1–2; Jubier-Galinier 2012, 275–281 fig. 2; Junker and Strohwald 2012, 71–76 fig. 58; Brinkmann 2013, 243, 246, 334 no. 47 fig. 268; Vollmer 2014, 434–436, 443–444, 446–447 n. 156 pl. 60.3–4; Isler-Kerényi 2015, 182–183 fig. 97.

16 For the position of the tree see Berger 1977, 126–128; Palagia 2005, 246.

17 Plut. *Mor.* 741a: Athens, Delphi, Argos (cf. Paus. 2.15.5; Nonnus *Dion.* 36.127–29), Aegina, Naxos. For Corinth see Paus. 2.1.6. For Troizen see Paus. 2.30.6. – Parker 1987, 198–200; Simon 1998[4], 66–68; Mylonopoulos 2003, 409–410.

and defeat serve to characterize the god. He, the divinity of natural forces, loses against gods who are protectors of civilized life and social organization. This is also a statement concerning the right choice of guidance and governance.[18] The logic behind this motif is indicative of a tradition which is much older than the sources which attest it. A *terminus ante quem* for the adoption of the *eris* motif in Athens is provided by the poet Simonides, who mentions an *eris* of Demeter and Hephaistos for Sicily.[19] This is obviously a derivation of the story commonly told for Poseidon. Simonides had lived in Athens before he went to Sicily after the Persian wars and died in 468/465 BCE.[20]

Two divinities quarreling over the possession of a land, with a winner and a loser, is definitely a narrative. However, I would like to differentiate between tales that are told for one specific set of agents and those which can be told for variable agents, like the *eris*. For the sake of communication I speak of the *eris* motif (and not of the *eris* myth).

Herodotus links the *eris* to physical marks on the Acropolis. Quite a number of authors try to include the *martyria* in a narrative of the *eris* – and none of them presents a coherent story.

The literary sources for the *eris* do not even agree on the reason for the strife.[21] In his play *Erechtheus*, performed shortly before 421 BCE and preserved only in fragments, Euripides says it was about the main cult in Attica.[22] Late authors say it was about founding a city[23] or giving a name to it[24] or producing something.[25] Plutarch states that the early kings used the *eris* to make the Athenians turn from seafaring to agriculture.[26]

18 Mylonopoulos 2003, 87–88, 409–411; Parker 2011, 90.

19 Page 1962, 288 fr. 552 (Schol. Theoc. 1.65–66).

20 Robbins 2001, 573.

21 An *eris* is attested by: Hdt. 8.55; Pl. *Menex.* 237c-d; Callim. *Ia.* 4.66–71; D. H. 14.2.1; Apollod. 3.14.1; Paus. 1.24.5; Plut. *Them.* 19.3; Plut. *Mor.* 489B; 741B; Aristid. *Or.* 1.40 (Panath.); Luc. *Salt.* 39; Himer. *Or.* 21.2; Ov. *Met.* 6.71 (*lis*). – *amphisbetesis:* Isoc. 1.193 (Panath.); Paus. 1.26.5. – *krisis:* Xen. *Mem.* 3.5.10; Himer. *Or.* 21.2; Synkell. *Ecl. chron.* p. 179 (ed. Mosshammer). – *agon:* Paus. 1.27.2. – *hamilla:* Himer. *Or.* 6.7. – *certatio:* Hyg. *Fab.* 164.

22 E. *Erechtheus* F 360, 43–49 (eds. Collard and Cropp 2008). For the date of the play: Collard, Cropp and Lee 1995, 155; Tiverios 2005, 313 with n. 94; Collard and Cropp 2008, xxxi, 366; Sonnino 2010, 27–34; Calame 2011, 3; Connelly 2014, 129, 146 with n. 6; 214. Cf. Apollod. 3.14.1: the strife was about *timai.*

23 Hyg. *Fab.* 164.

24 Varro *De gente pop. Rom.* fr. 17 (F 222 ed. Salvadore 1999) *apud* August. *De Civ. D.* 18.9; cf. Aristid. *Or.* 1.43 and Himer. *Or.* 6.7.

25 Schol. Aristid. *Or.* 1.40–42 (Panath.). Dindorf 1829, 106.11; 106.15.

26 Plut. *Them.* 19. Marx 2011, 33. Cf. Plut. *Mor.* 489B: the Athenians are said to have absurdly invented (ἀτόπως πλάσαντες) the myth about the *eris.*

The most elaborate account is in Apollodorus' *Bibliotheke:* In Cecrops' time the gods wanted cities that would worship them. Poseidon was the first who came to Attica and, with his trident, he produced a *thalassa* on the Acropolis. Athena came later and, after having made Cecrops the witness of her possession, she planted an olive-tree. Then there was an *eris* about the land, and the twelve gods gave it to Athena because Cecrops testified that she had been the first to plant the olive.[27] This account does not really make sense. If the victory was due to the first olive, why is a witness needed? Who would doubt Athena's merit? Why is the time of arrival an issue? Cecrops did not act as a false witness – he did not pretend that Athena had been the first to arrive; he testified what did not need testifying in the first place. The author obviously combines the common motif of *protos heuretes* – first finder – with a competition. The idea of *protos heuretes,* however, is to praise an invention as an independent, outstanding achievement, not the better performance in comparison to the one by someone else. Euripides, accordingly, credits Athena with the first olive without any reference to Poseidon.[28]

Athena's creation of the olive is the only fairly consistent trait in the tradition of the *eris.* Ten authors mention it in this context. Callimachus explicitly acknowledges the goddess' role als *protos heuretes:* "Who found the olive tree? Pallas, when she was striving for Attica with the sea-weed dweller ...".[29] Five of these writers as well as Himerios give the creation of the olive as the reason for the goddess' victory.[30]

As Poseidon's salt water cannot be a match for Athena's olive the *thalassa* is mentioned only rarely.[31] Three Latin authors substitute it for a horse. This might pass as a plausible variant given the god's connection with the horse and his epithet Hippios. However, this version turns out to be a derivation from the common tradition: The god is said to have produced the horse by hitting the rock with his trident (Ovid, Vergil)[32] – an idea either taken from the emergence of

27 Apollod. 3.14.1.
28 Eur. *Tro.* 801–803; Eur. *Ion* 1433–1434.
29 Callim. *Ia.* 4.66–67 (my translation). D. H. 14.2.1; Apollod. 3.14.1; Hyg. *Fab.* 164; Paus. 1.24.3; Ov. *Met.* 6.82; Plut. *Them.* 19.3; Aristid. *Or.* 1.41; Plin. *HN* 16.240; Lact. Schol. Stat. *Theb.* 12.632–634. Synkell. *Ecl. chron.* p. 179 (ed. Mosshammer) mentions the growing of the first olive tree and the *krisis* of Poseidon and Athena among the events of the time of Cecrops but does not connect them.
30 Apollod. 3.14.1; Hyg, *Fab.* 164; Ov. *Met.* 6.82; Plut. *Them.* 19.3; Aristid. *Or.* 1.40–45; Himer. *Or.* 6.7; *Or.* 21.2.
31 Poseidon produced the *thalassa* by hitting the rock with his trident: Apollod. 3.14.1; Paus. 1.26.5. He made it appear: Paus. 1.24.3; cf. Aristid. *Or.* 1.41 (Panath.): a wave appeared.
32 Ov. *Met.* 6.70–82; Verg. *G.* 1.12–14 (without reference to the olive tree or the *eris*).

water out of the rock or inspired by visual representations of the god with a raised trident. Lactantius Placidus says Poseidon gave a horse and Athena gave the olive *pacis insigne.*[33]

Only three authors do present the gods' products in a situation of direct competition and thus interpret Athena's triumph as the victory of the olive over whatever Poseidon has to give.[34] In Ovid, his product is a horse (hit out of the rock, see above). In Himerios and Aelius Aristides it is a wave. Himerios stresses the fact that the gods did not raise their weapons against each other but that the contest was decided between the olive twig and the wave.[35] Aelius Aristides harmonizes the conflict: The Athenians based their decision on the *symbola*, the wave and the olive twig. After Athena had won Poseidon retreated but continued to love the Athenians. They owed their victory in *sophia* to Athena, but their victories at sea to Poseidon (and so did their allies).[36]

This utterly divergent tradition about the reasons and the course of the *eris* leads to the conclusion that there never was a coherent narrative. There was just what Herodotus said; the *eris* and the *martyria*. The *eris* can be presented as a narrative (both gods aspired to be the main divinity, their rivalry was decided by Zeus or by a judge, with a possible involvement of witnesses), but it was not actually meant to be one: The *eris* was a means of conveying the antagonism of the forces both gods stood for: Civilization versus nature.

The pediment shows why any attempts to include the *martyria* in a tale of the *eris* must fail: The *martyria* have an attributive function, not a narrative one. Their existence or creation is telling for each of the gods, but not for a specific situation.

In the image, the *martyria* were added to the *eris*, the subject matter of the pediment. As for the pre-existing physical marks on the Acropolis, I claim that the *eris* motif was added to these *martyria*. This argument might require a profounder reasoning than I can offer here.[37]

According to my reconstruction the idea of Poseidon challenging Athena was adopted in Athens in order to give a new dimension to a traditional Athenian tale of an invasion: Foreigners attacked the city, and Erechtheus, the Athenian king, successfully repelled them. Erechtheus' fight against the Eleusinians and Eumol-

---

33 Lact. Schol. Stat. *Theb.* 12.632–634.

34 Varro (see n. 24) makes only the *martyria* compete (and does not even mention the gods). – Plut. *Them.* 19.3 and Hyg. *Fab.* 164, listed by Pollitt 2000, 222 in this context, do not present the *martyria* in direct confrontation.

35 Himer. *Or.* 6.7; *Or.* 21.2.

36 Aristid. *Or.* 1.41–42.

37 I will discuss these issues in my forthcoming book on the cult of Athena on the Acropolis.

pus[38] is first mentioned by Thucydides as the only example of a war in Attica in the times before the land was united by Theseus.[39] A war between Athens and Eleusis cannot have been invented in the 5th century. This myth must be much older. I call it the invasion myth. Euripides' play *Erechtheus* is the earliest source for the plot. Eumolpus' aim was to introduce his father Poseidon as the main god in Athens; he was to replace Athena. When Eumolpus is killed by Erechtheus Poseidon revenges his son by crushing Erechtheus into the rock of the Acropolis. In order to settle the hostilities, Athena tells the Athenians to install a cult and invoke Erechtheus by the name of Semnos Poseidon because the god killed him.[40]

The joint cult of Erechtheus and Poseidon was not the poet's invention. It is attested by a dedication on the Acropolis made a generation before Euripides wrote his play.[41] I think that it was founded when Erechtheus became an eponymous hero in 508/507 BCE and his *persona* was split into a king for whom the invasion myth continued to be told and the foster child of Athena that became known by the new name Erichthonius. The combination of the *eris* motif with the traditional Athenian invasion myth occurred at the same time. Originally a confrontation of mortals, this tale received a much larger dimension by the combination with the antagonism of two immortals.

An olive tree as Athena's sacred tree will have stood on the Acropolis ever since the goddess had a cult statue made of olive wood.[42] A cavity with salt water in the rock of the Acropolis can only be an artificial installation, given the geology of the site.[43] I think that Poseidon's *thalassa* used to be an ancient cult mark for Erechtheus.[44] The very existence of the olive tree and of a cavity which could be ascribed to Poseidon might have instigated the idea of presenting them as *martyria* for the strife when the *eris* motif was adopted in Athens and connected with the invasion myth.

---

38 The invader Eumolpus must not be confused with the founder of the mysteries (mentioned towards the end of Euripides' play *Erechtheus*), see Sourvinou-Inwood 2011, 59, 112–123.

39 Thuc. 2.15.1–2.

40 Eur. *Erechtheus* fr. 349–370 (ed. Collard and Cropp 2008); see especially fr. 360, 43–49 (intended replacement of Athena); fr. 370, 90–95 (foundation of cult).

41 Marble louterion, Athens, EM 6319 (now in the Acropolis Museum): IG I³ 873; Raubitschek 1949, 412–413 no. 384; Kron 1976, 48–49, 53; Christopoulos 1994, 123 no. 1; Luce 2005, 148–161; Sourvinou-Inwood 2011, 68; Eleftheratou 2014, 256–257 fig. 315. One of the dedicants, Epiteles, died as *strategos* in 447 BCE, see IG I³ 1162, 4–5.

42 Schol. D. *Or.* 22.13; Athenagoras, *Leg. pro Christ.* 17.3–4.

43 I owe thanks to the geologist Erich Draganits (Vienna) for the information.

44 Apollod. 3.14.1 asserts that the *thalassa* was still called the Erechtheid one.

In the pediment there is no allusion to a fight between mortals. The image is about the gods, and it is about placing Athena. Let's have a look at its setting.

The east front of the Parthenon showcases Athena in three contexts: In the pediment she appears as the newborn goddess, sprung from her father's head;[45] in the metopes she is one of the fighters in the gigantomachy,[46] and in the frieze she sits in an assembly of the Olympian gods.[47] These images have in common that they present Athena in the company of other Olympian gods and thus celebrate her as a member of the Olympian family. The east front prominently places the city goddess in a Panhellenic context. At the time the Parthenon was built, both the myth of Athena's birth and the gigantomachy could be regarded as traditional tales, known and visualized outside Attica, too.[48]

The west front, facing the entrance to the sanctuary, presents two local events: the strife for Attica in the pediment, and the so-called Attic amazonomachy in the metopes,[49] a version of an amazonomachy which takes the fight to Athens. These tales are not only explicitly located in Athens (as opposed to the Panhellenic myths on the east side), they are also recent innovations (as opposed to the traditional myths on the east side). The tale of the Amazons' invasion in Attica and their successful repulsion by Theseus and the Athenians arose only after the Persian wars, providing a mythical paradigm for the Athenians' collective *aristeia* in this historical event.[50] Until then, Eumolpus' attack had been the standard Athenian invasion myth. Only its recent extension (the conflict

45 Brommer 1963, 3–29, 112–157 pl. 1–62, 132–138, 151; Berger 1976, 122–123, 128–141 pl. 30–32; Berger 1977, 124–126, 134–140 fig. 6 pl. 30–34 fold-out II (reconstruction); Palagia 1993, 7–39, 60 fig. 1–2, 6, 7b; 12–21, 23–59, 61–70; Berger 1996, 348–352; Mostratos 2004, 114–149 fig. 5.1–4; 5.7–22 (reconstruction of figures G to K); Palagia 2005, 225–242 fig. 76a-b; 78, 81–85, 87–88; Williams 2013.

46 Praschniker 1928, 142–223 pl. 14–27; Tiverios 1982, 227–229 pl. 29; Berger 1986, 55–76 pl. 37–72; Schwab 2005, 161, 167–173 fig. 49–51; Schwab 2014, 19, 36–44 cat. 1–16 with fig. (O 1 – O 14).

47 Berger and Gisler-Huwiler 1996, 160–161, pl. 134–135; Neils 2001, 161–166 fig. 123; Fehr 2011, 93–103 fig. 75–81.

48 Both myths appear in Athenian imagery since ca. 560 BCE: Parker 1987, 190–192; Shapiro 1989, 38–40; Neils 2001, 227–229; Muth 2008, 271, 761; Pala 2012, 92–102 fig. 41–45. The gigantomachy myth – as a common enterprise of the Olympian gods – was invented for the first celebration of the Great Panathenaia in 566 BCE, see Giuliani 2000, 263–277 fig. 1–2 and my forthcoming book.

49 Praschniker 1954, 5–53 fig. 1–28; Wesenberg 1983, 203–208 fig. 1–3; Berger 1986, 99–107 pl. 1, 113–139; Schwab 2005, 166, 178–183 fig. 54; Arrington 2015, 133–141, 144, 147–153 fig. 4.6–8.

50 Aesch. *Eum.* 685–690 (458 BCE); cf. Hdt. 9.27; Plut. *Thes.* 27; Paus. 1.2.1. Castriota 1992, 5, 43–58; Bloedow 2005, 30–36; Meyer 2005, 288–289 with n. 61–62; Arrington 2015, 146–147.

of the immortals) was used for the imagery of the Parthenon: Athena's victory over Poseidon. The Athenians' victory over foreign invaders was, in this context, more forcefully visualized by the Attic amazonomachy.[51]

There is another aspect to consider, the issue of time: Four authors connect the *eris* with priority.

Apollodorus combines the notion of first finder with a notion of first arrival in order to provide Poseidon's claim with an argument (see above). Hyginus makes the gods quarrel about who should be the first to found a city in Attica. Both writers thus refer to relative priority, suggesting a contest, and both apply a different criterion for explaining Athena's victory: She planted the olive,[52] an undisputed *aristeia* which is independent of the sequence of arrival or founding.

Isocrates in his speech *Panathenaikos* makes Eumolpus justify his attack with the pretense that his father Poseidon had taken possession of Athens before Athena had.[53] In this case, the claim for priority is made by a mortal, not the god himself. As Isocrates' version of the myth depends on Euripides' play *Erechtheus*,[54] I would suggest that it was this poet's idea to present Eumolpus basing his intention to install Poseidon as main divinity on Poseidon's earlier rights.[55] Maybe this was Euripides' subtle way of conveying the pretentiousness and futility of Eumolpus' claim. Hyginus is the only author who follows this line of reasoning – and who combines the war of the mortals with the strife of the gods.[56] All the other authors concentrate on the *eris* alone.

51 By "this context" I specifically mean the metopes: All the four sides of the building show battles against enemies who either epitomize 'the other' (the giants, the centaurs, the amazons) or recall a severe war against Eastern enemies (the Trojans who only after the Persian wars began to be seen as foreigners). In addition to n. 50 see Castriota 1992, 134–174; Giuliani 2000, 263–287; Hölscher 2000, 287–320; Schneider and Höcker 2001, 144–147 fig. 154; Meyer 2005, 280–312.

52 Apollod. 3.14.1; Hyg. *Fab.* 164. In *Fab.* 46 Hyginus presents a different version (with Eumolpus' claim that Attica belonged to Poseidon), see below.

53 Isoc. 1.193.

54 See Roth 2003, 212.

55 Cf. Eur. *Erechtheus* fr. 360, 46–49 (ed. Collard and Cropp 2008): "... nor shall Eumolpus or his Thracian folk replace the olive and the golden Gorgon by planting a trident upright in the city's foundations ..." (transl. Collard and Cropp).

56 Hyg. *Fab.* 46: Eumolpus attacked *quod patris sui terram Atticam fuisse diceret* ("because he said the Attic land had been his father's"). Hyginus is the only author in addition to Euripides and Isocrates to combine the strife of the gods with the invasion myth. His dependence on Euripides' *Erechtheus* is revealed by the fact that in his tale the sisters of the girl who was sacrificed join her voluntarily, too (although Hyginus disagrees with Euripides in saying that it was Poseidon who demanded the sacrifice).

The fourth author to mention relative priority as a criterion is a scholiast who was trying to find a plausible reason for competing. According to him, the strife was about who would be the first one to produce something.[57]

Although these authors' references to priority (of arriving, founding, possessing, producing) are obvious attempts to expound the *eris* motif, some scholars concluded that the gods' competition was actually thought to be about priority or speed.[58] The chariots were taken as a clue that the pediment showed a race.[59] As I hope to have shown, the chariots were included in the composition for a double function: as attributes of the gods and as means of visualizing their vivid movement and its sudden change.

I would suggest, however, that the idea of Poseidon having possessed Athens *proteron* (before it became Athena's city, supposedly Euripides' invention) might have been inspired by the presence of the chariots and the well-known use of chariots for races – races of mortals, of course. In the world of mortals, first arrival in a chariot entitled one to victory. Eumolpus transferred this reasoning to the realm of the gods.

The *eris* motif – the idea of Poseidon challenging an established divinity – is not about speed or priority, and it does not reflect traditions of cult and Poseidon's seniority either. It reveals a specific trait of the god's character, and it explains why Athena was preferable as a city goddess. The composition in the pediment concentrates on the visualization of the antagonism of Poseidon and Athena, insinuating the goddess' victory by prominently placing the olive tree near the center. As it is a specific quality of images to inspire multiple connotations, single elements as well as their combination could lead to further readings and new interpretations.

This contribution, with its specific focus on two particular aspects (space and time) of a very complex image, tries to exemplify potentials and limits of images and visual communication. It serves its purpose if it does fuel further scholarly debates.

---

57 Schol. Aristid. *Or.* 1.40 (Panath.); Dindorf 1829, 106.11 (p. 58).

58 Parker 1987, 198: "It looks as if in the classical legend the issue was merely one of priority". Palagia 2005, 243: "All variants of the myth seem to agree that Athena stole the victory".

59 Binder 1984, 15–22; Palagia 1993, 40; Stewart 1990, 153–154; Shear 2001, 735–736; Palagia 2005, 243–253; Schultz 2007, 66–69 fig. 9.

Bibliography

Arrington, N.T. 2015. *Ashes, Images and Memories. The Presence of the War Dead in Fifth-Century Athens.* Oxford.

BAPD: Beazley Archive Pottery Database, http://www.beazley.ox.ac.uk.

Berger, E. 1976. Parthenon-Studien: Erster Zwischenbericht. *AK* 19: 122–142.

Berger, E. 1977. Parthenon-Studien: Zweiter Zwischenbericht. *AK* 20: 124–141.

Berger, E. 1986. *Der Parthenon in Basel. Dokumentation zu den Metopen*, Mainz.

Berger, E. 1996. Review of O. Palagia, The Pediments of the Parthenon. *Gnomon* 68: 348–356.

Berger, E. and M. Gisler-Huwiler. 1996. *Der Parthenon in Basel. Dokumentation zum Fries.* Mainz.

Binder, J. 1984. The West Pediment of the Parthenon: Poseidon. In *Studies Presented to Sterling Dow*, ed. A.L. Boegehold, 15–22. Durham.

Bloedow, E.F. 2005. Aspects of Cimon's Cultural Legacy. In *Omni pede stare. Saggi architettonici e circumvesuviani in memoriam Jos de Waele*, eds. S.T.A.M. Mols and E.M. Moormann, 20–38. Naples.

Brinkmann, V. (Hrsg.) 2013. *Zurück zur Klassik. Ein neuer Blick auf das alte Griechenland*, (exhibition catalog Frankfurt). Munich.

Brommer, F. 1963. *Die Skulpturen der Parthenongiebel.* Mainz.

Calame, C. 2011. Myth and Performance on the Athenian Stage: Praxithea, Erechtheus, their Daughters, and the Etiology of Autochthony. *CPh* 106: 1–19.

Castriota, D. 1992. *Myth, Ethos and Actuality. Official Art in Fifth-Century B.C. Athens*, Madison.

Christopoulos, M. 1994. Poseidon Erechtheus and *Erechtheis thalassa*. In *Ancient Greek Cult Practice from the Epigraphical Evidence*, ed. R. Hägg, 123–130. Stockholm.

Collard, C. and M. Cropp. 2008. *Euripides VII. Fragments.* Cambridge.

Collard, C., M.J. Cropp and K.H. Lee. 1995. *Euripides. Selected Fragmentary Plays I.* Warminster.

Connelly, J.B. 2014. *The Parthenon Enigma.* New York.

De Rycke, J.-P. 2007. Arnould de Vuez, auteur des dessins du Parthénon attribués à Carrey. *BCH* 131: 721–753.

Dindorf, W. 1829. *Aristides III.* Leipzig.

Drougou, S. 2000. Krieg und Frieden im Athen des späten 5. Jhs. v. Chr. Die rotfigurige Hydria aus Pella. *MDAI*(A) 115: 147–216.

Eleftheratou, S. (ed.) 2014. *Acropolis Museum.* Athens.

Fehr, B. 2004. 'Kinder, die den Eltern gleichen' – ein Beitrag zur Deutung des Parthenon-Ostgiebels. In *Bildergeschichte. Festschrift Klaus Stähler*, eds. J. Gebauer, E. Grabow, F. Jünger and D. Metzler, 125–150. Möhnesee.

Fehr, B. 2011. *Becoming Good Democrats and Wives. Civic Education and Female Sozialisation on the Parthenon Frieze.* Zürich.

Giuliani, L. 2000. Die Giganten als Gegenbilder der attischen Bürger im 6. und 5. Jh. v. Chr. In *Gegenwelten zu den Kulturen Griechenlands und Roms in der Antike*, ed. T. Hölscher, 263–286. Munich and Leipzig.

Giuliani, L. 2003. *Bild und Mythos.* Munich.

Greco, E. (ed.) 2010. *Topografia di Atene I.* Athens and Paestum.

Harrison, E.B. 2000. Eumolpos Arrives in Eleusis. *Hesperia* 69: 267–291.

Hölscher, T. 2000. Feindwelten – Glückswelten: Perser, Kentauren und Amazonen. In *Gegenwelten zu den Kulturen Griechenlands und Roms in der Antike*, ed. T. Hölscher, 287–320. Munich and Leipzig.
Hurwit, J.M. 2004. *The Acropolis in the Age of Pericles*. Cambridge.
Isler-Kerényi, C. 2015. *Dionysos in Classical Athens. An Understanding through Images*. Leiden and Boston.
Jubier-Galinier, C. 2012. Athéna et Poséidon en conflit: adaptations céramiques à l'ombre de l'Acropole. In *La pomme d'Eris. Le conflit et sa représentation dans l'antiquité*, eds. H. Ménard, P. Sauzeau and J.-F. Thomas, 271–292. Montpellier.
Junker, K. and S. Strohwald. 2012. *Götter als Erfinder. Die Entstehung der Kultur in der griechischen Kunst*. Mainz.
Kreuzer, B. 2005. 'Geschlossene Gesellschaft' im Parthenon-Westgiebel. In *Otium. Festschrift für Volker Michael Strocka*, eds. Th. Ganschow and M. Steinhart, 193–200. Remshalden.
Kron, U. 1976. *Die zehn attischen Phylenheroen. Geschichte, Mythos, Kult und Darstellungen*. Berlin.
Luce, J.-M. 2005. Erechthée, Thésee, les tyrannoctones et les espaces publiques athéniens. In *Teseo e Romolo. Le origini di Atene e Roma a confront*, ed. E. Greco, 143–164. Athens.
Marx, P.A. 2011. Athens NM Acropolis 923 and the Contest Between Athena und Poseidon for the Land of Attica. *AK* 54: 21–40.
Meyer, M. 2005. Bilder und Vorbilder. Zu Sinn und Zweck von Siegesmonumenten Athens in klassischer Zeit. *JÖAI* 74: 279–314.
Mostratos, G. 2004. A Reconstruction of the Parthenon's East Pediment. In *The Parthenon and its Sculptures*, ed. M.B. Cosmopoulos, 114–149. Cambridge.
Mylonopoulos, J. 2003. *Πελοπόννησος οἰκητήριον Ποσειδῶνος. Heiligtümer und Kulte des Poseidon auf der Peloponnes*. Liège.
Neils, J. 2001. *The Parthenon Frieze*. Cambridge.
Neils, J. 2005. 'With Noblest Images on All Sides': The Ionic Frieze of the Parthenon. In *The Parthenon. From Antiquity to the Present*, ed. J. Neils, 198–223. Cambridge.
Neils, J. 2013. Salpinx, Snake, and Salamis. The Political Geography of the Pella Hydria. *Hesperia* 82: 595–613.
Page, D.L. 1962. *Poetae melici Graeci*. Oxford.
Pala, E. 2012. *Acropoli di Atene. Un microcosmo della produzione e distribuzione della ceramica attica*. Rome.
Palagia, O. 1993. *The Pediments of the Parthenon*. Leiden.
Palagia, O. 2005. Fire from Heaven: Pediments and Akroteria of the Parthenon. In Neils 2005, 225–259.
Parker, R. 1987. Myths of Early Athens. In *Interpretations of Greek Mythology*, ed. J. Bremmer, 187–214. London.
Parker, R. 2011. *On Greek Religion*. Ithaca and London.
Pollitt, J.J. 2000. Patriotism and the West Pediment of the Parthenon. In *Periplous. Papers on Classical Art and Archaeology Presented to Sir John Boardman*, eds. G.R. Tsetskhladze, A.J.N.W. Prag and A.M. Snodgrass, 220–227. London.
Praschniker, C. 1928. *Parthenonstudien*. Augsburg and Vienna.
Praschniker, C. 1954. Neue Parthenonstudien. *JÖAI* 41: 5–54.
Raubitschek, A.E. 1949. *Dedications from the Athenian Akropolis*. Cambridge.

Robbins, E. 2001. Simonides [2]. In *Der Neue Pauly 11*, 573–575. Stuttgart.
Roth, P. 2003. *Der Panathenaikos des Isokrates*. Munich and Leipzig.
Schneider, L. 2010. Der Parthenonfries – Selbstbewußtsein und kollektive Identität. In *Die griechische Welt. Erinnerungsorte der Antike*, eds. E. Stein-Hölkeskamp and K.-J. Hölkeskamp, 259–279. Munich.
Schneider, L. and C. Höcker. 2001. *Die Akropolis von Athen*. Darmstadt.
Schultz, P. 2007. The Iconography of the Athenian Apobates Race: Origins, Meanings, Transformations. In *The Panathenaic Games*, eds. O. Palagia and A. Choremi-Spetsieri, 59–72. Oxford.
Schwab, K.A. 2005. Celebrations of Victory: The Metopes of the Parthenon. In *The Parthenon. From Antiquity to the Present*, ed. J. Neils, 159–197. Cambridge.
Schwab, K.A. 2014. *An Archaeologist's Eye. The Parthenon Drawings of Katherine A. Schwab* (exhibition catalog). Fairfield, CT.
Shear, J.L. 2001. *Polis and Panathenaia: The History and Development of Athena's Festival* (Ph.D. Diss.). University of Pennsylvania.
Simon, E. 1980. Die Mittelgruppe im Westgiebel des Parthenon. In *Tainia. Festschrift für Roland Hampe*, eds. H.A. Cahn and E. Simon, 239–255. Mainz.
Simon, E. 1998⁴. *Die Götter der Griechen*. Munich.
Sonnino, M. 2010. *Euripidis Erechthei quae exstant*. Florence.
Sourvinou-Inwood, C. 2011. *Athenian Myths and Festivals*. Oxford.
Stewart, A. 1990. *Greek Sculpture. An Exploration*. New Haven.
Tiverios, M.A. 1982. Observations on the East Metopes of the Parthenon. *AJA* 86: 227–229.
Tiverios, M.A. 2005. Der Streit um das attische Land. Götter, Heroen und die historische Wirklichkeit. In *Meisterwerke. Internationales Symposion anläßlich des 150. Geburtstags von A. Furtwängler*, ed. V.M. Strocka, 299–319. Munich.
Tiverios, M. 2009. Bild und Geschichte. In *An Archaeology of Representations. Ancient Greek Vase-Painting and Contemporary Methodologies*, ed. D. Yatromanolakis, 159–199. Athens.
Valavanis, P. 2013. *The Acropolis Through Its Museum*. Athens.
Vollmer, C. 2014. *Im Anfang war der Thron. Studien zum leeren Thron in der griechischen, römischen und frühchristlichen Ikonographie*. Rahden.
Wesenberg, B. 1983. Wesenberg, Perser oder Amazonen? Zu den Westmetopen des Parthenon. *AA:* 203–208.
Williams, D. 2013. *The East Pediment of the Parthenon: From Perikles to Nero*. London.
Yalouris, N. 1984. Das Ketos des Parthenon-Westgiebels. In *Parthenon-Kongress Basel*, ed. E. Berger, 281–283. Mainz.

Cecilia Nobili

Cattle-raid Myths in Western Peloponnese

In *Odyssey* 21.11–41 Odysseus recounts that when he was a boy he travelled to Ortilochus' house in Pherai in Messenia[1] to recover some cattle that the Messenians had stolen from the Ithacans. There he met Iphitus, who was visiting Messenia for the same reason: he was looking for some mares and suspected that the Messenians had abducted them.

> τὼ δ' ἐν Μεσσήνῃ ξυμβλήτην ἀλλήλοιϊν
> οἴκῳ ἐν Ὀρτιλόχοιο δαΐφρονος. ἦ τοι Ὀδυσσεὺς
> ἦλθε μετὰ χρεῖος, τό ῥά οἱ πᾶς δῆμος ὄφελλε
> μῆλα γὰρ ἐξ Ἰθάκης Μεσσήνιοι ἄνδρες ἄειραν
> νηυσὶ πολυκλήϊσι τριηκόσι' ἠδὲ νομῆας.
> τῶν ἕνεκ' ἐξεσίην πολλὴν ὁδὸν ἦλθεν Ὀδυσσεύς,
> παιδνὸς ἐών· πρὸ γὰρ ἧκε πατὴρ ἄλλοι τε γέροντες
> Ἴφιτος αὖθ' ἵππους διζήμενος, αἵ οἱ ὄλοντο
> δώδεκα θήλειαι, ὑπὸ δ' ἡμίονοι ταλαεργοί.

> The two of them came on one another in Messene, in the house of the war-minded Ortilochus. Odysseus had come after a debt that the whole people owed him, for the men of Messene had stolen three hundred sheep from Ithaca in their ships with many benches, and the herders along with them. On their account Odysseus had come a long distance on embassy while still a youth, for his father and the other elders had sent him forth. Iphitus, for his part, had come in search of twelve mares that he had lost, with mules at the teat.[2]

Iphitus was wrong, since the theft had actually been committed, according to Homer, by Heracles, and, according to other versions, by Autolycus, Odysseus' grandfather.[3] However, both Odysseus and Iphitus are convinced about the guilt of the Messenians, who seem to have a reputation for being cattle-raiders. In *Iliad* 5.541–560, a passage devoted to the offspring of river Alpheus, Ortilochus, the king of Pherai, and his brother Kreton are compared to a couple of

---

1 In Mycenaean times, Pherai was the capital of the so-called 'Pylos eastern province', (Chadwick 1961) and was in some way subjected to the greater kingdom of Pylos. See Kiechle 1960, 56–63; Brillante 1996; Musti and Torelli 1991, 248–249.

2 The adapted translations of passages from *Iliad* and *Odyssey* are by B.B. Powell; from the *Homeric Hymn to Hermes* by M.L. West.

3 Cf. Apollod. *Bibl.* 2.6.2 and Eust. *Comm. ad Hom. Od.* 2.246, 37–39. On the Odyssean passage see Marcozzi 1994; Crissy 1997. Autolycus had the power of stealing without being caught and was a famous cattle-raider: he also stole Sisyphus' cows with a tricky device (Hyg. *Fab.* 201; Polyaen. *Strat.* 6.52; schol. Soph. *Aj.* 190; Tzetzes in Lycophr. 344, p. 134.1–5 Scheer).

DOI 10.1515/9783110535150-011

lions who abduct cows and ram herds, thus confirming their attitude. Furthermore, the theft perpetrated against the Ithacans was on a large scale, given that Ithaca was a small and rocky island, scarcely suitable for cattle breeding.[4] The organization that such an operation requires attests to the experience of Messenians in this regard.

This is confirmed by other episodes in Greek myth which connect Messenia (and more generally Western Peloponnesian areas) with cattle raids. As we shall now see, Pylos in particular seems to be extremely relevant for cattle-raid myths.[5] As Zanetto explained,[6] the location of this town shifts between Messenia and Elis, but it is now generally recognized that a local saga originally rooted in Messenia was later transferred to northern areas;[7] in any case we are always dealing with the western Peloponnese and its fertile plains. This saga is usually called 'Pylian epos' and traces of it may still be detected in some passages of Homeric and Hesiodic poems, one of the longest of which is Nestor's long speech in *Iliad* 11, where he recounts the war between the Pylians and the inhabitants of Elis.

The reason for the struggles was a long series of cattle-raids carried out by the Eleians against the Pylians: the Eleian king Augeias first stole the four-horse chariot which Neleus (Nestor's father) had sent to Elis for a competition (11.697–702);[8] later the Eleians repeatedly attacked Pylos, taking advantage of the weakness of the city after Heracles' assault (11.692–695).[9] The Pylians took their revenge by stealing the cattle of the Eleian Itymoneus (11.670–681), a rich breeder, and shared it among the inhabitants of Pylos, as compensation for past losses

4 According to *Od.* 4.634–637 and 14.96–102, the Ithacans used to keep most of their cattle on the mainland, especially in Elis. See Nobili 2009a, 177–179.

5 See Nobili 2011, 23–70, for a fuller account.

6 See Zanetto ("Fighting on the River: The Alpheus and the *Pylian Epic*") in this volume.

7 On the shifting position of Pylos in ancient sources see Meyer 1951; Bölte 1934; Kiechle 1959 and 1960; Frame 2009, 651–686. Brillante 1993 and Vetta 2003, followed by Aloni 2006, assume that the inhabitants of the Mycenaean kingdom of Pylos in Messenia (whose existence is confirmed by the discovery of the palace of Ano Englianos) later moved to Triphylia, carrying with them their mythic heritage.

8 An allusion to an early experience of *agones* by Elis seems to be implied here, as in *Il.* 23.629–642, where Nestor recounts that when he was a boy he went to Elis to compete in a chariot race organized for the funerals of the local king Amarynceus. Although the traditional date for the institution of the Olympic games is 776 BC and archaeological finds substantially confirm, some sort of local games may had been connected with the sanctuary from Mycaenaean times. See Hermann 1972, 37–59, 64–65; Morgan 1990, 43–56; Maddoli and Saladino 1995, xv; Valavanis 2004, 39–44.

9 See *Il.* 5.395–402; Hes. fr. 33a-35 M.-W; Apollod. *Bibl.* 1.9.9, 2.7.3; Ov. *Met.* 12.549–572. On the Hesiodic passages see Nobili 2009b and c.

(11.687–688, 11.703–705).[10] The episode ends with the open war between the Eleians and the Pylians, which comes to a close with the victory of the Pylians (11.711–752).

> εἴθ' ὣς ἡβώοιμι βίη δέ μοι ἔμπεδος εἴη,
> ὡς ὁπότ' Ἠλείοισι καὶ ἡμῖν νεῖκος ἐτύχθη
> ἀμφὶ βοηλασίῃ, ὅτ' ἐγὼ κτάνον Ἰτυμονῆα,
> ἐσθλὸν Ὑπειροχίδην, ὃς ἐν Ἤλιδι ναιετάασκε,
> ῥύσι' ἐλαυνόμενος· ὁ δ' ἀμύνων ᾗσι βόεσσιν
> ἔβλητ' ἐν πρώτοισιν ἐμῆς ἀπὸ χειρὸς ἄκοντι,
> κὰδ δ' ἔπεσεν, λαοὶ δὲ περίτρεσαν ἀγροιῶται.
> ληΐδα δ' ἐκ πεδίου συνελάσσαμεν ἤλιθα πολλήν,
> πεντήκοντα βοῶν ἀγέλας, τόσα πώεα οἰῶν,
> τόσσα συῶν συβόσια, τόσ' αἰπόλια πλατέ' αἰγῶν,
> ἵππους δὲ ξανθὰς ἑκατὸν καὶ πεντήκοντα,
> πάσας θηλείας, πολλῇσι δὲ πῶλοι ὑπῆσαν.
> καὶ τὰ μὲν ἠλασάμεθα Πύλον Νηλήϊον εἴσω
> ἐννύχοι προτὶ ἄστυ· γεγήθει δὲ φρένα Νηλεύς,
> οὕνεκά μοι τύχε πολλὰ νέῳ πόλεμόνδε κιόντι.

> Would that I were young and my strength were as when a quarrel broke out with the Eleians and ourselves over some stolen cattle – then I killed Itymoneus, the noble son of Hypeirochos, who lived in Elis, when I was driving off booty seized in reprisal. He was defending his cows and got hit among the foremost by spear thrown from my hand. He fell, and the country folk around him fled in terror. We took booty aplenty from the plain – fifty herds of cows, as many flocks of sheep, as many herds of pigs, as many herds of roving goats, and one hundred-fifty tawny horses, all mares, and there were many with foals at the teat. We drove them toward Neleian Pylos during the night, toward the city. Neleus rejoiced in his heart when he learned that I had been successful in going to war, though still a youth.

The theft committed by the Pylians is even greater than that committed by the Messenians against the Ithacans: we are dealing with fifty herds of cows and the same number of herds of sheep, goats and pigs, in addition to 150 horses. Such a rich booty, in fact, had to be divided among all the inhabitants, since Neleus kept for himself one herd of cows and one of sheep. The result of such a massive cattle-raid was a war between the two powers, in which Pylos regained the power it had enjoyed before Heracles' assault. The herds of Pylian kings were famous in antiquity: Pausanias, on his visit at Messenian Pylos (4.36.2–4) describes a cave, which was credited with being the stable where Neleus and Nestor

---

10 On this episode see Cantieni 1942; Bölte 1934; Frame 2009, 105–130. The same story was reported by Pherecydes (Schol. *Il.* 11.674 = Pherec. *FGrH* 3 F 118), with some differences.

kept their cattle. Mycenaean tablets found in the palace of Ano Englianos confirm the importance of cattle for the economy of the community.[11]

Another myth connected to the Pylian epos deals with a cattle-raid: the seer Melampus, son of Amitaon, wanted to marry Neleus' daughter Pero, but her father intended to concede her only to the man who would manage to steal the cows of Iphiclus, son of Phylakus, the Thessalian king of Phylake.[12] Melampus was the only one who accepted the challenge but was imprisoned by Phylakus' guards. He was released only when he revealed his prophetic powers and cured Iphiclus of his sterility. As a reward, he obtained the much desired herd and carried it back to Pylos. Homer recounts Melampus' deeds in two passages.[13] The first is *Odyssey* 11.287–297, a section in the heroines' catalogues devoted to Tyro, Neleus' mother, and her descendants:

> τοῖσι δ' ἐπ' ἰφθίμην Πηρὼ τέκε, θαῦμα βροτοῖσι,
> τὴν πάντες μνώοντο περικτίται· οὐδέ τι Νηλεὺς
> τῷ ἐδίδου ὃς μὴ ἕλικας βόας εὐρυμετώπους
> ἐκ Φυλάκης ἐλάσειε βίης Ἰφικληείης
> ἀργαλέας· τὰς δ'οἶος ὑπέσχετο μάντις ἀμύμων
> ἐξελάαν· χαλεπὴ δὲ θεοῦ κατὰ μοῖρα πέδησε,
> δεσμοί τ' ἀργαλέοι καὶ βουκόλοι ἀγροιῶται.
> ἀλλ' ὅτε δὴ μῆνές τε καὶ ἡμέραι ἐξετελεῦντο
> ἂψ περιτελλομένου ἔτεος καὶ ἐπήλυθον ὧραι,
> καὶ τότε δή μιν λῦσε βίη Ἰφικληείη,
> θέσφατα πάντ' εἰπόντα· Διὸς δ' ἐτελείετο βουλή.

> In addition she bore Pero, a marvel to men. All those men who lived nearby sought Pero's hand in marriage, but Neleus would give her only to the man who rustled from Phylake the obstinate cattle with curly horns and broad faces of powerful Iphiclus. The prophet Melampus undertook to drive them off, but a cruel decree of the gods ensnared him – herdsmen carted him off in grievous chains. But when the months and days were complete as the year rolled onward, and the seasons came around, then powerful Iphiclus freed Melampus after he told of all the gods had decreed. Thus was the will of Zeus fulfilled.

Melampus is not explicitly mentioned because the passage only refers to a μάντις ἀμύμων; his identification with Melampus is proved by the other Odyssean passage, 15.225–255, which narrates the same episode of the theft of Iphiclus' cattle, with some significant changes:

---

11 McInerney 2010, 63–68.

12 Melampus' father, Amitaon, was Neleus' step-brother; they left Thessaly together and moved to Pylos (Apollod. *Bibl.* 1.9.11; Diod. Sic. 4.68.3).

13 On which see Frame 2009, 227–241, 277–300.

... ἀτὰρ γενεήν γε Μελάμποδος ἔκγονος ἦεν,
ὃς πρὶν μέν ποτ' ἔναιε Πύλῳ ἔνι, μητέρι μήλων,
ἀφνειὸς Πυλίοισι μέγ' ἔξοχα δώματα ναίων·
δὴ τότε γ' ἄλλων δῆμον ἀφίκετο, πατρίδα φεύγων
Νηλέα τε μεγάθυμον, ἀγαυότατον ζωόντων,
ὅς οἱ χρήματα πολλὰ τελεσφόρον εἰς ἐνιαυτὸν
εἶχε βίῃ. ὁ δὲ τεῖος ἐνὶ μεγάροις Φυλάκοιο
δεσμῷ ἐν ἀργαλέῳ δέδετο, κρατέρ' ἄλγεα πάσχων
εἵνεκα Νηλῆος κούρης ἄτης τε βαρείης,
τήν οἱ ἐπὶ φρεσὶ θῆκε θεὰ δασπλῆτις Ἐρινύς.
ἀλλ' ὁ μὲν ἔκφυγε κῆρα καὶ ἤλασε βοῦς ἐριμύκους
ἐς Πύλον ἐκ Φυλάκης καὶ ἐτείσατο ἔργον ἀεικὲς
ἀντίθεον Νηλῆα, κασιγνήτῳ δὲ γυναῖκα
ἠγάγετο πρὸς δώμαθ'· ὁ δ' ἄλλων ἵκετο δῆμον,
Ἄργος ἐς ἱππόβοτον· τόθι γάρ νύ οἱ αἴσιμον ἦεν
ναιέμεναι πολλοῖσιν ἀνάσσοντ' Ἀργείοισιν.

He was a prophet, a descendant of Melampus, who used to live in Pylos, the mother of flocks. Melampus was rich among the Pylians, and he lived in a fancy house. Then he went to the land of strangers, fleeing his fatherland and fleeing great-hearted Neleus, the nobles man alive, who for a full year had been taking much of Melampus' wealth by force. Then he was bound in tight bonds in the house of Phylakus, suffering pain on account of the daughter of Neleus, and due to a heavy blindness of heart that the terrible Erinys laid upon him. But he avoided fate and drove off the deep-lowing cattle from Phylake to Pylos. He took vengeance on godlike Neleus for his awful deed and brought the woman home to be his brother's wife. Melampus himself went then to land of strangers, to horse-pasturing Argos. For it was his destiny to live there as king over the many Argives.

One difference between this passage and the passage in Book 11 is that in the latter Melampus was forced to leave Pylos due to a struggle with Neleus. Another is that he went on his mission on behalf of his brother Bias, who wanted to marry Pero, and after his return to Pylos Melampus moved to Argos.[14] We get the impression that the poet is expanding the details given in the previous passage, in accordance with a well-known tradition concerning Melampus. The seer was a popular figure: a poem in the Hesiodic corpus was called *Melampodia*, and an extensive section of the *Catalogue of women* was devoted to him. As we can see from fr. 37 M.-W., the presence of Melampus' brother Bias and his move to Argos were part of the tradition concerning the seer.

---

14 On these divergences cf. Harrauer 1999. Other sources preserving this story include Hes. fr. 37 M.-W.; Apollod. *Bibl.* 1.9.11–12; Schol. *Od.* 11.287 (= Pherekyd. *FGrH* 3 F 33); Paus. 4.36.2–4; 2.18.4–8; 8.18.7–8; Theoc. *Id.* 3.43–45; Hdt. 2.49; Schol. Ap. Rhod. 1.118–121 (= Hes. fr. 261 M.-W.); Schol. *Od.* 15.225 (= Pherekyd. *FGrH* 3 F 114); Stat. *Theb.* 3.450–455.

In any case, the cattle-raid appears as an instrument through which the leading families of Amitaon and Neleus reinforce their kinship. A common element is shared by the episodes of Odysseus, Nestor and Melampus: the young age of those who commit the theft and the leading role of an older figure. Odysseus is defined as a παιδνός (11.21), Nestor as a νέος (11.684) and Melampus is still unmarried; Melampus and Odysseus are sent on mission by the Laertes and Neleus respectively, and even Nestor acts on behalf of his father Neleus. However, the latter tries to prevent him from fighting and abducts his horses, thus putting him in a dangerous and unfavourable position. In all cases, an older figure forces a young member of the family to undertake a dangerous and hard endeavour, whose fulfillment will enable him to enter the community of adults. The cattle-raid thus assumes a clear initiatic function and represents the first *aristeia* that a young member of a royal family must carry out to gain power and reward.[15]

To the same geographical area belongs the saga of the Apharetidai, Idas and Lynkeus, who became Messenian national heroes at the time of Messene's new foundation in 370 BC.[16] Aphareos was in fact considered to be the first inhabitant of the region, who received Neleus fleeing from Iolkos.[17] In earlier times the myths concerning the Apharetidai were included in Spartan mythologies and their destinies connected to those of the Dioscuri; their figures, however, are strictly integrated into Western Peloponnesian mythical traditions, as confirmed by the important role played by cattle-raids in their stories. The sons of Aphareos quarreled with the Dioscuri and died in the mortal duel described by Pindar in *Nemean* 10 (54–79). As is well known, the only one who survived was Pollux, who obtained from Zeus the privilege of sharing his own immortality with his brother Castor; they could alternatively spend one day on the Olympus and one in the Hades. According to the most ancient sources, the reason for the struggle was the theft of the cows of the Dioscuri by the Apharetidai.[18] But Hellenistic poets later attributed it to the abduction of the Leukippides by the Dioscuri, even if the girls were already betrothed to the Apharetidai.[19] The antiquity of the cattle-raid motif is demonstrated by a metope from the treasure of the Sykionans at

15 See Walcot 1979, 334–343; Vallebella 2000–2002, 24–27.

16 They appeared on the temple of Messene together with Nestor and Antilochus, Leukippos and his daughters; Arsinoe and Asklepios (Paus. 4.31.11–12).

17 Paus. 4.2.5.

18 Pind. *Nem.* 10.60: Ἴδας ἀμφὶ βουσίν πως χολωθείς; see also Paus. 4.3.1

19 Theoc. 12.145–151; Lykoph. 544–561. See Gengler 2003.

Delphi (dating from around 570 BC), which represents the four heroes with a cows herd.[20]

The concentration of myths of cattle-raid in the Messenia, and particularly in connection with Pylos, may serve to explain an awkward passage in the *Homeric Hymn to Hermes.* As is well known, baby Hermes steals the cows of his brother Apollo in Pieria and leads them all across continental Greece, up to the banks of the river Alpheus, where he shuts them in a high-ceiling stable (95–104):

πολλὰ δ' ὄρη σκιόεντα καὶ αὐλῶνας κελαδεινοὺς
καὶ πεδί' ἀνθεμόεντα διήλασε κύδιμος Ἑρμῆς.
ὀρφναίη δ' ἐπίκουρος ἐπαύετο δαιμονίη νὺξ
ἡ πλείων, τάχα δ' ὄρθρος ἐγίγνετο δημιοεργός
ἡ δὲ νέον σκοπιὴν προσεβήσατο δῖα Σελήνη
Πάλλαντος θυγάτηρ Μεγαμηδείδαο ἄνακτος,
τῆμος ἐπ' Ἀλφειὸν ποταμὸν Διὸς ἄλκιμος υἱὸς
Φοίβου Ἀπόλλωνος βοῦς ἤλασεν εὐρυμετώπους.
ἀδμῆτες δ' ἵκανον ἐς αὔλιον ὑψιμέλαθρον.
καὶ ληνοὺς προπάροιθεν ἀριπρεπέος λειμῶνος

Many were the shadowed mountains and echoing valleys and flowered plains that glorious Hermes drove through. His ally, the dark divine night, was coming to an end, the greater part, and soon it would be lightening and arousing people to work; the lady Moon had reached her height, daughter of Megamedes' son, lord Pallas. Then it was that Zeus' brave son drove Phoevus Apollo's braod-browed cattle to the river Alpheus, and they came, still innocent of the yoke, to the high-roofed steading and the water troughs in front of the magnificent meadow.

Later on, at lines 397–402, the poet confirms that the place is actually Pylos, which in the epic tradition was often located on the banks of the Alpheus.[21] Why does Hermes lead the cows to Pylos?[22] The journey is long and unnecessary;

20 Delphi, Mus. 1322.

21 Pylos' location is shifting also in the *Hymn to Hermes.* At lines 101, 139, 398 it is located in Triphylia, near the Alpheus, but an allusion to Messenian Pylos may also been detected: at line 402 the poet says that the cows were shut in a high-ceiling cave, where Hermes also hangs the skins of the sacrificed cows. Müller 1833 identified it with a cave at Cape Coryphasios, where several cults are attested and where some huge stalactites and stalagmites possibly recalled the cows' skins. On Pylos and the Alpheus see Zanetto, in this volume.

22 Burkert 1984 tries to connect the episode of the sacrifice with institution of the cult of the twelve gods at Olympia, located near the river Alpheus (Pind. *Ol.* 1.92 defines it Ἀλφεοῦ πόρῳ). Such an interpretation is in contrast with the official Olympic version which assigned the institution of the cult to Heracles (Pind. *Ol.* 5.5–6, 10.24–59), and any reference in the *Hymn* to Olympia itself of the gods' altars is absent.

it would have been more reasonable to bring them on mount Kyllene, to his own cave (as in Sophocles' *Ichneutai*), or to other places where Hermes was the object of a cult, like Pheneos, in Arcadia, or Delos. No special connection exists between Pylos and Hermes.[23] The only plausible explanation is to consider the role played by Pylos (and more generally by Messenia) in the epic tradition the *Hymn* belongs to: the poet of the *Hymn* locates the cattle-raid episode at Pylos because he was influenced by several local traditions on the same theme which were connected to this place.

We may now wonder if there is any historical explanation for the special connection between these areas and cattle-raids. As has long been established, this phenomenon was important in pastoral Indo-European societies because it served to their enrichment. Cattle in fact are "the ultimate measure of wealth" and "cattle wealth favors a range of complimentary human institutions: bride price, gift exchange, and raiding are all integral parts of a cattle culture".[24] This is the reason why, far from being a shameful action, cattle-raid represented an important test for the young men who wanted to be admitted into the adults' society and supports exchanges and relations among neighbour tribes.[25] Even in Greece it played a fundamental role in Mycenaean society and is present in several traditions: Hesiod for example mentions a war which took place in Thebe over Oedipus' cattle, Achilles' sacks against Troy's allied cities included cattle-raids, and it is also part of the dynastic struggle for the throne of Mycenae among the Pterelaus' and Elektryon's children.[26]

Nevertheless, in no other tradition does it appear with the same insistence as in Western Peloponnesian myths. Elis and Messenia were well known for the abundance of their grazing lands and livestock; the Alpheus' region, in particular, represented the main point of passage for the transhumance of cattle from the northern to the southern Peloponnese. The plain territory, the abundance of water resources and the proximity to the sea made these areas a junction for herds, herdsmen and caravans, which also played an important role in the birth of the Olympian sanctuary.[27] Iron age votives found in the sanctuary contain with unparalleled concentration bronze figurines of oxen and sheep; this

---

23 The cult of Hermes in Mycenaean Pylos is occasionally attested by some tablets found in the Ano Englianos palace, but the god seems to be rather marginal in the local pantheon: see Sbardella 2009, 160–167.

24 McInerney 2010, 30. On cattle wealth in the Homeric epic see also pp. 80–96.

25 Lincoln 1976; Walcot 1979; Vallebella 2000–2002, 5–11; McInerney 2010, 97–112.

26 Hes. *Op.* 161–165; Hom. *Il.* 6.414–424, 20.89–92; Apollod. *Bibl.* 2.4.6. A cattle raid is also depicted on the ships fresco from Thera: see Lincoln 1976, 329–330; Immerwahr 1977.

27 Taita 2007, 19–30, 96–98.

suggests that farmers and breeders from neighbouring regions (Messenia) constituted the greatest part of the visitors of the cult site at this time.[28] In this social and economic context, cattle-raids represented an important sustenance for local populations; it is therefore unsurprising that it also played a major role in its poetic traditions and in the sagas developed in these areas.

Bibliography

Aloni, A. 2006. *Da Pilo a Sigeo. Poemi cantori scrivani al tempo dei Tiranni.* Alessandria.

Bölte, F. 1934. Ein pylisches epos. *RhM* 83: 319–347.

Brillante, C. 1993. Pilo e i Neleidi in un frammento di Mimnermo. In *Tradizione e innovazione nella cultura greca da Omero all'età ellenistica. Scritti in onore di Bruno Gentili*, ed. R. Pretagostini, 267–278. Rome.

Brillante, C. 1996. Nestore Gerenio: le origini di un epiteto. In *Atti e memorie del secondo congresso internazionale di Miceneologia*, I, eds. E. De Miro, L. Godart and A. Sacconi, 209–219. Rome.

Burkert, W. 1984. Sacrificio-sacrilegio: il trickster fondatore. *StudStor* 25: 835–845.

Cantieni, R. 1942. *Die Nestorerzählung im XI. Gesang der Ilias (v. 670–762).* Zürich.

Chadwick, J. 1961. The Two Provinces of Pylos. *Minos* 7: 125–141.

Crissy, K. 1997. Heracles, Odysseus and the Bow: "Odyssey" 21.11–41. *CJ* 93: 41–53.

Frame, D. 2009. *Hippota Nestor.* Washington.

Gengler, O. 2003. Héritage épique et lyrique dans la poésie alexandrine: les Dioscures et les Apharétides d'Homère à Lycophron. In *Forme di comunicazione nel mondo antico e metamorfosi del mito: dal teatro al romanzo*, eds. M. Guglielmo and E. Bona, 135–147. Alessandria.

Harrauer, C. 1999. Die Melampus-Sage in der "Odyssee". In *Euphrosyne. Studies in Greek Epic and its Legacy in Honor of Dimitris N. Maronitis*, eds. J.N. Kazakis and A. Rengakos, 132–142. Stuttgart.

Hermann, H.V. 1972. *Olympia. Heiligtum und Wettkampfstätte.* Munich.

Immerwahr, S.A. 1977. Mycenaeans at Thera: Some Reflections on the Paintings from the West House. In *Greece and the Eastern Mediterranean in Ancient History and Prehistory*, ed. K.H. Kinzl, 173–191. Berlin and New York.

Kiechle, F. 1959. *Messenische Studien. Untersuchungen zur Geschichte der messenischen Kriege und der Auswanderung der Messenier.* Kallmünz.

Kiechle, F. 1960. Pylos und der pylische Raum in der antiken Traditionen. *Historia* 9: 1–67.

Luraghi, N. 2008. *The Ancient Messenians. Constructions of Ethnicity and Memory.* Cambridge.

Maddoli, G. and V. Saladino 1995. *Pausania. Guida della Grecia, V: L'Elide e Olimpia.* Milan.

Marcozzi, D. 1994. Connotazioni messeniche nella leggenda di Eurito. *RCCM* 36: 79–86.

McInerney, J. 2010. *The Cattle of the Sun: Cows and Culture in the World of the Ancient Greeks.* Princeton.

Meyer, E. 1951. Pylos und Navarino. *MH* 8: 119–136.

28 Morgan 1990, 30–9, 61–99; Valavanis 2004, 32–38.

Morgan, C. 1990. *Athletes and Oracles. The Transformation of Olympia and Delphi in the Eighth Century BC.* Cambridge.
Müller, K.O. 1833. Die Hermes-Grotte bei Pylos. In *Hyperboreisch-römische Studien für Archäologie*, ed. E. Gerhard, 310–316. Berlin.
Musti, D. and M. Torelli 1991. *Pausania. Guida della Grecia, IV: La Messenia.* Milan.
Nobili, C. 2009a. L'Odissea e le tradizioni peloponnesiache. *Pasiphae* 3: 171–185.
Nobili, C. 2009b. La sezione pilia del Catalogo delle donne (frr. 30–37 M.-W.). Parte prima. I frammenti. *Prometheus* 35.1: 11–33.
Nobili C., 2009c. La sezione pilia del Catalogo delle donne (frr. 30–37 M.-W.). Parte seconda. Data e luogo di composizione. *Prometheus* 35.2: 105–124.
Nobili, C. 2011. *L'Inno Omerico a Ermes e le tradizioni locali.* Milan.
Nobili, C. 2013. I carmi di Bacchilide per Sparta. In *La cultura spartana in età classica*, ed. F. Berlinzani, 31–69. Trento.
Sbardella, L. 2009. Nascita di un "dio minore": l' "Inno omerico a Ermes" tra letteratura, mito e culto. *Paideia* 54: 153–178.
Taita, J. 2007. *Olimpia e il suo vicinato in epoca arcaica.* Milan.
Valavanis, P. 2004. *Games and Sanctuaries in Ancient Greece. Olympia, Delphi, Isthmia, Nemea, Athens.* Los Angeles.
Vallebella, M.G. 2000–2002. Razzia di bestiame e iniziazione virile nei poemi omerici. *Sandalion* 23–25: 5–38.
Vetta, M. 2003. L'epos di Pilo e Omero. Breve storia di una saga regionale. In *Ῥυσμός. Studi di poesia, metrica e musica greca offerti dagli allievi a Luigi Enrico Rossi per i suoi settant'anni*, ed. R. Nicolai, 13–33. Rome.
Walcot, P. 1979. Cattle Raiding, Heroic Tradition and Ritual: The Greek Evidence. *HR* 18: 326–351.

Ezio Pellizer

Time and Space in Argolic Traditions: From Ocean to Europe[1]

Chronotopes

During the 1970s and 1980s, in an attempt to work out a practical analysis of stories (mythical and other) likely to offer an effective method and a certain objectivity of results, some structuralist scholars, inspired by the research of V. Propp (also Eco, Brémond, Todorov, Greimas,[2] Courtés, the "Groupe of Entrevernes" and others),[3] produced studies that employed a "segmentation of the utterances" (*segmentation des enoncés*) of a narrative text. In this kind of analysis it was important to highlight every single trace, implicit or explicit, of utterances that would enable the insertion of the narrative process into the context of spatial and temporal actualization. If this practice was relatively simple in the case of stories of small size, it could easily prove to be quite complicated when it came to analyzing entire mythical cycles which, like the Greek, could cover many generations of time, and from the point of view of space, entailed geographical movements of vast dimensions, like the most famous travel stories (Heracles, the Argonauts, Odysseus), which could extend to a scale as vast as the entire Mediterranean, from the Pillars of Hercules to the eastern shores of the Black Sea. But this kind of attention to narrative still seems useful to this writer, even in times of post-structuralism, because it allows one to get an idea of both the 'internal time' of the stories and of the position of the enunciator ('*enonciateur*') in relation to the stories that are told. We shall examine if it is possible from this perspective to study one of these mythical cycles, the story of Inachus and his descendants, with a particularly close look at the chronological processes and the highly complicated spatial movements of its protagonists.

In particular, we shall examine the movement in space of a few female (and male) characters, connected by kinship, who represent an important example of how mythical discourse can describe, within a complex structure, a series of characters ranging from a virgin transformed into a heifer, with its vagabond ad-

1 Translated into English by Lowell Edmunds.

2 These names are grouped in the famous special number of *Communications* of 1966, with an introduction by R. Barthes, translated into Italian by Bompiani (Milan 1969).

3 Compare Edmunds 2014, 280–281; Edmunds 2016, 44–47. The concept of 'chronotope' was developed, for narrative and the novel, by Bachtin 1937/1938.

DOI 10.1515/9783110535150-012

ventures, to the vicissitudes of an abandoned girl (Ariadne), who in the end reaches the apotheosis of a wedding with a divine groom. These women include the homicidal Danaids, Europa abducted by a bull, Pasiphaë, who hid in a fake cow to join with another bull, and also lesser-known characters, like the mother of the eponymous hero of Byzantium (now Istanbul) the lady of the Golden Horn.

The God of the River

Argos; Phoroneus; Io the cow; children of a minor god: the god of a river (called today Panitsa, west of Argos, apparently),[4] to which Sophocles dedicated one of his plays, entitled *Inachus*, probably a satirical drama, of which there remains very little. Fuller of myths than of water,[5] this river leaves a semblance of history, at least a succession of generations, a genealogy.[6] What will happen after this primordial union of a river *daimon* and a nymph (Melia) is an impressive demonstration of how there can develop, in a mythical tradition, a chrono-topical (spatial-temporal) synthesis that concerns not only Greece or the Argolid, but a large portion of the south-eastern Mediterranean, as far as Egypt and Libya.

From the point of view of time ('temporal situation', '*circonstant temporel*') internal to the story, this river-god, son of Oceanus (primordial god), begets a maiden – eventually raped by Zeus – and a son called Phoroneus, a name that in turn expresses a very old collocation, from the time of the origins: grandson of Oceanus and Tethys, he is considered the inventor of fire, the first man, or at least the founder of Argive society and its organization.

The variants sometimes make it difficult even to reconstruct a genealogical series (that is, of generations). We know that this is an exercise that we are not the only ones to attempt. The ancients practised it, too.[7]

4 Brewster 1997, 59–60.

5 An *aition* for the scarcity of water in the region of thirsty (*polydípsios*) Argos is not lacking. The rivers Inachus, Cephisus, and Asterion decided that the region of Argos, contested between the two gods, would belong to Hera, and Poseidon, to get revenge, brought it about that the rivers of the Argolid would have a torrential character (Paus. 2.15.4; Apollod. 2.1.4, *13*).

6 On the genealogies and the catalogue there was held, in September 2005 in Brussels, a colloquium (CIERGA – POLYMNIA), the acta of which were published in *Kernos* 19 (2006). See in particular the contributions of Papadopoulou 2006; Calame 2006b; Moutsopoulos 2006. On these themes, cf. also Jacob 1994.

7 For example, late information collected by Paus. 2.16.1 and Apollod. 2.1.3 (dubious; different versions are given) situate the young Io some generations later and call her the daughter not of Inachus but of Iasus (Inachus – Phoroneus – Argos – Phorbas – Triopas – Iasus – Io). On the whole question see the excellent Dowden 1989, 149–184 and 185–208 of the Italian edition

Of course, the 'spatial situation' ('*circonstant spatial*'),[8] i.e. the geographical (topological, spatial) collocation of this myth, or set of myths, builds a structure, a coherent system, which can be displayed as a ring, or rather a spiral[9] that goes from Argos to the Ionian Sea and Dodona, to Euboea, then further, from Scythia and the Black Sea to Byzantium, and from there, beyond the Bosphorus, to North Africa (Phoenicia-Lebanon, Egypt, Libya); then, a return journey follows with many branches and points of arrival: a branch that reaches Crete (Minos, Rhadamanthys), giving Europe its name, and then on to Thebes and from there again to the northwest to Istria (foundation of Polai). Another branch carries out a nostos to Argos,[10] which is not yet the city of Abas, Proetus, Danae and Perseus (and still less the Mycenae of Atreus and Agamemnon),[11] but it is one that Aeschylus will make famous for a massacre committed by forty-eight terrible virgins who butchered their husbands on their wedding night: the Danaids.[12]

Things are therefore very complicated. Before the Perseids (Abantids) and Pelopids we have in Argos four generations, Abas (– Acrisius – Danae – Perseus) descended from a Danaid (therefore from Egyptus and Danaus, themselves descendants of Epaphus, the son of the cow), called Hypermestra, the only one who did not want to end her wedding night in a pool of blood, but let herself be seduced by her cousin Lynceus, generating Abas.[13]

The Theban branch is even more complicated, and involves a descendant of Epaphus (a grandson of the dark-skinned son of the heifer; Aesch. *PV* 451: τέξεις κελαινὸν Ἔπαφον), named Agenor (or Phoenix), who begets – in addition to famous Europa, ancestor of the Cretan kings – Cadmus, who married the daughter of Ares and Aphrodite, Harmonia, and founded the legendary city of Thebes after following – he, too – a fatal heifer, and became the grandfather of a god, before going into exile with Harmonia, and – perhaps – before he was buried with her in Istria;[14] that is, a journey from Phoenicia to Thebes (foundation of Cadmea), finishing in Illyria (perhaps in Polai).

(very bad translation of Guano), Calame 2000, chapter 4, "Iô, les Danaïdes, l'extérieur et l'inflexion tragique" (reprinted in Calame 2015, chapter 5).

8 Calame 1985, 144.

9 Aeschylus (*PV* 838) speaks of παλίμπλαγκτοι δρόμοι ("back-driven courses"), and, as is well known, has Io arrive also in the Caucasus, where Prometheus is chained.

10 Apparently passing from Rhodes. See Apollod. *Bibl.* 2.1.4 (12).

11 See Brillante 2004.

12 On the Danaids, see – among others – Sissa 1987 and Pellizer 2004.

13 An ample volume of studies has been dedicated to the Argive traditions, see Angeli Bernardini 2004, from which I limit myself to citing the contribution of Brillante 2004, 35–56. See also West 1997, 443–446.

14 On the Theban traditions see the studies collected in Angeli Bernardini 2000.

In the Cretan space, a son of Europa named Minos – direct cousin of Agave, of Ino, and especially of Semele – will have the good fortune to become, so to speak, the stepfather of Dionysus, who, on the island of Naxos, will celebrate his wedding with his daughter Ariadne.

One of the temporal paradoxes (of which the ancient Greeks themselves were sometimes aware) is therefore the relative proximity of the birth of some of the gods to the time of some 'historical' characters who are relatively recent.

Oceanus = Tethys
|
Inachus
|
Io – (Phoroneus)
|
Epaphus – Ceroessa
|
Libya = Poseidon
|
Agenor = Telephassa
|
Cadmus = Harmonia, – Europa
| |
Semele Minos
| |
Dionysus = Ariadne

We know that myths need to be interesting, symbolic, sometimes terrible, always evocative of deep realities of the human soul, not to respect chronology – eternal misunderstanding of those who read them in the naive hope of recovering "historical" data or realities hidden under the veil of fantasy, or the marvelous, because myth is also "the ability to discover new things": Strabo 1.2.8: αἴτιον δ', ὅτι καινολογία τίς ἐστιν ὁ μῦθος.

Ariadne, daughter of Minos, is also the daughter of Pasiphae, who conceived Asterion, the monster of the labirinth, entering a fake cow to join with, to make love with, a willing bull, and so she followed in the footsteps of her aunt Europa, the virgin seduced by a white bull.[15]

15 Zeus seduces and abducts her as a bull but in order to unite with her he re-transforms himself – as we imagine and as the poets make clear – into human form. Plutarch *Thes.* 19 reports a nice 'interpretation' of the myth of Pasiphaë: it is not a matter of a bull but of a handsome official in service in the Palace, who by chance was called Tauros!

Ariadne, the sister of Asterion, the man-bull, will finally marry an elusive god that no doubt is the grandson of Cadmus, and in turn loves to manifest himself in the form of bull (*phanethi tauros*, Eur. *Bacch.* 1017).

The time of myth thus connects Dionysus, the ancient Mycenaean god of Lerna and of Crete, to Ariadne, a granddaughter of the virgin bull-rider who gave her name to us Europeans, and bore to Zeus the famous kings of Crete, Minos and Rhadamanthys.

A Sophistic (and Sophisticated) Critique

The "second-degree modeling systems"[16] devised by the narrators of myths – for the most part poets – are built from the works that centuries and millennia do not tire of reading (and representing, narrating or reproducing).

It seems a Borgesian irony, the fact that precisely the one who took the step to History (after all, it is a word "found" by him) in order to tell "stories for today's enjoyment", that is, to make a good impression in (oral) communications, opens his famous preface with a scathing *Kritik der mythischen Vernunft*, which explores the vastest geographical spaces of Asia and Europe, linking Egypt (Io), Lebanon (Europa), Georgia (Medea), and Phrygia (Helen) in a single, vertiginous and multi-century debate over kidnapped women.

The events cover a history that, for the author himself, goes roughly from the seventeenth century (1600 or earlier if Dionysus was born about a thousand years before him [Hdt. 2.45], thus around 1450) down to 1280 – 1270, the date – for Herodotus – of the war at Troy. For Christian scholars of the third century (Clem. Alex. *Strom.* 1.21.136, approx. 150 – 215 CE), Inachus was a contemporary of Moses, of the Mosaic Exodus, and of the Pharaoh Amosis, around the fourteenth century. In other authors, too, like Polemon of Ilium, the Mosaic Exodus was contemporary with Ogyges, son of Apis, grandson of Phoroneus, thus two or three generations after Inachus and Io.

As concerns the position of the author-enunciator (*enonciateur*), it is known that Herodotus, who could also read Hecataeus, Acusilaus, Pherecydes, Damastes and many more, in these famous "rationalist" pages summons, as if to shun all responsibility, not only the rather vague Persian *logioi*, but also, and for the most ancient history, that is, for Io, certain other no less vague Phoenicians, who even interpret the rape and violation of the virgin Io as a guilty eva-

16 Lotman 1970, 15 of the Ital. transl. of Mursia 1976: myth and religion are communicative structures that develop at the level of natural language.

sion of her having been impregnated by a handsome ship captain and Phoenician merchant. Then the main enunciator says: but *I know* that all this had no importance, in the enmity between Persians and Greeks, between Asians and Europeans.[17]

Therefore, regarding chronology, Herodotus ranges from the Assyrian epoch to the time of Croesus. He shows a tendency not to take a stand on the mutual accusations between the Persians and the Greeks (1.5.3), and then says that the Persians took it as a serious fault (or rather, as a stupid thing), on the part of the Argives, to make war for an abducted woman (and furthermore a consenting one), and that since then (they did not yet call themselves Persians) they regarded the Greeks as enemies, and Anatolia as their possession.[18]

Needless to say, the Trojans, somewhat hellenized Phrygians, cannot in any way be considered Persians. So the whole Herodotean discourse is to be understood on the basis of a general spatial opposition Asia / Europe (and later, East and West). Herodotus tries to be plausible with time (counting the years that separate him from the time in which a god was born), but a total confusion of spaces and nations is allowed.

These interpretative assumptions, that Herodotus pretends not to share, must have emerged from the sophistic circles that he frequented (he was certainly in relations with Protagoras around the year 444 BCE). The total "demythologizing" of entire centuries of events of heroic times (from Phoroneus to the rape of Europa, from the Argonautica finally to the Trojan War) is not only subjected to a sort of "rationalization", but *also had to be subject of hilarity*, though a few years later (in 425) Aristophanes could "quote" this funny interpretation in order to make Olympian Pericles look foolish and describe him as an irresponsible leader, since he had waged an entire war for "three whores" (*Ach.* 528–529): κἀντεῦθεν ἀρχὴ τοῦ πολέμου κατερράγη / Ἕλλησι πᾶσιν ἐκ τριῶν λαικαστριῶν. In these lines of the *Acharnians* a reference to the Herodotean prooemium is fairly evident.[19]

17 In this regard, Calame 2000, 191–194 of the Ital. ed., § 2.3., *Erodoto arbitro e promotore della storia* (151–153 of the orig. ed.). It is worthwhile to point out a small detail that seems to me even comical: Herodotus' mention of Paris/Alexander, who, "having heard of the misadventures of Medea, and of the Greeks' refusal to respond to the diplomatic envoy sent by Aeetes to bring them to account for the rape of Medea, decides as a prank (act of spite, revenge, cf. Hdt. 1.3.1–2) to abduct Helen in turn and not to restore her".

18 Philologists and historians have often discussed these pages of the Herodotean prooemium and it is rather strange that it is not mentioned in Veyne 1983.

19 See Asheri's commentary to Herodotus (1988, lxiii and 263); see also Olson 2002 ad loc.

Herodotus seems to wish to dissociate himself from these witty hearsay, but at the same time he does not fail to report it, whilst hiding behind a reasoning that is being attributed to the barbarians.

The fact remains that the author of the treatise *de Herodoti malignitate* 856d-e can reproach him precisely for having covered with *psogos* a "heroine", i. e. Io, the daughter of Inachus:

> ... ἣν πάντες Ἕλληνες ἐκτεθειῶσθαι νομίζουσι ταῖς τιμαῖς ὑπὸ τῶν βαρβάρων καὶ καταλιπεῖν ὄνομα πολλαῖς μὲν θαλάτταις, πορθμῶν δὲ τοῖς μεγίστοις ἀφ' αὐτῆς διὰ τὴν δόξαν, ἀρχὴν δὲ καὶ πηγὴν τῶν ἐπιφανεστάτων καὶ βασιλικωτάτων γενῶν παρασχεῖν
>
> ... whom all the Greeks consider to have been divinized with great honors by the Barbarians and that through her fame she left her name to many seas and to the most important straits and that she was the beginning and the source of the most illustrious royal lines.

Ceroessa (the Horned One): In the Sign of the Heifer

The daughter of the river god (Inachus) will have descendants who return from Egypt to their place of origin, Argos, the city of the Mycenaean peoples. But in particular, a great-grandson of Inachus, the son of a sister of Epaphus called Ceroessa (The Horned) and the sea god Poseidon, will found, on the European side of the Sea of Marmara, a city that was to have a great destiny. This colony of the Megareans, according to a different legend, was actually founded by a son of this sister of Epaphus, called Byzes.[20]

The name of this sister (or probably step-sister) of Epaphus, whose very birth appears suspicious, Ceroessa, suggests that it may be a matter of a heroine *invented* to explain what from century to century will be the Golden Horn (*keras*), even if it is a name that is entirely plausible for the daughter of a cow. However it may be, this son of hers was called Byzas (or Byzes), and the city was per force called Byzantium but in time will have also other names: Second or New Rome, Constantinople, now Istanbul. A city on the Bosphorus, full of history, which can well serve as a symbol of the exchanges and the relations, for the most hostile and warlike, that affected, down the centuries, the East and the West.[21]

---

20 Steph. Byz. s.v. Byzes; there seems to be an anachronism in Diodorus Siculus: a Byzas (perhaps another Byzas?) was ruling on the Bosphorus in the time in which the Argonauts crossed it when they were returning from Colchis with Medea and the Golden Fleece, Diod. Sic. 4.49.
21 See Pellizer 2006.

Aelian, among others, observes that in antiquity rivers were represented in human form, and still more often in bovine form. As we have seen, a deep structure based on a "bovine" model presides over the whole group of myths that, from the time of Inachus (contemporary of Moses) to the wedding of Dionysus and Ariadne, covered not Argos or Thebes, or Memphis or Sidon, but the entire Mediterranean mythology.

Did the Greeks believe in their myths? This is the effective title of a well-known book by P. Veyne. Obviously, it is a meaningless question, no matter how challenging it may be. It would be like asking: do the Italians believe in the recently sanctified Padre Pio of Pietrelcina? E. Dodds says that the poet Sophocles hosted in his house – he was a priest – a sacred serpent, which represented or embodied the god Asclepius in transit to Athens. Then citing Burkhardt, he recalls that according to the Swiss scholar, in the nineteenth century religion was "rationalism for the few and magic for the many". Later, J. Assmann will be back on this rather obvious subject (1998), showing with abundant evidence how popular religion could be used by the wielders of power as a means of controlling the masses, while the dominant figures (in religion) shared more or less secretly a much more abstract (counter-) religion.

We are little interested in knowing just how many (and which) of the Greeks, in the times of Homer, of Herodotus, of Plato or of Plutarch believed in the voyage of Io, of a girl with the horns of a cow, transformed into a cow, which swam across a strait of the sea (Ox-ford). No more than the visitors who admire Michelangelo's Moses in Rome are interested in explaining why the biblical figure has horns (which seem to derive from a reading of Hebrew KRN as KeReN, "horn" and not as KaRaN, "ray": *Ex.* 34:29 – 30). What makes these materials interesting, and offers many of us today an extraordinary field of reflection and creative imagination (as it offered to the Greeks who went to the theatre in the fifth century, to readers of Dante Alighieri,[22] to the spectators of the theatre of Syracuse, or the one in Epidaurus), and also a great and inexhaustible intellectual pleasure, is precisely the study of Greek myth, to follow the countless stories that ancient Greek culture has left us as a legacy: the pleasure of cultivating the second "Great Codex" of the cultural history of our planet (or at least a good half of it, and certainly not only of Europe), in the infinite creative reproduction of these ancient stories, as long as men, in Europe, the Americas, Australia or China, are sensitive to the ancient legends, and to their living and inexhaustible beauty.

---

22 Dante Alighieri called the Minotaur "son of the fake cow" (*Inf.* XII, 12 – 13); cf. Langdon's translation: "the infamy of Crete / who in the seeming heifer was conceived".

Bibliography

Angeli Bernardini, P. (ed.) 2000. *Presenza e funzione della città di Tebe nella cultura greca* (Atti del Convegno Internazionale, Urbino, 7–9 luglio 1997). Pisa and Rome.

Angeli Bernardini, P. (ed.) 2004. *La città di Argo. Mito, storia, tradizioni poetiche* (Atti del Convegno Internazionale, Urbino, 13–15 giugno 2002). Rome.

Asheri, D. 1988. *Erodoto. Le storie*, vol. I (Coll. Valla). Milan.

Assmann, J. 1998. *Moses der Ägypter. Entzifferung einer Gedächtinsspur.* Munich and Vienna.

Bachtin, M. 1937/1938. Formy vremeni i chronotopa v romane. Očerki po historičesko poetike. (cited from Italian transl.: Forme del tempo e del cronotopo nel romanzo. Saggi di poetica storica. In *Michail Bachtin. Estetica e Romanzo*, eds. R. Platone and C. Strada Janovic, 231–405. Turin)

Brewster, H. 1997. *The River Gods of Greece. Myths and Mountain Waters in the Hellenic World.* London and New York.

Brillante, C. 2004. Genealogie argive: dall'*asty phoronikon* alla città di Perseus. In Angeli Bernardini 2004, 35–56.

Calame, C. 1985. *La formulation de quelques structures sémio-narratives ou comment segmenter un texte.* In *Exigences et perspectives de la sémiotique. Recueil d'hommages pour Algirdas Julien Greimas,* vol. 1, eds. H. Parret and H.-G. Ruprecht, 135–147. Amsterdam.

Calame, C. 2000. *Poétique des mythes dans la Grèce antique.* Paris 2000. (Italian transl. Lecce 2011)

Calame, C. 2004. Le funzioni di un racconto genealogico: Acusilao di Argo e la nascita della storiografia. In Angeli Bernardini 2004, 229–243 (French transl. in *Europe* 945/946, 2008: 87–108).

Calame, C. 2006a. *Pratiques poétiques de la mémoire. Représentations de l'espace-temps en Grèce ancienne.* Paris.

Calame, C. 2006b. Logiques catalogales et formes généalogiques: mythes grecs entre tradition orale et pratique de l'écriture. *Kernos* 19: 23–29.

Detienne, M. 1994. *Transcrire les mythologies.* Paris.

Dodds, E. 1951. *The Greeks and The Irrational.* Berkeley, LA. (Italian transl. Florence 1959)

Edmunds, L. 2002. Oral Story-Telling and Archaic Greek Hexameter Poetry. In López Férez 2002, 17–33.

Edmunds, L. 2009. Interpreting Greek Myth. *SIFC* (special issue) 7: 75–78.

Edmunds, L. 2014² (ed.). *Approaches to Greek Myth.* Baltimore (review by N. Serafini, in GRiMM, grmito.units.it/recensioni)

Edmunds, L. 2016. *Stealing Helen. The Myth of the Abducted Wife in Comparative Perspective.* Princeton, NJ and Oxford.

Jacob, C. 1994. *L'ordre généalogique entre le mythe et l'histoire.* In Detienne 1994, 160–202.

López Férez, J.A. (ed.). 2002. *Mitos en la literatura griega arcaica y clásica.* Madrid.

Lotman, J. 1970. *La struttura del testo poetico*, Moskow (Italian transl. Milan 1972, 1976²).

Miralles, C. and J. Pórtulas (eds.). 1987. *La dona en l'antiguitat.* Barcelona.

Moutsopoulos, E. 2006. Généalogies et structures de parenté dans la mythologie grecque. *Kernos* 19: 31–34.

Olson, S.D. 2002. *Aristophanes Acharnians.* Oxford.

Papadopoulou, I. 2006. Hésiode, Homère, Hérodote: forme catalogique et classifications génériques. *Kernos* 19: 79–95.

Pellizer, E. 2004. Épreuves d'amour, épreuves de mort. In *Europe* 904–905: 103–118.
Pellizer, E. 2006. Nel segno della giovenca. Miti e culture tra Oriente e Occidente. In *DYNASTHAI DIDASKEIN. Studi Càssola*, eds. M. Faraguna and V. Vedaldi, 343–350. Trieste.
Sissa, G. 1987. La giara delle ingrate. In Miralles and Portulas 1987, 85–95.
Veyne, P. 1983. *Les Grecs ont-ils cru à leur mythes? Essay sur l'imagination constituante.* Paris 1983 (Italian transl. Bologna 1984).
West, M.L. 1997. *The East Face of Helicon. West Asiatic Elements in Greek Poetry and Myth.* Oxford.

Paolo Daniele Scirpo

About the Boeotian Origin of the Emmenidai's Genos: An Indication from Gela

Introduction

From the recent analysis of literary sources carried out by Gianfranco Adornato,[1] it is clear that we must distinguish two strands of tradition about Akragas: one concerning the origin of the polis (sub-colony of Gela, founded around 580 BC) and on the other hand, the origin of the Emmenidai's genos. To celebrate Theron's victory in the chariot race which took place in Olympia in 476 BC, Pindar would have created (or saved) in his 2nd Olympic ode the family tree that starts from Cadmus and ends up to the tyrant of Akragas and whose tormented stories are being mentioned in the Pindaric ode Scholia. However, a passage of Theron's ancestors from Gela, Akragas's metropolis, remains doubtful.[2]

The Traditions about the Genealogical Tree of Emmenidai

We can thus distinguish two different genealogical traditions, reflected in *Scholia*, mentioned by Hellenistic historians: the *scholium* 82(d) shows the complete genealogy of Theron from Laius (and Cadmus) through Polynices (and his connection to the family of Argive Adrastus), his son Thersander (connected instead to Amphiaraus), the stop at Thera (with an oecistic legend and ties to the family of Heraclides of Sparta) and the arrival in Sicily where Telemachus, grandfather (or great-grandfather) of Theron, participated in the conspiracy against the tyrant Phalaris.

The *scholia* 15(a-d) and 16(c) and 29(d), however, in which the historian Timaeus is being quoted, in addition to the acceptance of their Theban ascendancy

---

1 See Adornato 2011. For some observations, see Scirpo 2013. I express my warmest thanks to Anita Leontopoulou and Anna Manesioti-Jones for the translation of this paper in English. I would thank also the anonymous referees who have enabled me with their comments to clarify some points of discussion expressed perhaps too 'hermetically' and using a language not mine.
2 See van Compernolle 1959; Caserta 1995; Caserta 2000; Schneider 2000.

DOI 10.1515/9783110535150-013

from Cadmus, speak about a mysterious Haemon, son of Polydor and brother of Labdacus (not to be confused with the namesake son of Creon) who, after an act of violence during a hunt, is being forced to take refuge in Athens. From there, his group which joined the Argives, went to Rhodes and from there they went to colonize Akragas.

In this second current, a controversy among Hellenistic historians was created, where Artemon of Pergamum believed the Emmenidai's genos tied in Gela and, on the other end, Hippostratos, Aristarchus and Menecrates claimed, that they were of Rhodian origin, without the intermediate step of Gela.

Three hypotheses (by Musti, Fileni, and Buongiovanni respectively[3]) have been formulated to explain the reasons why Timaeus (and his successors) emphasized the link of Akragas with Rhodes.

Cristina Caserta believes that this recall is not due to the wish to separate Akragas from Gela, but rather to separate it from Syracuse, bringing it closer to the metropolis of Rhodes, an ally of Athens during the Peloponnesian War, on the age of the Sicilian expedition (422 BC).[4]

Thus following the path indicated by Pindar about the Theban ancestry of Emmenidai, the Athenian propaganda claimed an oecistic tradition that bound Akragas to Rhodes (via Athens), the same way it was running pro-Argive, the praise that Pindar himself composed in honour of Diagoras for his Olympic victory in 464 BC.[5]

And a Rhodian oecist, Heraclid Tlepolemos, whose cult on the island had from then a civilian national valor,[6] returns as 'supporting actor' in the subsequent history of Poemander.

A Hero-Cult of Boeotian Origin in Gela

In the Regional Archaeological Museum of Gela is being kept a cylindrical marble base of *louterion*, which was published in 1899 by Paolo Orsi, and was delivered to him by a citizen, that unearthed it during the excavation of a well in his private residence in the center of the city[7] (fig. 1a-b). On the upper surface of the

3 See Musti 1992; Fileni 1993; Buongiovanni 1995.
4 See Caserta 1995; Caserta 2000.
5 Pind. *Ol.* 7.
6 Scirpo forthcoming (c).
7 See Orsi 1900, 50, n. 19.

base, around the cavity for the insertion of the lustral basin, it has been engraved a dedication[8] of Leukon (or Teukon) to the hero *Chaeresilaus.*[9]

Figure 1a: Marble cylindrical base of louterion (from Panvini 1998)

Λεύκōν {Τεύκōν(?)} Χαιρēσίλεōι.

Provided that the inscription is authentic,[10] it dates between the late 6th and early 5th century BC and testifies to the cult of this Boeotian hero.[11]

M. Guarducci analyzing the inscription believed that his cult had come to Gela, along with a core of Boeotians by Megara Hyblaea, because the alphabet is Megarian.[12] As evidence of this, she reports that the name of Mnasithales, devoting offer to oecist Antiphemos, is of Boeotian origin and he was attested in Orchomenos.[13]

Also on a burial stone, that was found in the ancient necropolis of Gela and dated to the mid-6th century BC, a series of graffiti with ten names (*nomina de-*

8 *IGASMG II*, n. 63 (450 BC). *IGDS*, n. 131 (500 BC).
9 Stoll 1884/1890.
10 Against its authenticity see Adamesteanu 1959, 432–434.
11 See Tümpel 1899, c. 2029.
12 See Guarducci 1949/1951, 109–110, fig. 4; Orlandini (1956, 153, n. 104) doesn't agree.
13 *IG*, VII, n. 3175.

Figure 1b: Marble cylindrical base of louterion (from Panvini 1998)

votorum) was engraved,[14] half of which can also be found in Boeotia and precisely in Tanagra.[15]

But also the name of the one who devoted to Chaeresilaus, e. g. Leukon, is attested in Boeotia (Mycalessos, Plataea, Thespiae).[16] Not to forget the discovery on the acropolis of Gela of a crock with written the name of Ainesidamos on it, dating from the mid-5th century BC.[17] Also, between lead tables (*pinakes*), that

14 See Gentili 1946.

15 *LGNA* IIIB.

16 *LGNA* IIIB.

17 See Orlandini 1956, 143 – 144. He dated it at the beginning of the 5th century, recalling Ainesidamos, son of Pataikos, "doryphoros" of Hippocrates with Gelo and later for a short time, the tyrant of Leontini, mentioned by Herodotus and Aristotle. He is also considered an Emmenides, but not to be confused with the father of Theron.

were found in a *cista* at the temple of Athena in Camarina, with the list of civilian soldiers enrolled at the time in Punic painful siege of 406/5 BC, is mentioned the ability of the hoplite Thrasys, Emmenidas' son.[18]

So, an alternative hypothesis would tie this hero-cult (clearly of private character) with the participation of Boeotian *gene*, as the Emmenidai from Akragas. Two passages of Pausanias inform us in fact, about the divine origins of Chaeresilaus and his connection to the environment of Boeotia.[19]

> Ταναγραῖοι δὲ οἰκιστήν σφισι Ποίμανδρον γενέσθαι λέγουσι Χαιρησίλεω παῖδα τοῦ Ἰασίου τοῦ Ἐλευθῆρος, τὸν δ' Ἀπόλλωνός τε καὶ Αἰθούσης εἶναι τῆς Ποσειδῶνος. Κορίννῃ δέ ἐστιν ἐς αυτὴν πεποιημένα Ἀσωπού παῖδα εἶναι.
>
> φασὶ δὲ καὶ Ἐλευθῆρα ἀνελέσθαι Πυθικὴν νίκην μέγα καὶ ἡδὺ φωνοῦντα, ἐπεὶ ᾄδειν γε αὐτὸν οὐχ αὐτοῦ τὴν ᾠδήν.
>
> The people of Tanagra say that their founder was Poemander, the son of Chaeresilaus, the son of Iasius, the son of Eleuther, who, they say, was the son of Apollo by Aethusa, the daughter of Poseidon. It is said that Poemander married Tanagra, a daughter of Aeolus. But in a poem of Corinna she is said to be a daughter of Asopus.
>
> They said that Eleuther won a Pythian victory for his loud and sweet voice, for the song that he sang was not of his composition (transl. by W.H.S. Jones, 1969).

Chaeresilaus was the son of Iason and grandson of Eleuther (the eponymous founder of Eleutherae and winner of the musical contests in Delphi, the first to have erected a statue of god Dionysus). Eleuther, now, was the son of Apollo and Aithousa, daughter of Poseidon and the Pleiad Alcyone. The latter was the daughter of the Titan Atlas and the Oceanid nymph Pleione. Atlas was the son of the Titans Iapetus and the Oceanid Clymene while Pleione was the daughter of Oceanus and Tethys.[20]

Chaeresilaus, marrying Stratonice, is the father of Poemander who founded *Poimandria* (later *Tanagra*) who married the naiad Tanagra (daughter of Aeolus or the river god Asopos). She had two children, Leucippus and Ephippus.[21]

By refusing to participate in the Trojan War, Poemander was besieged by Achilles who kidnapped his mother Stratonice and killed his nephew Acestor, son of Ephippus. Wanting to fortify the citadel, Poemander quarreled with the architect Polycritus, and wanting to hit him with a stone, killed by mistake his other son Leucippus. To ward off the *miasma*, he was obtained by Achilles to

---

18 Cordano 1992; *IGDS*, II, n. 114–116; Manganaro 2011, 35.
19 Paus. 9.20.1 and 10.7.3.
20 Roscher 1884/1937; Grimal 1987; Parada 1993.
21 Schultz 1884/1890.

go to Chalkis to be purified by Elephenor, escorted also by the Heraclid Tlepolemus and Peneleos (son of Hippalcimus/Hippalmos and Asterope) regent of Thebes, in the name of the minor Tisamenus, son of Thersander. In memory of this concession, Poemander founded a sanctuary to Achilles near Tanagra. [22]

This mythological precedent may have also been used by Leukon as a thanksgiving for a received vote or for an obtained purification. Guarducci had keenly observed that given the location of the epigraph, hardly on the pedestal, rested a lustral basin, but a votive gift (*pinax* or sculpture) that was stuck into the nut.

Oecistic and Heroic Cults in Gela and Akragas

However, the pedestal, both would hold a *louterion* or a *pinax*, had to be certainly within the temenos of a sanctuary in Gela. A likely candidate would be that of Pythian Apollo, Chairesilaos's ancestor, and deity linked to the health sphere.[23]

Taking the existence of a *Heroon* dedicated to the oecists of Gela (Antiphemos and Entimos) for granted,[24] why not believe that there was also a *Heroon* of Chaeresilaus? It being led by Megarians or Boeotians that fled with Autesion, a noble worship might have well survived in the difficult years of the formation of the Geloan *polis*. The *stasis* in the middle of the 6th century BC would give a new social balance with the dominance of the Rhodian (and from Dodecanese in general) ethnic component, primarily of the Dinomenidai's genos, native of Telos.

From the famous passage of Thucydides,[25] who probably served the historiographical work of Antiochus of Syracuse, we know that Akragas was founded in 580 BC by settlers from Gela, in command of two oecists, Aristonoos and Pistylos, and that the new *polis* acquired the *Geloa nomima*.[26]

But if in Gela, the presence of *Heroa* dedicated to oecists is confirmed (at least for Antiphemos) by archaeological evidence, there are currently no tracks about sanctuaries for its two oecists in Akragas. The mixed contingent led by the two men, would certainly include people coming from other areas of the Hellenic world (the same Phalaris was however born in Cos).[27]

22 Plut. *Quest. Gr.* 37. About the cult of Achilleus in Tanagra, in the Bronze age, see Sakellariou 2009, 101–104.
23 See Scirpo forthcoming (c).
24 See Scirpo forthcoming (c).
25 Thuc. 6.4.4.
26 Scirpo forthcoming (b).
27 Vaglio 2000.

After the fall of Phalaris, a conspiracy in which a member of Emmenidai had also participated (whether Emmenes or Telemachus), we should question if a *Heroon* dedicated to Emmenidai ever existed in Akragas.[28]

We know from Pindar, that *Theoxenia* was celebrated in Akragas, in honour of *Dioscuri.*[29] The "sons of Zeus" were the model of harmony and brotherly love, as also the prototype of the perfect athlete.[30] Nothing prevents us from thinking that Pindar wanted in his verses to allude to the harmony between the brothers, Theron and Xenocrates.[31] Although lacking of date track of a place of worship dedicated to them in Akragas, the temple F (the so-called 'Temple of Concordia') could be ascribed to them, dating from 440–430 BC, work commissioned by the democratic government of the *polis.*[32]

When Theron took power in 488 BC, his family had enough money and power to participate in the Panhellenic games.[33]

The Emmenidai and The Panhellenic Games

But what was the relationship between the latter and the Emmenidai that Pindar mentions in his verses?

In chronological order, the Boeotian poet dedicated an ode to Xenocrates, brother of Theron, for his victory in a chariot race in August 490 BC in Delphi.[34] Shortly thereafter was the "Praise" about Thrasybulus, son of Xenocrates. After the battle of Himera (480 BC), Pindar dedicated two odes (*Ol.* 2 and 3) to the same Olympic victory in the chariot race won by Theron in August 476 BC and finally, he composed custom to Thrasybulus, probably in 474, an ode dedicated to the victory in the chariot race won by Xenocrates in April 476 BC in *Isthmia.*[35]

---

28 After his death (472 BC), Theron was honored as a hero (Diod. 11.53.7 et 13.86).

29 Pind. *Ol.* 3.72.

30 See Brelich 2010^2, 86–93.

31 See Athanassaki 2014, 209.

32 See Berve and Gruber 1961, 256.

33 See Scirpo forthcoming (a). The Emmenidai would also sponsored the victory of Midas in *aulos*'s race, in the same Pythian games, which is also sung by Pindar (*Pyth.* 12). Since in Boeotia were widespread musical contests, in which the locals excelled in playing the *aulos*, favored by the presence in nearby Lake Copais, of cane required for its construction, it could be argued that the poet has drawn inspiration from here for the theme of his ode. See Gentili, Bernardini, Cingano and Giannini 2012^5, 311.

34 Pind. *Pyth.* 6.

35 Pind. *Isth.* 2. In a *scholium* on Pindar a passage of Aristotle is cited (fr. 617 Rose), which mentions the double victory of Xenocrates in the Pythian (490/489 BC) and Isthmian games (476 BC).

But if these Olympic victories can explain the Panhellenic cult of Olympic Zeus, in whose honour Theron designed the largest and most mysterious temples (482/472 BC), how can we explain the participation in the Pythian and Isthmian games? And especially, what reflection of the theronian propaganda can we find in Akragas?

Temple A (*Apollonion*), which was built in the late 6th century BC[36] as the first peripteral Doric like a sacred chain around the *polis* of Akragas, may have been dedicated not to the hero-god Heracles[37] but to the Delphic god who predicted the coming of the genos in Sicily. One may wonder, however, whether it was wanted by the *aisymnetes* (Alcamenes and Alcandrus) before the tyranny of Theron or was funded by the money of the rich Emmenidai? One can give them at least one of the remakes that affected the coverage of the temple, with the first of the two series of lion gutters examined by Pirro Marconi[38] and dated to the years around 470/460 BC.

The first victory in the Pythian Games would still have brought honour to the family, so to gain the favour of Apollo.

The oldest evidence of the heroic cult of Melicertes[39] at Isthmia,[40] is being deduced from a fragment of Pindar,[41] where the Panhellenic games in honour of Poseidon were founded, under the aegis of Corinth.[42]

It is self explanatory that Melicertes was Ino's son and Cadmus' grandson.[43] Even this genealogy goes back to the House of Labdacides, which is *Emmenidai*'s descend.

Their pro-Rhodian policy would also obscure the Cretan component of Akragas, almost erasing its memory, both in civilian life and in the cults of the *polis*. The conquest of *Minoa*,[44] sub-colony of Selinus in 497 BC, since it was the first

---

36 De Waele 1971; De Waele 1980; De Waele 1992 dates the temple to the tyranny of Theron (488–480 BC) due to the architectural similarities with the temple of Apollo, funded by Alcmeonids, in Delphi. Gullini 1985 dates it instead to the post-Himera years, as René van Compernolle does (1959, 70–71).

37 See Adornato 2011, 108 n. 23. For the cult of Heracles in Akragas, see Scirpo 2014.

38 See Marconi 1929, 56, n. 1. The second set, however, is dated around 450/440 BC.

39 See Vikela and Vollkommer 1992.

40 See Gebhard and Dickie 1998; Gebhard and Reese 2005.

41 To Burkert (1972, 197) due to the discovery of the passage of Pindar [fr. 6.5 (1) Snell]: "They ordered Sisyphos, son of Aiolos, to raise up a far-shining honor for his dead son, Melikertes" (translation of W.H. Race, 1997).

42 At least until its destruction by Romans (146 BC), when the organization went to Sicyon, to return again at its peak during the Imperial period.

43 See n. 21.

44 De Miro 2014.

step for the hegemony in Sicania, is linked to the symbolic return to the Cretans (probably from Knossos) of Minos' bones, found in the tomb built by Daedalus, inside the temple of Aphrodite.[45]

After the restoration of 'democracy' in Akragas was a fact, the cult of the ancient Cretan god Velchanos[46] was finally reintroduced to the civil pantheon with the construction of a new Doric temple that incorporated the archaic *naiskos*.[47]

Conclusions

The story of Akragas' early years is yet to be written: in Adornato's praiseworthy attempt, the combination of the ancient sources (already known and analyzed through time) and the recent archaeological finds, had not allowed to not even touch the half-mythical figure of tyrant Phalaris, who, ten years after the founding of Geloan *apoikia*, took control and then traced the expansionist policy that the successive dynasties of Emmenidai only partly succeeded to achieve.[48]

The building of Temple A (maybe an *Apollonion*) in thanks to the God who blessed their arrival in Sicily, and the victories in the Panhellenic contests (mainly in Delphi), are being sung in the verses of the Boeotian poet Pindar, creating a dense network of connections intended to glorify the past and the present of Emmenidai.[49]

If, as we believe, their arrival on the island has taken place at first in the city of Gela, this inscribed base, that was found at the end of the 19th century by Paolo Orsi, could be the missing link to the correct reconstruction of the Emmenidai's *genos* eventful past, confirming also the traditional fondness of the Family for the Olympic, Pythian and Isthmian games.

Bibliography

Adamesteanu, D. 1959. Le iscrizioni false da Licata e Gela. In *Atti del III Congresso Internazionale di Epigrafia Greca e Latina (Roma, 4–8/9/1957)*, 425–434. Rome.

Adornato, G. 2011. *Akragas arcaica. Modelli culturali e linguaggi artistici di una città greca d'Occidente.* Milan.

45 Fontana 1978.
46 Capdeville 1995.
47 Marconi 1933; De Miro 1965.
48 La Torre 2011, 76–79.
49 Scirpo forthcoming (a).

Arena, R. (ed.) 1992 (2002[2]). *Iscrizioni greche arcaiche di Sicilia e Magna Grecia. Iscrizioni di Sicilia II. Iscrizioni di Gela e Agrigento.* Milan (= *IGASMG*).
Athanassaki, L. 2014. The Creative Impact of the Occasion: Pindar's Songs for the Emmenids and Horace's Odes 1.12 and 4.2. In *Defining Greek Narrative* (Edinburgh Leventis Studies 7), eds. D.L. Cairns and R. Scodel, 197–225. Edinburgh.
Berve, H. and G. Gruber. 1961. *Griechische Tempel und Heiligtümer.* Munich.
Brelich, A. 2010[2]. *Gli eroi greci. Un problema storico-religioso.* Milan.
Buongiovanni, A.M. 1985. Una tradizione filo-emmenide sulla fondazione di Acragas. *AnnPisa* 15: 493–499.
Burkert, W. 1972. *Homo Necans. Interpretationen altgriechischer Opferriten und Mythen.* Berlin. (Italian transl. by F. Bertolini: *Homo Necans. Antropologia del sacrificiocruento nella Grecia antica.* Turin 1981.)
Capdeville, G. 1995. *Volcanus. Recherches comparatistes sur les origines du culte de Vulcain.* Rome.
Caserta, C. 1995. *Gli Emmenidi e le tradizioni poetiche e storiografiche su Akragas fino alla battaglia di Himera* (SEIA 12). Palermo.
Caserta, C. 2000. Le genealogie mitiche degli Emmenidi negli Scholia Vetera all'Olimpica II di Pindaro. *Hormos* 2: 5–42.
Cordano, F. 1992. *Le tessere pubbliche dal tempio di Atena a Camarina.* Roma.
De Miro, E. 1965. Terracotte architettoniche agrigentine. *CronA* 4: 39–78.
De Miro, E. 2014. *Heraclea Minoa: mezzo secolo di ricerche* (Sicilia Antiqua 9 2012). Pisa and Rome.
De Waele, J. 1971. *Acragas Graeca. Die historische Topographie des griechischen Akragas auf Sizilien, 1. Historischer Teil.* Rome.
De Waele, J. 1980. Der Entwurf der dorischen Tempel von Akragas. *AA:* 180–241.
De Waele, J. 1992. I grandi templi. In *Akragas 1. "Agrigento e la Sicilia greca". Atti della Settimana di studio (Agrigento, 2–8/5/1988)*, eds. L. Braccesi and E. De Miro, 157–205. Rome.
Dittenberger, G. (ed.) 1892. *Inscriptiones Megaridis et Boeotiae.* Berlin (= *IG* VII).
Dubois, L. 1989. *Inscriptions grecques dialectales de Sicile. Contribution à l'étude du vocabulaire grec colonial.* Paris (= *IGDS*).
Fileni, M.G. 1993. Una pagina di storia agrigentina: Pindaro frr. 118e 119 Sn.-Maehl. In *Tradizione e innovazione nella cultura greca da Omero all'età ellenistica. Scritti in onore di Bruno Gentili*, ed. R. Pretagostini, 427–440. Rome.
Fontana, M.J. 1978. Terone e il τάφος di Minosse. Uno squarcio di attività politica siceliota. *Kokalos* 24: 201–219.
Gebhard, E.R. and M.W. Dickie 1998. Melikertes-Palaimon, Hero of the Isthmian Games. In *Ancient Greek Hero Cult. Proceedings of the Fifth International Seminar on Ancient Greek Cult, Göteborg University, 21–23 April 1995* (*ActaAth* 16), ed. R. Hägg, 159–165. Stockholm.
Gebhard, E.R. and D. Reese 2005. Sacrifices to Poseidon and Melikertes-Palaimon at Isthmia. In *Greek Sacrificial Ritual, Olympian and Chthonian, Proceeding of the Sixth International Seminar on Ancient Greek Cult, Göteborg University, 25–27 April 1997* (*ActaAth* 18), eds. R. Hägg and B. Alroth, 125–154. Stockholm.
Gentili, B., P.A. Bernardini, E. Cingano and P. Giannini (eds.) 2012[5]. Pindaro. *Le Pitiche.* Milan.

Gentili, B., C. Catenacci, P. Giannini and L. Lomiento (eds.) 2013. Pindaro. *Le Olimpiche*. Milan.
Gentili, M.V. 1946. Iscrizione arcaica sul coronamento di cippo gelese del Museo di Siracusa. *Epigraphica* 8: 11–18.
Grimal, P. 1987. *Dizionario di mitologia greca e romana*. Brescia.
Guarducci, M. 1949/1951. Note di epigrafia siceliota. *ASAtene* 27–29: 103–140.
Gullini, G. 1985. L'architettura. In *Sikanie. Storia e civiltà della Sicilia greca*, ed. G. Pugliese Carratelli, 415–492. Milan.
Jones, W.H.S. 1965. *Pausania. Description of Greece*, IV. London and Cambridge.
La Torre, G.F. 2011. *Sicilia e Magna Grecia. Archeologia della colonizzazione greca d'Occidente*. Rome and Bari.
LGPN IIIA = Fraser, P.M. and E. Matthews (eds.). 1997. *Lexicon of Greek Personal Names*, vol. IIIA: *Peloponnesus, Western Greece, Sicily and Magna Graecia*. Oxford.
LGPN IIIB = Fraser, P.M. and E. Matthews (eds.). 2000. *Lexicon of Greek Personal Names*, vol. IIIB: *Central Greece: From the Megarid to Thessaly*. Oxford.
Manganaro, G. 2011. Il sistema anagrafico nella Sicilia in epoca ellenistica: a proposito di due tabelle di piombo erroneamente attribuite ad area siracusana e di due nuove iscrizioni, una di Kale akté ed una di Halaisa. In *Da Halaesa ad Agathyrnum: studi in memoria di Giacomo Scibona* (*Contributi alla conoscenza del territorio dei Nebrodi* 5), 33–68. Sant'Agata di Militello.
Marconi, P. 1929. *Agrigento. Topografia ed Arte*. Florence.
Marconi, P. 1933. *Agrigento arcaica. Il santuario delle divinità ctonie e il tempio di Vulcano*. Rome.
Musti, D. 1992. Le tradizioni ecistiche di Agrigento. In *Akragas 1. "Agrigento e la Sicilia greca". Atti della Settimana di studio (Agrigento, 2–8/5/1988)*, eds. L. Braccesi and E. De Miro, 27–45. Rome.
Orlandini, P. 1956. Storia e topografia di Gela dal 405 al 282 a.C. alla luce delle nuove scoperte archeologiche. *Kokalos* 2: 158–176.
Orsi, P. 1900. Frammenti epigrafici sicelioti. *RStorAnt*, n.s. 5.1: 39–66.
Panvini, R. (ed.) 1998. *Gela. Il Museo Archeologico. Il catalogo*. Gela.
Parada, C. 1993. *Genealogic Guide to Greek Mythology* (*SiMA*, CVII). Jonsered.
Privitera, G.A. (ed.) 2009[5]. *Pindaro. Le Istmiche*. Milan.
Race, W.H. 1997a. *Pindar I (Olympic Odes, Pythian Odes)*. Cambridge, MA and London.
Race, W.H. 1997b. *Pindar II (Nemean Odes, Isthmian Odes, Fragments)*. Cambridge, MA and London.
Roscher, W.H. 1884–1937. *Ausführliches Lexikon der griechischen und römischen Mythologie*, I-VI. Leipzig.
Sakellariou, M. 2009. *Ethne grecs à l'âge de Bronze*, I–II (*Meletemata* 47). Athens.
Schneider, K. 2000. De Cadmos aux Emménides. *Kentron* 16.1–2: 65–81.
Schultz, X. 1884/1890. s.v. *Ephippos*. In Roscher, I: c. 1281.
Scirpo, P.D. 2013. Review of Adornato 2011. *Thiasos* 2: 3–8.
Scirpo, P.D. 2014. Η ροδο-κρητική προέλευσις της λατρείας του Ηρακλέους στην Γέλα και τον Ακράγαντα. *Electra* 3: 65–87.
Scirpo, P.D. forthcoming (a). 'Violence' and 'State' of the Tyrants in Sicily. In *Violence and Politics: Ideologies, Identities, Representations. Proceedings of the 3rd Colloquium of*

Young Researchers in Memory of Prof. Nikos Birgalias (Athens, January 14–16, 2016). Cambridge.

Scirpo, P.D. forthcoming (b). Αποσπάσματα κρητικής λατρείας στις εν Σικελία ροδοκρητικές αποικίες ($7^{ος}$–$6^{ος}$ αι. π.Χ.). In *Πρακτικά του $12^{ου}$ Διεθνούς Κρητολογικού Συνεδρίου (Ηράκλειο, 21–25/9/2016)*, Herakleion.

Scirpo, P.D. forthcoming (c). *Η προέλευσις και η εξέλιξις των ροδο-κρητικών λατρειών στη Γέλα και τον Ακράγαντα ($8^{ος}$–$5^{ος}$ αι. π.Χ.)*. Athens.

Stoll, H.W. 1884/1890. s.v. *Chairesilaos*. In Roscher I: c. 868.

Tümpel, X. 1899. s.v. *Chairesilaos*. In *RE* III^2: c. 2029.

Vaglio, M. 2000. La patria perduta di Falaride. *Anemos* 1: 151–155.

Van Compernolle, R. 1959. *Etude de chronologie et d'historiographie siceliotes*. Brussels and Rome.

Vikela, E. and R. Vollkommer, 1992. s.v. *Melikertes*. In *LIMC* 6.1–2: cc. 436–444.

Giuseppe Zanetto

Fighting on the River: The Alpheus and the 'Pylian Epic'

In the *Iliad* the old Pylian king Nestor talks on several occasions about his past glorious deeds.[1] Many scholars think that these passages are what survives of a 'Pylian Epic', i.e. of the epic songs which in Mycenaean age celebrated the lords of Pylos.[2] It is assumed in fact that there was Mycenaean epic poetry, which was performed at the courts of the kings on the occasion of festive or funeral banquets;[3] and that this poetry has been almost completely lost, the only exception being the Pylian songs which were incorporated into our text of the *Iliad*. The survival of the 'Pylian Epic' must probably be connected with the important role that many families of Pylian origin played in archaic Greece. In archaic Athens some of the most powerful γένη claimed descent from Neleus and the Neleids. In the cemetery of the Ceramicus the excavations have brought to light geometric funerary vases, produced at the end of the VIII century BC, which reproduce the fight between Nestor and the Moliones.[4] We can hypothesise that the mythical episode which was depicted on the vases containing the ashes of the deceased corresponded to a poem performed by a singer during the funeral. If this is so, this means that in geometric Athens the epic tradition originating from Pylos still survived, was still performed and was a key-element in the creation of identity.[5]

Now, if we look at the content of Nestor's speeches, we see that he praises victories in battles against neighbouring peoples (Epeans, Arcadians), successful cattle raids, victories in athletic contests and in particular in chariot races. Let us have a short review of these narratives.

Iliad 11.670–762 offers the most extended passage of 'Pylian Epic'. Nestor tells a very complicated story which can be divided in two main sections.[6] In the first one he reports how he seized a huge number of cattle, fighting against the Epeans and killing Itymoneus, and how he drove this booty down to Pylos, where on the day after it was divided among the Pylians. Neleus takes the major

1 Hom. *Il.* 7.132–156, 11.670–761, 23.629–642.
2 Bölte 1934; Hainsworth 1993, 296–298; Vetta 2003; Nobili 2009, 105–110.
3 Vetta 2001, 23–30.
4 Ahlberg-Cornell 1992, 32–35.
5 Vetta 2003, 22–27; Aloni 2006, 67–75; Nobili 2009, 110–113.
6 Hainsworth 1993, 296.

DOI 10.1515/9783110535150-014

portion for himself, in retaliation for the loss of his four race-horses, stolen from him by the Epean king Augeas. In the second section Nestor tells how the Epeans came down towards Pylos to take revenge and how the Pylians, warned by Athena, left their city and marched out against the enemies: here the tale is very rich in topographic details, so that it is not difficult to reconstruct the geographic setting.

1. The Epeans move to the river Alpheus, which marks the border of the Pylian territory, and attack Thryoessa.

Iliad 11.711–713:

ἔστι δέ τις Θρυόεσσα πόλις αἰπεῖα κολώνη
τηλοῦ ἐπ' Ἀλφειῷ, νεάτη Πύλου ἠμαθόεντος
τὴν ἀμφεστρατόωντο διαρραῖσαι μεμαῶτες.

Now there is a city Thryoessa, perched on a steep cliff, overlooking the Alpheus, on the far border of sandy Pylos, and there they camped, aiming to destroy it.[7]

2. The Pylians set out; the cavalry in a few hours reaches the river Minyeios (a minor stream, presumably flowing about 20 kilometres south of the Alpheus).[8] Once the infantry has also arrived (shortly after daybreak), they all move to the Alpheus, where they offer sacrifices to the gods, take their evening meal and bivouac under arms.

Iliad 11.722–732:

ἔστι δέ τις ποταμὸς Μινυήϊος εἰς ἅλα βάλλων
ἐγγύθεν Ἀρήνης, ὅθι μείναμεν Ἠῶ δῖαν
ἱππῆες Πυλίων, τὰ δ' ἐπέρρεον ἔθνεα πεζῶν.
ἔνθεν πανσυδίῃ σὺν τεύχεσι θωρηχθέντες
ἔνδιοι ἱκόμεσθ' ἱερὸν ῥόον Ἀλφειοῖο.
ἔνθα Διὶ ῥέξαντες ὑπερμενεῖ ἱερὰ καλά,
ταῦρον δ' Ἀλφειῷ, ταῦρον δὲ Ποσειδάωνι,
αὐτὰρ Ἀθηναίῃ γλαυκώπιδι βοῦν ἀγελαίην,
δόρπον ἔπειθ' ἑλόμεσθα κατὰ στρατὸν ἐν τελέεσσι,
καὶ κατεκοιμήθημεν ἐν ἔντεσιν οἷσιν ἕκαστος
ἀμφὶ ῥοὰς ποταμοῖο.

A river, Minyeios, meets the sea near Arene, and there the chariots waited for the dawn, and then the infantry arrived. From that point, travelling armed and at speed, by noon we reached Alpheus' holy stream. There we sacrificed fine victims to mighty Zeus, bulls to Al-

7 The English translations from the *Iliad* are taken from the on-line text of A.S. Kline.
8 Hainsworth 1993, 302.

pheus and Poseidon, and a heifer to bright-eyed Athena. Then each company ate supper, and we slept in battle-gear on the bank.

3. On the following day there is the battle. The text is not explicit on this point, but we must think that the two armies fight on the northern shore of the Alpheus, because the Pylians, after defeating the Epeans, chase them until Bouprasion, well to the north of the river. This means that Nestor and his men have crossed the Alpheus, immediately after their arrival and before the night which they spend bivouacking on the shore: the sacrifices to the gods, and in particular to the god of the river, are διαβατήρια.[9]

Iliad 7.132–156 is the account of another exploit of young Nestor, this time in an one on one fight against the Arcadian champion Ereuthalion. The context is a battle between the Pylians and the Arcadians; Nestor doesn't explain the reasons for the conflict, he only says that the two armies gathered together near the wall of Pheia and along river Iardanus. It is not clear to which part of the western Peloponnesian shore these geographic markers point;[10] we don't know, in particular, which river is meant by the name Iardanus (and the same goes for the other stream mentioned just before, the Keladon);[11] but Pheia should probably be identified with the town which in other epic passages[12] is called Pheai, about fifteen kilometres north of the mouth of the Alpheus, in southern Elis.[13] If it is so, then this struggle between Pylians and Arcadians takes place not far from the location where Nestor kills Itymoneus and seizes his cattle in *Iliad* 11: the setting of both episodes is the plain along the northern shore of the Cyparissian Gulf, north and south of river Alpheus; and we can argue that the Alpheus – in this context too – is meant to mark the northern border of the Pylian territory.

Nestor's opponent is Ereuthalion, whose favourite weapon is an iron mace; we are told that in the past it was the property of Areithoos, who used to massacre his enemies with this club; but Lycurgus killed him by trapping him in a narrow place where he had no room to swing it; then it was Lycurgus who wore Areithoos' armour, until he grew old and gave it to his friend Ereuthalion. It is very likely that Areithoos was originally an Arcadian brigand (or an Arcadia-related brigand):[14] from a fragment of Pherecydes (fr. 158) we learn that he came to Arcadia and accumulated a huge booty, until he was killed by Lycurgus, who

9 Hainsworth 1993, 303.
10 Robert 1920, 191.
11 Kirk 1990, 252 ("the rivers remain mysterious").
12 Hom. *Od.* 15.297; *Hom. Hymn. Ap.* 427 (φεράς codices, Φεάς editores).
13 Aloni 2006, 38–39.
14 Nobili 2011, 34–36; see also Kirk 1990, 253.

led his Arcadian troops against him to recover their stolen property. Lycurgus' attack against Areithoos has therefore much in common with the killing of Itymoneus by Nestor. And because Ereuthalion uses the same iron mace (that is to say, the typical weapon of a brigand), the fight between him and Nestor can also be seen as an episode of a story of cattle raids: we can imagine that the Pylians are attacking the Arcadians of Ereuthalion in response to their incursions.

Iliad 7.132–156:

αἳ γὰρ Ζεῦ τε πάτερ καὶ Ἀθηναίη καὶ Ἄπολλον
ἡβῷμ' ὡς ὅτ' ἐπ' ὠκυρόῳ Κελάδοντι μάχοντο
ἀγρόμενοι Πύλιοί τε καὶ Ἀρκάδες ἐγχεσίμωροι
Φειᾶς πὰρ τείχεσσιν Ἰαρδάνου ἀμφὶ ῥέεθρα.
τοῖσι δ' Ἐρευθαλίων πρόμος ἵστατο ἰσόθεος φὼς
τεύχε' ἔχων ὤμοισιν Ἀρηϊθόοιο ἄνακτος
δίου Ἀρηϊθόου, τὸν ἐπίκλησιν κορυνήτην
ἄνδρες κίκλησκον καλλίζωνοί τε γυναῖκες
οὕνεκ' ἄρ' οὐ τόξοισι μαχέσκετο δουρί τε μακρῷ,
ἀλλὰ σιδηρείῃ κορύνῃ ῥήγνυσκε φάλαγγας.
...
ἀλλ' ἐμὲ θυμὸς ἀνῆκε πολυτλήμων πολεμίζειν
θάρσεϊ ᾧ· γενεῇ δὲ νεώτατος ἔσκον ἁπάντων
καὶ μαχόμην οἱ ἐγώ, δῶκεν δέ μοι εὖχος Ἀθήνη.
τὸν δὴ μήκιστον καὶ κάρτιστον κτάνον ἄνδρα
πολλὸς γάρ τις ἔκειτο παρήορος ἔνθα καὶ ἔνθα.

Oh, Father Zeus, Athena and Apollo, if only I were young again as when our Pylian host was fighting the Arcadian spearmen by swift-running Keladon, under Pheia's walls, at the streams of Iardanus. Ereuthalion was their champion. Like a god he was, clad in the armour of noble King Areithoos whom men and fair women called the Mace-man, because he ignored long-spear or bow, and shattered the lines with his iron mace.
...
Though the youngest there, in my boldness my doughty heart spurred me to fight him, and Athena granted me glory. He was the tallest and strongest I ever slew: yet he lay sprawling there in all his mighty breadth and height.

So these two narratives of 'Pylian Epic' which we find in *Iliad* 7 and 11 seem to refer to the same situation and to the same geographic environment. In both cases the Pylians are engaged in military actions in the northern area of their land, in proximity to the Alpheus. The Alpheus is explicitly referred to in Book 11, whereas it is only evoked through the mention of Pheia and of the river Iardanus (perhaps one of its tributaries) in Book 7; but its presence is a core element in both stories. The role of the river is double. First, it is the boundary line that cattle raiders cross as they drive their booty from the foreign land to their own land: crossing the river is something that must necessarily happen, but is also the symbolic image of a successful raid. Second, the river is the setting of the battle that is a consequence of the raid, as the victims come to rescue. The report of Book 11 is a complete one, as the two segments of the story (raid and battle) are narrated at lenght. The narrative of Book 7 is shorter, because the purpose of Nestor is to encourage the Achaean warriors to accept Hector's challenge: so he focuses his report on the scene of himself coming out to fight against the Arcadian champion. But the scenery of the fluvial battle is clearly alluded to: Pylians and Arcadians are fighting beside the swift-flowing Keladon and on the stream of the Iardanus. Thus in this case too the river is the line of contact between the two armies and defines the battlefield.

We may suppose that exactly this topic, i.e. the abduction of cattle and the heroic actions connected with this (fights during the raids and struggles in consequence to the raids) was a standard theme (a kind of 'typical scene') of the 'Pylian Epic'. As Cecilia Nobili has shown, there is a very large number of myths in which stories of cattle raids have Messenian heroes as their protagonists or are located in Messenia.[15] The existence of such mythical accounts is of course very significant and can be explained as a survival of a local epic tradition. We can therefore argue that in the songs performed in the so called 'Palace of Nestor' the lords of Pylos were celebrated for their bravery[16] in collecting booty with successful raids or in defending their property from hostile incursions. I would suggest – and this is a core point of my contribution – that in these songs a river was often the background of the heroic action.

One of the wall paintings which decorated Hall 64 of the palace of Ano Englianos seems to confirm this idea: it depicts a battle between two armies, the one in the typical Mycenaean armour and the other dressed in animal skins; in the background a curvilinear decoration brings to mind the meandering stream of a river.[17] The archaeologist Nikolaos Yalouris suggests that this painting refers exactly to the struggle between Pylians and Arcadians which is narrated by Nestor in *Iliad* 7:[18] the Arcadians were well known in antiquity for wearing sheepskins, so Yalouris thinks that the visitors of the palace immediately associated this painting with that famous battle which was a favourite object of poetic performances. The style of the narration is in fact close to the epic mode, because the battle is split into individual fights, to focus on the bravery of the champions. And it is a fluvial battle, if Yalouris is right in interpreting the curvilinear shapes on the background as a river.

15 Nobili 2011, 23–70; see also her contribution in this volume.

16 Vetta 2001, 21–24.

17 Lang 1969, pls. 16, 117.

18 Yalouris 1989.

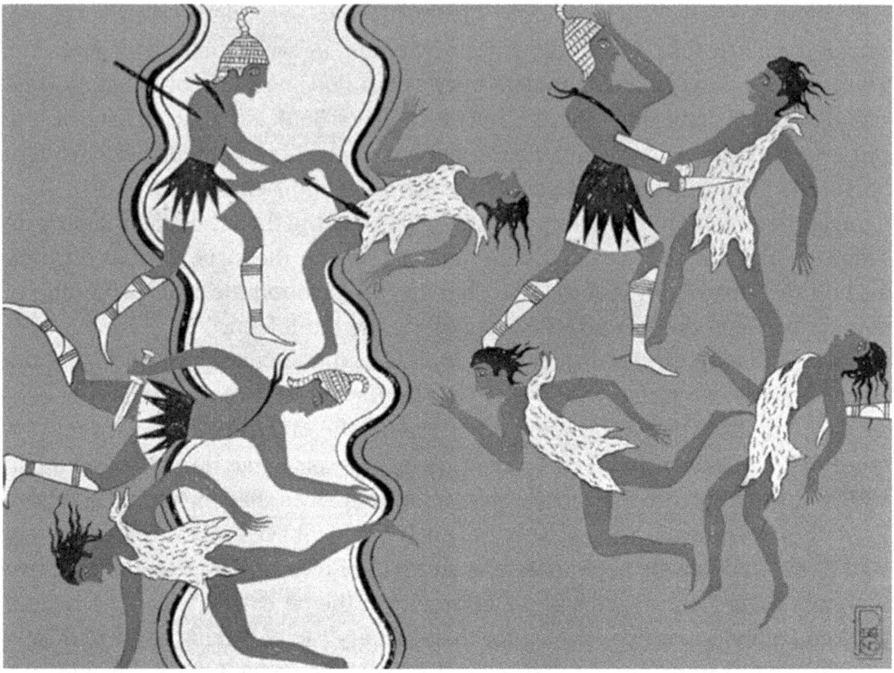

Palace of Nestor, 'The battle' – Fresco from Hall 64 (Watercolour by Piet de Jong)

Against this interpretation there is an obvious argument: the plain of the Alpheus (at which the mention of Pheia and of the Iardanus points) is very far (about 100 kilometres) from the Messenian 'Palace of Nestor'. Our fresco has been dated by Reinhard Jung and Mathias Mehofer to 1250 BC;[19] so it was *in situ* on the wall when the final destruction of the palace occurred. Why should a poet, performing for the lords of the Palace, have set the battle in a geographical area which was outside their authority (and presumably outside their usual military activity)? From the documents written in Linear B we can infer that the northern boundary of the Pylian Kingdom was marked by the river Neda, well to the south of the Alpheus.[20]

The same can be said also in relation to the other episode of 'Pylian Epic' in *Iliad* 11. Here too – as we have seen – the setting of Nestor's exploits (the fight with Itymoneus and the battle against the Epeans) is the plain of the Alpheus, well outside of the Pylian Kingdom. From Nestor's account, however, it is clear

19 Jung and Mehofer 2008, 121.

20 Kelder 2010, 8–9; see also Niemeier 1991, 126–132; Bennet 1998, 117; Bennet 2011, 151–155.

Reconstruction of the Pylian Kingdom according to the Linear B Texts (from Kelder 2010)

that the Alpheus is only a few hours' march from Pylos, because the Pylian cavalry leaves at night and arrives at the river before daybreak. This means that the poet is not thinking of a Messenian Pylos. The problem of the position of the Homeric Pylos was discussed already in antiquity (by Strabo, for example)[21] and has been discussed also in modern times.[22] A brilliant and influential paper on this issue has been published by Massimo Vetta.[23] In Vetta's opinion the Homeric (or at least the Iliadic) Pylos is not the Messenian one, but the other town with the same name which the Pylians founded and inhabited in Triphylia, near modern Kakovatos, after the fall of Mycenaean society.

21 Strab. 8.3.7 p. 339c.

22 Meyer 1951; Kiechle 1960; Brillante 1993; Frame 2009, 651–686.

23 Vetta 2003.

The Pylian colonists, as it was the rule in ancient Greece, brought with themselves in the new site not only the goods that they had saved from the destruction but also their habits and their cultural identity, as it was registered in their traditional songs. But the setting of these songs had to be adapted to the new geographical context. This is the reason why the 'Pylian Epic' was – in Vetta's words – re-located, so that it could fit with the new Triphylian location.

The narratives of the 'Pylian Epic' needed a river as their setting. The relocation forced the poets to look for a river that could be an eligible option for the new context. The Alpheus, which marks the border between Triphylia and Elis, only 30 kilometres north of Kakovatos, became the new setting of the Pylian exploits. Nestor's accounts in *Iliad* 7 and 11 are clearly the result of this re-location and refer therefore not to the primary version but to the sub-Mycenaean re-styling of the 'Pylian Epic'. The Arcadians are still there, because they are plausible opponents for the Pylians of Kakovatos too. But now the standard competitors for Nestor and the Neleids become the Epeans, who live beyond the Alpheus and can make raids against the Pylians or be the target of Pylian raids.

But once the Alpheus has been incorporated into the 'Pylian Epic' (and the 'Pylian Epic' has been absorbed into the Panhellenic poetry), the Alpheus plays an active role in also 'attracting' narratives which are not originally 'Pylian' but have to do with cattle theft or cattle abduction. This is the reason why, for example, in the *Homeric Hymn to Hermes* the Alpheus is the final destination of the god's journey, after the abduction of Apollo's cattle.[24]

One last point. The proximity of Kakovatos to Elis also makes possible the attraction of the Pylians into narratives which are specifically Eleian. The new 'Pylian Epic' develops stories in which the Neleids turn out to be brilliant charioteers. We know that chariot races were a typical Eleian tradition: the competition between Pelops and Oenomaus is the mythical counterpart of horse races that took place in the plain of the Alpheus long time before the foundation of the Olympic games. In our 'Pylian Epic' there are two passages in which the Pylians seem to share the Eleian love for chariot races. In *Iliad* 11 there is the mention of the four race horses send by Neleus to Elis and abducted by Augeas (699–702). In *Iliad* 23 Nestor remembers another glorious episode of his youth, as he went to Bouprasion to take part in the splendid funeral games for king Amarynceus and won all the competitions. His only defeat

24 *Hom. Hymn. Herm.* 101; Nobili 2011, 28–29.

was in the chariot race, where the twins Moliones had the great advantage of being two on the same chariot.

Iliad 23.629–642:

εἴθ' ὣς ἡβώοιμι βίη τέ μοι ἔμπεδος εἴη
ὡς ὁπότε κρείοντ' Ἀμαρυγκέα θάπτον Ἐπειοὶ
Βουπρασίῳ, παῖδες δ' ἔθεσαν βασιλῆος ἄεθλα
ἔνθ' οὔ τίς μοι ὁμοῖος ἀνὴρ γένετ', οὔτ' ἄρ' Ἐπειῶν
οὔτ' αὐτῶν Πυλίων οὔτ' Αἰτωλῶν μεγαθύμων.
πὺξ μὲν ἐνίκησα Κλυτομήδεα Ἤνοπος υἱόν,
Ἀγκαῖον δὲ πάλῃ Πλευρώνιον, ὅς μοι ἀνέστη·
Ἴφικλον δὲ πόδεσσι παρέδραμον ἐσθλὸν ἐόντα,
δουρὶ δ' ὑπειρέβαλον Φυλῆά τε καὶ Πολύδωρον.
οἴοισίν μ' ἵπποισι παρήλασαν Ἀκτορίωνε
πλήθει πρόσθε βαλόντες ἀγασσάμενοι περὶ νίκης,
οὕνεκα δὴ τὰ μέγιστα παρ' αὐτόθι λείπετ' ἄεθλα.
οἳ δ' ἄρ' ἔσαν δίδυμοι· ὃ μὲν ἔμπεδον ἡνιόχευεν,
ἔμπεδον ἡνιόχευ', ὃ δ' ἄρα μάστιγι κέλευεν.

I wish I were as young and strong as that time when the Epeans were interring King Amarynceus at Bouprasion, and his sons held funeral games in his honour. Then no man proved himself my equal, Epeans, Pylians or proud Aetolians. I beat Clytomedes, the son of Enops, in the boxing and Ankaios of Pleuron, who took me on in the wrestling. In the foot race I outran Iphiclus, good as he was, and my spear out-threw Phyleys and Polydorus. Only in the chariot race did the two Moliones beat me, by their combined superior strength, forcing their team to the front, begrudging me the victory since the race carried the best prize. They were twins, and one could drive with a sure hand, while the other plied the whip.

To sum up, the Alpheus of the renewed 'Pylian Epic' is a real river, because it is the boundary of the 'New Pylos' territory; but it is at the same time a literary river, because it plays a pivotal role in defining and suggesting stories.[25]

Bibliography

Ahlberg-Cornell, G. 1992. *Myth and Epos in Early Greek Art. Representation and Interpretation.* Jonsered.
Aloni, A. 2006. *Da Pilo al Sigeo. Poemi, cantori e scrivani al tempo dei Tiranni.* Alessandria.
Bennet, J. 1998. The Linear B Archives and the Kingdom of Nestor. In *Sandy Pylos: An Archaeological History from Nestor to Navarino*, ed. J.L. Davies, 111–133. Austin.

25 Zanetto 2004, 151.

Bennet, J. 2011. The Geography of the Mycenaean Kingdoms. In *A Companion to Linear B. Mycenaean Greek Texts and their World*, vol. II, eds. Y. Duhoux and A. Morpurgo Davies, 137–168. Louvaine-La-Neuve and Walpole, MA.
Bölte, F. 1934. Ein pylisches Epos. *RhM* 83: 319–347.
Brillante, C. 1993. Pilo e i Neleidi in un frammento di Mimnermo. In *Tradizione e innovazione nella cultura greca da Omero all'età ellenistica*, ed. R. Pretagostini, 267–278. Rome.
Frame, D. 2009. *Hippota Nestor*. Washington.
Hainsworth, A. 1993. *The Iliad: A Commentary*, vol. iii: Books 9–12. Cambridge.
Jung, R. and M. Mehofer. 2008. A Sword of Naue II Type from Ugarit and the Historical Significance of Italian-type Weaponry in the Eastern Mediterranean. *Aegean Archaeology* 8: 111–136.
Kelder, J.M. 2010. *The Kingdom of Mycenae. A Great Kingdom in the Late Bronze Age Aegean*. Bethesda, MA.
Kiechle, F. 1960. Pylos und der pylische Raum in der antiken Tradition. *Historia* 9: 1–67.
Kirk, G.S. 1990. *The Iliad: A Commentary*, vol. ii: Books 5–8. Cambridge.
Lang, M. 1969. *The Palace of Nestor at Pylos in Western Messenia*. Princeton.
Meyer, E. 1951. Pylos und Navarino. *MH* 8: 119–136.
Niemeier, W.-D. 1991. La struttura territoriale della Grecia micenea. In *Geografia storica della Grecia antica. Tradizioni e problemi*, ed. F. Prontera, 123–149. Rome and Bari.
Nobili, C. 2009. La sezione pilia del "Catalogo delle donne" (frr. 30–37 M.-W.). *Prometheus* 35: 11–33, 105–124.
Nobili, C. 2011. *L'Inno omerico a Ermes e le tradizioni locali*. Milan.
Robert, C. 1920. *Die Griechische Heldensage*, Berlin.
Vetta, M. 2001. Prima di Omero. I luoghi, i cantori, la tradizione. In *La civiltà dei Greci. Forme, luoghi contesti*, ed. M. Vetta, 19–58. Rome.
Vetta, M. 2003. L'epos di Pilo e Omero. Breve storia di una saga regionale. In *Ῥυσμός. Studi di poesia, metrica e musica greca offerti dagli allievi a Luigi Enrico Rossi per i suoi settant'anni*, ed. R. Nicolai, 13–33. Rome.
Yalouris, N. 1989. Ein Schlachtengemälde im Palast des Nestors. *MDAI(A)* 104: 162–184.
Zanetto, G. 2004. Il Peloponneso nella tradizione epica. In *Il Peloponneso di Senofonte. Giornate di Studio del Dottorato di Ricerca in Filologia. Letteratura e Traduzione classica*, eds. G. Daverio Rocchi and M. Cavalli, 143–154. Milan.

Nereida Villagra

Time and Space in the Myth of Byblis and Caunus[1]

The plot of the myth of Byblis can be summarized as follows: Byblis and Caunus are sister and brother and one of them falls in love with the other – in some versions it is Caunus, whereas in others it is Byblis. Either way, the result is the departure of Caunus, who ends up founding a city, and Byblis' death originating a water source that will bear her name. The first treatment of the myth appears, in a fragmentary form, in Hellenistic poetry, but it must have been known at least since the fourth century BCE, when Aristotle uses the expression Καύνιος ἔρως to refer to illicit loves.[2] The tale must have become quite popular from Hellenistic times onwards, for we can find it in the poetic works of Apollonius Rhodius, Nicaenetus, Nicander, Ovid and Nonnus of Pannopolis,[3] and it is also narrated by three Imperial mythographers, Parthenius, Conon and Antoninus Liberalis.[4] The myth is also mentioned by Stephanus Byzantinus, Diogenianus and the scholia to Theocritus.[5]

In her book on Hellenistic literature, Kathryn Gutzwiller points out that tales about passion between sister and brother were a usual topic in Hellenistic literature, a likely reaction to the Ptolemaic marriage between siblings, either against or in favour of it.[6] She reckons that the story on Biblys and Caunus, who were linked to Miletus, was meant to justify the sibling marriage by Mausolus, the Carian satrap who ruled between the 377/376 and the 353 and who had married his sister Artemisia.[7] However, I will try to show in this paper that several elements in the myth are meant to disapprove of incest rather than to validate it.[8] In order to do this, I will focus on the symbolism of time and especially of space in the

1 This paper has been written with the support of the FCT Postdoctoral scholarship SFRH/BPD/90803/2012, at the Centro de Estudos Clássicos of the Universidade de Lisboa.
2 Arist. *Rh.* 1402b3. See also Diog. Laert. 5.71.1; Suda κ 1138; Eust. in Dionys. Per. 533.9. Cf. Hsch. κ 1915.
3 Artistocritus *FGrH* 493 F 1; Ap. Rhod. fr. F5 P.; Nicaenetus *Coll. Alex.* F1 P.; Nicaenetus fr. F46 G-S.
4 Conon, *Narr.* 2; Parth. *Amat. narr.* 11; Ov. *Met.* 9.450 – 665; Ant. Lib. 30; Nonnus *Dion.* 13.548 – 561.
5 Schol. Theoc. *Id.* 7.115c; Diogenian. 5.71; Steph. Byz. s.v. Καῦνος.
6 Gutzwiller 2007, 126 – 127.
7 On Mausolus see Hornblower 1982.
8 On incest in the Greek culture see Bremmer 1987, 41 – 59; in this myth see Nagle 1983, 301 – 315.

DOI 10.1515/9783110535150-015

version of Parthenius of Nicaea's *Erotica Pathemata*, for it provides a text that integrates different versions of the myth, although I will of course also refer to other versions when necessary.

The *Erotica Pathemata*, from the first century BCE, is a collection of stories about unfortunate passions, which are usually classified as Hellenistic mythographical prose, and provides a set of short narrations on erotic myths organised in independent chapters.[9] These narrations cannot be seen as mere summaries of former material and scholars have recently vindicated the aesthetic value of this prose and its intimate relation to poetry.[10] Indeed, Parthenius, a poet himself, dedicated the work to his friend Cornelius Gallus, a poet as well, and declared that the stories are meant to provide inspiration for poetical composition.[11] As a matter of fact, several lines of two different poems are quoted in the chapter on Byblis and Caunus. The first poet to be cited is Nicaenetus and the transmitted verses probably belong to the hexametric poem titled *Lyrcus*.[12] Parthenius also quotes some lines from his own poetic treatment. The two fragments are quoted in order to illustrate two different versions of the myth, as Parthenius himself points out. Since it is a short text, I quote the complete chapter:[13]

Περὶ Βυβλίδος
Ἱστορεῖ Ἀριστόκριτος περὶ Μιλήτου καὶ Ἀπολλώνιος ὁ Ῥόδιος Καύνου κτίσει.
Περὶ δὲ Καύνου καὶ Βυβλίδος, τῶν Μιλήτου παίδων, διαφόρως ἱστορεῖται. Νικαίνετος μὲν γάρ φησι τὸν Καῦνον ἐρασθέντα τῆς ἀδελφῆς, ὡς οὐκ ἔληγε τοῦ πάθους, ἀπολιπεῖν τὴν οἰκίαν καὶ ὁδεύσαντα πόρρω τῆς οἰκείας χώρας πόλιν τε κτίσαι καὶ τοὺς ἀπεσκεδασμένους τότε Ἴωνας ἐνοικίσαι. λέγει δὲ ἔπεσι τοῖσδε·
αὐτὰρ ὅ γε προτέρωσε κιὼν Οἰκούσιον ἄστυ
κτίσσατο, Τραγασίην δὲ Κελαινέος ἤγετο παῖδα
ἥ οἱ Καῦνον ἔτικτεν ἀεὶ φιλέοντα θέμιστας.
γείνατο δὲ ῥαδαλῆς ἐναλίγκιον ἀρκεύθοισι
Βυβλίδα. τῆς ἤτοι ἀέκων ἠράσσατο Καῦνος.

9 On Parthenius see Lightfoot 1999; Biraud-Voisin-Zucker 2008; Cuartero 1982; Calderón Dorda 1988. On the *Erotica Pathemata* as a mythographical work see Pellizer 1993, 291–292; Lightfoot 1999, 215–282; Higbie 2007, 237–254.

10 Gallé Cejudo 2013, 247–275.

11 The epistolographical introduction is most probably a rhetorical exercise than an actual letter, as Lightfoot warns (1999, 223–224), for epistolary prefaces were more honorific than utilitarian at their time. However, it is still adressed to Cornelius Gallus, a poet.

12 For an edition of Nicaenetus' fragments see Powell 1925 (1970), 1–2. The manchette in the margin of Parthenius' manuscript quote Aristocritus and Apollonius Rhodius but not Nicaenetus. On the manchetes see Papathomopulos 1968, the first to use that term for the notes on authorities in the manuscript which transmits Parthenius' and Antoninus Liberalis' works.

13 I follow Lightfoot's text and translation, except for a passage from Nicaenetus' quotation (lines 13–15), for which I follow White's edition and interpretation (1982, 192).

βῆ δὲ φερένδιος φεύγων ὀφιώδεα Κύπρον,
καὶ Κάπρος ὑλιγενὲς καὶ Κάρια ἱρὰ λοετρά.
ἔνθ' ἤτοι πτολίεθρον ἐδείματο πρῶτος Ἰώνων.
αὐτὴ δὲ γνωτὴ ὀλολυγόνος οἶτον ἔχουσα
Βυβλὶς ἀποπρὸ πυλῶν Καύνου ὠδύρατο νόστον.

οἱ δὲ πλείους τὴν Βυβλίδα φασὶν ἐρασθεῖσαν τοῦ Καύνου λόγους αὐτῷ προσφέρειν καὶ δεῖσθαι μὴ περιιδεῖν αὐτὴν εἰς πᾶν κακὸν προελθοῦσαν· ἀποστυγήσαντα δὲ οὕτως τὸν Καῦνον περαιωθῆναι εἰς τὴν τότε ὑπὸ Λελέγων κατεχομένην γῆν, ἔνθα κρήνη Ἐχενηΐς, πόλιν τε κτίσαι τὴν ἀπ' αὐτοῦ κληθεῖσαν Καῦνον. τὴν δὲ ἄρα ὑπὸ τοῦ πάθους μὴ ἀνιεμένην, πρὸς δὲ καὶ δοκοῦσαν αἰτίαν γεγονέναι Καύνῳ τῆς ἀπαλλαγῆς, ἀναψαμένην ἀπό τινος δρυὸς τὴν μίτραν ἐνθεῖναι τὸν τράχηλον. λέγεται δὲ καὶ παρ' ἡμῖν οὕτως·

ἡ δ' ὅτε δή <ῥ'> ὀλοοῖο κασιγνήτου νόον ἔγνω,
κλαῖεν ἀηδονίδων θαμινώτερον, αἵ τ' ἐνὶ βήσσῃς
Σιθονίῳ κούρῳ πέρι μυρίον αἰάζουσιν.
καί ῥα κατὰ στυφελοῖο σαρωνίδος αὐτίκα μίτρην
ἀψαμένη δειρὴν ἐνεθήκατο· ταὶ δ' ἐπ' ἐκείνῃ
βεύδεα παρθενικαὶ Μιλησίδες ἐρρήξαντο.
φασὶ δέ τινες καὶ ἀπὸ τῶν δακρύων κρήνην ῥυῆναι ἀίδιον τὴν καλουμένην Βυβλίδα.

Various stories are told about Caunus and Byblis, the children of Miletus. Nicaenetus says that Caunus fell in love with his sister, and that when the passion did not abate he left his home and travelled far from his native land, founding a city and settling there the scattered Ionians. He says in the following hexameters:

But faring further on, the town of Oecous
He founded, took to wife Celaeneus' daughter
Tragasia, who bore him justice-loving Caunus.
But like the slender poplars was her girl –
Byblis, whom Caunus loved against his will.
And he went enduring midday, in flight to snaky Cyprus,
And to wooded Caprus, and to the holly streams of Caria.
There he, first of all the Ionians, built a city.
But Byblis shared the nightingale's sad fate:
Without the gates she mourned for Caunus gone.

Most, however, say that Byblis fell in love with Caunus and made overtures to him, begging him not to look on while she went through every sort of misery. But Caunus felt only loathing, and crossed over into the land at that time possessed by the Leleges, where there is a stream called Echeneis; and there he founded a city named Caunus after him. But as for her, her passion did not abate; and in addition, when she considered that she was the reason for Caunus' departure, she fastened her girdle to an oak tree and put her neck in it. Here is my own version of the story:

And once she knew her cruel's brother mind,
Her shrieks came thicker than the nightingales'
In woods, who ever mourn the Thracian lad.
Her girdle to a rugged oak she tied,
And laid her neck within. And over her

> Milesian maidens rent their lovely robes.
> Some also say that an everlasting stream flowed from her tears, and that the stream's name was Byblis.

The story on Byblis and Caunus is related several times, thus constituting an account that can be described as a manifold narrative. In fact, some scholars consider that the chapter should be seen as a practical example of Parthenius' poetics: he displays multiple versions of the same subject-matter, both poetical and prose, as a sort of exhibition of his working methodology.[14]

In the introduction – after the title and the manchette –,[15] the first information provided is the genealogy of the siblings: their father is Miletus, the founder of the homonymous city.[16] This brief mention sets the space and time of the narration.[17] Indeed, all other versions of the myth begin with some genealogical information as well, and all agree that they were the children of Miletus.[18] The quoted verses of Nicaenetus also include a reference to the origin of the siblings, although Parthenius had already mentioned it in the prose introduction. The poetic fragment, however, offers a variant, for it presents their father not as the founder of Miletus but as the founder of the city of Oecous.[19] Some sources, indeed, identify Oecous with Miletus. Therefore, the choice of this toponym could be simply a poetical, possibly metrical, choice. However, the fact that Parthenius

---

14 Biraud, Voisan and Zucker 2008, 22–23.

15 The manchettes are found in the margins of the manuscript. Papathomopoulos (1968, xi-xix) showed that they do not belong to the original text, but were inserted by the scribe of the manuscript.

16 Apollod. 3.2.1; schol. Ap. Rhod. 1.185; Ant. Lib. 30.

17 Genealogy usually provides a chronological and geographical frame to a mythical figure. It is a very common procedure in mythographical literature – Hecateus' work and some other fragmentary logographers' works were titled *Genealogies*. Genealogy also plays an important role in archaic epic – Homeric heroes state who their ancestors are when they meet each other; Hesiod uses it to order the cosmos and the divine forces. Indeed, it is a tool to organise the traditional material because it integrates the double aspects of space and time: it is, on one side, the expression of an individuals' kin relations projected on time and, on the other side, the expression of the location of a group – a family – within space, when many Greek cities' names are explained by the name of their founders. This is the case for Miletus, who is the mythical founder of the city in Asia Minor which bears his name, and the father of the siblings. On the genealogical order in doxography see Delattre 2006, 145–159. One can consider the use of genealogy as a chronotopic reference. On the concept of chronotope see Bakhtin 1981. On chronotopes in the Aeschylean tragedy see Seaford 2012.

18 Conon *Narr.* 2; Ov. *Met.* 9.450–453, where Caunus and Byblis are twins; Ant. Lib. 30.2, also twins. Nonnus *Dion.* 13.546–549, 557–559 gives an aberrant version where Caunus is said to be brother of Miletus as well as brother of Byblis. On Nonnus see Hollis 1976, 142–150.

19 Eponymy is a type of aetiology. On aetiology see Valverde 1989; Delattre 2009, 285–310.

includes in the quotation the verses on the settlement of Oecous and the marriage to Tragasia strongly suggests that, on one hand, he considers this background information as belonging to the myth of Byblis and, on the other hand, that the variant provides additional details. In fact, a scholion to Dionysius Periegetes mentions a tradition according to which the hero Miletus founded first the city of Oecous, where he dedicated a temple to Aphrodite, and then his son Celadon founded Miletus.[20] According to this, Lightfoot surmises that Nicaenetus' version could reflect Oecous' desire to assert his priority (1999, 438). Also a scholium to Theocritus links Oecous with a temple to Aphrodite.[21] The scholiast interprets a problematic point of Theocritus' *Idyll* 7 (115 – 117 Gow):[22]

> ὔμμες δ' Ὑετίδος καὶ Βυβλίδος ἁδὺ λιπόντες
> νᾶμα καὶ Οἰκοῦντα, ξανθᾶς ἕδος αἰπὺ Διώνας,
> ὦ μάλοισιν Ἔρωτες ἐρευθομένοισιν ὁμοῖοι
>
> But do you leave the sweet stream of Hyetis and Byblis,
> and Oecus, that steep seat of Golden-haired Dione,
> ye Loves so rosy as apples

In line 116 the manuscripts read οἰκεῦντες, but the scholiast understood it as a toponym, Οἰκοῦντα. This reading is probably also to be found in a 5th century papyrus, where the ending -ṿτα is to be read and Οικευ is restored by the editors.[23] Hecker and Gow accordingly correct Theocritus' manuscript accepting the scholiast interpretation.[24] If this is right, it would provide evidence of the fact that Oecous was known as the place of Aphrodite's temple in Theocritean times. Regardless of whether Miletus and Oecous are the same city or not, Huxley (1970, 253) pointed out that "from the testimonies it is evident that the spring Byblis, Oikous and the precinct of Aphrodite are all connected". One wonders, therefore, if the choice of Oecous in Nicaenetus was meant to evoke an associa-

---

20 Schol. Dionys. Per. 825: Μίλητος δὲ τῶν ἐπιφανῶν τις ἦν ἐν Κρήτῃ, ἀφ' οὗ καὶ πόλις ἐκεῖ Μίλητος, ὃς Μίνωος ἐπιστρατεύσαντος ἀπάρας τῆς Κρήτης κατάγεται εἰς Λυδίαν τῆς Ἀσίας, οὗ οἰκήσας Οἰκοῦντα τὸν τόπον ὠνόμασε, καὶ ἱερὸν Ἀφροδίτης ἱδρύσατο. Γαμεῖ δὲ Δοίην τὴν Μαιάνδρου, ἀφ' οὗ ποταμὸς ἐν Καρίᾳ, καὶ ποιεῖ Κελάδωνα, Καῦνον, Βυβλίδα. Ὧν ὁ Καῦνος οὐ φέρων τὸν ἔρωτα τῆς ἀδελφῆς μετῴκισται εἰς Λυκίαν. Κελάδων δὲ ἄρξας Οἰκοῦντος τὸν πατέρα εἰς τὴν πλησίον νῆσον ἔθαψεν, οὗ καὶ αὐτὸς μετῳκίσθη κατὰ χρησμόν, καὶ Μίλητον αὐτὴν ὠνόμασεν. Γέφυρα δὲ διορίζει τὰ νῦν Οἰκοῦντα καὶ Μίλητον.

21 Schol. Theoc. 7.115 – 118: Οἰκεῦντα: ἐν Μιλήτῳ τόπος, <ἔνθα> ἱερὸν Ἀφροδίτης.

22 Transl. Gow 1950, 65.

23 P. Oxy. 13.1618. Grenfell and Hunt 1919, 174.

24 Gow 1950, 64.

tion to Aphrodite,[25] for his audience would have been able to recognise this toponym as a reference to the goddess of love and sex. In other words, the variation might be aimed at bringing into the picture Aphrodite, whose power is exemplified within the story, since the myth of Byblis and Caunus is basically a story of unfortunate love. However, we must admit that the commentary of the scholiast is an explanation that can actually be inferred from the Idyll itself. Indeed, White (2007, 126) rejects the correction and favours the reading of the manuscript.[26] If we reject the correction together with White, the interpretation that Nicaenetus is alluding to Aphrodite loses strength, since the scholium could be of a much later composition. However, she does not take into account the fact that the papyrus reads also Οἰκοῦντα, which would show that at least in the 5th century this word would have been understood as a toponym. Furthermore, in the fragment of Nicaenetus, Aphrodite is once again alluded to by another toponym in line 12, Κύπρος, obelised by Lightfoot and corrected by Powell for φρικώδεα Κύπριν,[27] but accepted by White, whose interpretation of the passage I follow.[28] Indeed, Cyprus was an important centre of cult of the goddess and one of the islands which claimed to be her birth-land.[29]

Some attention should be given to another aspect of the toponym Oecous: the name of the city contains – or at least evokes – the same root found in the verb οἰκείω, "to inhabit", or the noun οἴκος, "home". Thus, the name of the father land of Byblis and Caunus would sound like 'Home', or something very close to it. As a matter of fact, in the different versions presented by Parthenius in this text, the idea of separation from the home land is expressed very sharply. Note that the wording of Parthenius' prose underlines that idea: ἀπολιπεῖν τὴν οἰκίαν καὶ ὁδεύσαντα πόρρω τῆς οἰκείας χώρας. Nicaenetus' version de-

25 On the fragment of Nicaenetus see White 1982, 185 – 192.

26 She interprets the passage as follows: "But leave the sweet stream of Hyetis and Byblis, although you inhabit the steep seat of golden-haired Dione, Loves as rosy as apples".

27 Following the idea that Caunus' was fleeing Aphrodite's wrath. Huxley (1970, 6) states that the incestuous passion was a punishment of the Goddess: "Kaunos was the victim of passion inspired by Aphrodite, who had a precinct at Oikous" (for the discussion see pp. 6 – 7). However, no version of the myth describes Caunus' love explicitly as Aphrodite's punishment, only this one suggests it.

28 Several editors had obelised the passage, interpreting that the mentioned places are accusatives depending of the participle φεύγων. This syntax poses a problem, since we do not know why Caunus would escape from Cyprus. White showed that this is a misinterpretation of the regent verb of these accusatives, which are direction accusatives depending of βῆ.

29 Aphrodite is already related to Cyprus in the Homeric poems (*Il.* 5.440, 458, 883) and in Hesiod (*Theog.* 188 – 206) and her name appears in early archaic Cypriot inscriptions (Cyrino 2010, 27 – 28). For an overview on the goddess see Pirenne-Delforge 2010, 3 – 16, esp. 9 – 12.

scribes Caunus' departure with the expression βῆ φεύγων ("went in flight") and he also describes Caunus' journey. The fact that Nicaenetus mentions the route, though in three synthetic verses,[30] slows down the narrative rhythm and increases the feeling of distance. Other versions of the myth also stress the importance of distance between the siblings. Indeed, in the second version provided by Parthenius (lines 19–21), Caunus also leaves (see infra). This is made extremely clear in Conon's version, which includes an episode on how Caunus, after erring for a while (πλανώμενος), arrived to Lycia, met Pronoe and was persuaded to marry her, and how one of their children founded the city called Caunus.[31] Hence, the episode on Caunus' departure is expanded in comparison to Nicaenetus or Parthenius' versions.

Nicaenetus' description of Byblis lamenting her brother employs the term νόστος (line 16). Lightfoot comments that the word should be interpreted as "journey", not "journey home". However, in my opinion, it should be interpreted precisely as the concept of "journey home", because Byblis is longing for her brother to come back to Oecous, "Home". It would also play ironically with the idea that he will find a new home to settle far away. Furthermore, the reference to νόστος allows Nicaenetus to draw from the rich poetic tradition of the return of the Greek heroes after the war of Troy. By using this term, the poet would be subtly characterising Caunus as a 'war hero'. Indeed, he is metaphorically fighting a war while being dominated by his passions. But in opposition to the Homeric heroes who, after winning the war, go back home – though finding many troubles in their way or even death – the only way Caunus can win his war is precisely by not going back home but instead leaving it.

The structure of the narrative, thus, follows the pattern of the foundation myths: a hero commits a crime and therefore has to leave his city.[32] The passion felt by Caunus is, therefore, equated to a crime. Certainly, the way Nicaenetus'

---

30 Editors have postulated several lacunae in the verses which refer to Caunus' parcours. However, White 1982 showed that it is not necessary to postulate them, for the text as it is follows perfectly the Hellenistic allusive poetics.

31 Καῦνος δὲ πλανώμενος εἰς Λυκίαν φθάνει, καὶ τούτῳ Προνόη (Ναῒς δ' ἦν αὕτη) ἀναδῦσα τοῦ ποταμοῦ τά τε συνενεχθέντα τῇ Βυβλίδι λέγει, ὡς ἐχρήσατο Ἔρωτι δικαστῇ, καὶ πείθει αὐτὸν αὐτῇ ἐπὶ τῷ τῆς χώρας λαβεῖν τὴν βασιλείαν, (καὶ γὰρ εἰς αὐτὴν ἀνῆπτο) συνοικῆσαι. For a commentary on Conon see Brown 2002.

32 Indeed, Francese (2001, 138) pointed out that Parthenius' treatment of the incest theme focuses on "the desire as an erotic pathology" and defended that this was an innovation of the Nicaean author. One of his arguments is that the former treatments of the myth of Byblis and Caunus, from which only Nicaenetus and Apollonius Rhodius' fragments have been transmited, would have presented the incest in a non-erotic way, describing it as a crime or a trangression and subordinating the narrative to local history interests.

verses refer to Caunus' passion suggests that it is seen as a transgression: Caunus is described as ἀεὶ φιλέοντα θέμιστας, and his passion as involuntary (ἀέκων). The origin of the incestuous passion is not specified in any version of the myth. In other traditions, though, love is a punishment sent by a divinity – and in fact, Huxley (1970, 6) interprets Nicaenetus' passage as meaning precisely that Caunus' love was sent by Aphrodite.[33] The indirect references to Aphrodite in the toponyms – Oecous, if we accept the correction, and Cyprus – suggest also the idea of divine punishment. We find this idea explicitly formulated in chapter 5 of the *Erotica Pathemata*, in which Parthenius says that Leucipus' desire for his daughter was a punishment from Aphrodite.[34] Be as it may, the important point is that the passion is depicted as negative and the consequence to such an inappropriate feeling is departure.

In the second version of the myth, Byblis' confession of her feelings justifies, from a narrative point of view, the fact that Caunus leaves the city (lines 17–18). Indeed, the pattern of the foundation myth no longer fits here, since the crime is now committed by Byblis and not by Caunus.[35] However, Caunus still has to leave. The brother's feelings in this version are described with the verb ἀποστυγήσαντα, "to hate deeply", precisely the opposite to what he felt in the former version.[36] As a consequence, he leaves. This is expressed by a verb formed on the root of πέραιος, "on the further side", expressing the idea of separation in a very definite and strong way. Thus, he crosses over to the land of the Leleges, a generic designation for early inhabitants of Asia Minor. The utter separation of Caunus is repeated again when Parthenius reformulates his departure from Byblis' point of view in his prose version (lines 21–22): she realises that she was the reason of Caunus' escape, expressed this time by the term ἀπαλλαγῆς, "release", as if he would be set free from her passion. The word can even be a synonym of "death".[37]

33 See n. 25. The desire that the Lemnian men felt for the Thracian women is described as a punishment from Aphrodite for neglecting her cult. See Apollod. 1.9.17–18 [114–115]; Hyg. *Fab.* 15; *BNJ* 12 F 14 (= schol. *Il.* 7.467). Phaidras' passion for Hyppolitus is also explained as a punishment sent by Aphrodite against the boy due to his rejection of sexual love: Eur. *Hipp.* 1–57.

34 Parth. 5.2: κατὰ μῆνιν Ἀφροδίτης εἰς ἔρωτα ἀφικόμενος τῆς ἀδελφῆς.

35 That the passion is viewed as crime is clear in Byblis' shame in the confession. Antoninus' version, mentioned above, clearly explains that she hid form the parents.

36 Lightfoot (1999, 437) notes that ἀποστυγήσαντα describes an "empahtic rejection".

37 Xen. *Cyr.* 5.1.13; Theophr. *Hist. pl.* 9.8.3. In Parthenius' poetic version (line 24) Byblis realises of Caunus' νόον, mind, purpose, but also heart or feelings, emphasising the pathetic aspect of the story. Byblis sorrow would be caused by the fact that he did not love her back, more than by his departure.

On the other hand, Byblis also leaves the house or the city in both versions of the myth, even though in the first version she is not guilty of anything. Her departure is not referred to in Parthenius' first part in prose, but it is explicitly mentioned in Nicaenetus' quotation: Βυβλὶς ἀποπρὸ πυλῶν Καύνου ὠδύρατο νόστον. She mourns Caunus "at the doors" of the city, we must infer. The image of the girl crying at the doors strongly reminds the Hellenistic *topos* of the *paraclausithyron*, the lament for the lover at the door of the loved one.[38] One wonders if this poem could have triggered the version of the myth in which it was Byblis the one in love with her brother, or Conon's version where the love is mutual.

Thus, Byblis leaves the πόλις, the civilised world, in a sort of social exile. Her separation from home is explicitly expressed in Conon's version, where she leaves the "paternal house" and "wanders through a very lonely place" or through "the wilderness".[39] Clearly, she is no longer within civilised society. This idea is also conveyed by the fact that she ends up giving place to a water source – mentioned in all versions – a geographical element which belongs to nature. Incorporation to the natural world would be also suggested by Nicaenetus' αὐτὴ δὲ γνωτὴ ὀλολυγόνος οἶτον ἔχουσα, which seems to imply that Byblis undergoes a metamorphosis into a nightingale.[40] This idea of integration in the wilderness is brought to an extreme by Antoninus Liberalis, who describes her as saved by nymphs and becoming herself a hamadryad nymph.[41]

In Parthenius' poetic and prose versions, the resolution for Byblis is expressed in a radical and suggestive way: she ties a girdle to a tree, which implies the idea of union, as if she was trying to tie Caunus.[42] However, the knot will actually not bind anything, but on the contrary it will release her from life. Death is a definite form of departure. The location of her suicide by hanging from a tree is not specified in the prose version. The mention of the δρύος and the stream that will appear in the spot, though, suggests a place outside the city. In Parthenius'

---

38 About the paraclausithyron see Canter 1920, 355–368; Copley 1940, 52–61; Copley 1956; Cummings 1996.

39 Conon *Narr.* 2: ἡ Βυβλὶς ἐκλείπει καὶ αὐτὴ τὴν πατρῴαν οἰκίαν, καὶ πολλὴν ἐρημίαν πλανηθεῖσα.

40 On the metamorphosis of Byblis see Forbes Irving 1990, 24, 31, 300; Buxton 2009, 199–200. Interestingly chapter 13 of Parthenius' mythographical work provides a parallel of a character, Harpalyce, involved in an incestuous passion which is transformed into a bird. On hamadryad nymphs see Larson 2001, 11 and 33.

41 On Antoninus Liberalis see Celoria 1992 and Del Canto Nieto 2003.

42 In erotic magic tying also plays a symbolic role.

verses a location outside her house and the city is suggested by the comparison to a crying nightingale – maybe a reminiscence of Nicaenetus' version.

Regarding time, the term φερένδιος in Nicaenetus' fragment (line 12), obelised by Lightfoot and other commentators, is, in my opinion, satisfactorily explained by Giangrande (1982, 81–82). This term places the escape of Caunus in the hottest moment of the day, with clear light, which posed a problem to other editors since broad daylight would make a stealthy escape more difficult. On the contrary, Giangrande reckons that escaping at noon, with bright light, would actually be the best way to hide, because in this moment everybody would be having the 'midday sleep', attested in several sources. On the other hand, I would like to point out that the apparent contradiction of hiding in the light fits nicely the Hellenistic poetics.[43] Besides, it contrasts with an element in the second version in prose (lines 17–18): Parthenius relates that Byblis confessed Caunus her love in a covert way by asking him not to look when she declared to him.[44] It is indeed logical that the notion of 'hiding' plays a role in the story of an illicit love. Remarkably, in Antoninus Liberalis' version, Byblis decides to commit suicide by night. The fact that daylight is associated to Caunus and night-time or hiding from sight are associated to Byblis, suggests a symbolic distribution in which each sibling occupies opposite extremes. Furthermore, in Antoninus' version day and night are also associated to hiding and escaping. In his narrative Caunus' departure plays no role at all and is not even mentioned, but the conflict revolves around Byblis. She is the one in love and she is depicted as trying to hide her feelings from her parents.[45] Antoninus' text continues saying ἐπεὶ δὲ καθ' ἡμέραν εἴχετο χαλεπωτέρῳ δαίμονι, νυκτὸς ἔγνω καταβαλεῖν ἐκ τῆς πέτρας ἑαυτήν, which is translated by Celoria (1992, 89) as "but daily she was being gripped by an even more unmanageable demon and one night she decided to throw herself from a rock". Besides the more frequent meaning of 'daily' chosen by the translator, the expression καθ' ἡμέραν can also mean "by day".[46] This would establish a contrast between the effort to hide the incestuous passion during the day, and the moment in which Byblis decides to commit suicide, νυκ-

43 See Durbec 2009, 21–24.

44 μὴ περιιδεῖν αὐτήν, line 18. Parthenius does not specify how she hides. Ovid also includes the detail of the declaration and resolves it by making her send a letter to her brother. The fact that their communication is indirect allows Byblis to hide from Caunus' sight. Ov. *Met*. 9.515–570.

45 Ant. Lib. 30.3: καὶ τὸ πάθος ἄχρι μὴν ἐδύνατο κρύπτειν ἐλελήθει τοὺς γονεῖς.

46 Cf. Aesch. *Cho*. 818 where it is opposed to night: νυκτὸς προὔμματων σκότον φέρει, / καθ' ἡμέραν δ' οὐδὲν ἐμφανέστερος.

τὸς. Again, we find an association of hiding, light, darkness and separation.[47] Thus, the fact that different versions specify the precise moments in which departure or suicide take place, both movements of separation, suggests that their symbolic value associated to hiding and prohibition, are connected precisely with the split-up or union of the siblings.

In conclusion, Parthenius provides a multiple account where prose and poetry are intertwined within different reformulations of a story about a forbidden feeling, which affects both of the individuals involved in the incestuous passion, and not only the one who feels it. In all versions, the idea of distance and separation appears as the way to resolve the conflict created by the incestuous desire. Both siblings have to leave their home city, Miletus or Oecous, a city called "Home", which is opposed to both the countryside or a new land far away. Caunus leaves and travels to a remote place where he founds a new city. Biblys stays in the χώρα, but outside the οἶκος and outside the πόλις. Thus, she suffers also an exilium, though indoors. Her integration into the wilderness symbolises the fact that she will be permanently excluded from the civilised world. Therefore, the variant on who is in love with whom seems secondary. There is a constant playing with the notion of near and far, of inside and outside. Passionate love for a sibling is too close, distance needs to be imposed to protect the home and the city. Both Caunus' and Byblis' destinies are movements of separation from the family. Death is the most extreme form of separation. On the contrary, exile is a separation with continuity through marriage and the foundation of a new city.

It could be surmised that the story warns about the dangers of prohibiting sibling marriage: the royal family would lose their heirs. However, as I argued above, the structure of the story following the foundation myths, the combination of the characterization of Caunus as a just man and his feelings as unwilling, Byblis' worries to hide during confession or to hide her feelings from her parents, reveal altogether a negative vision of a possible incestuous relation. On the other hand, Miletus, their father, who provides the link to the homonymous city, is not described as incestuous. In fact, Nicaenetus portraits him as a newcomer (προτέρωσε κιών, line 7 "faring further on") who obtained access to the power by marring a local woman, Tragasia, who is only mentioned here, and is said to be the daughter of a blurred Celaenus, which might be the eponym of the Κελαιναί, the territory in Phrygia (Lightfoot 1999, 438).

Seaford (1990, 76) showed that in tragic myth the imprisonment of a women is symbolically associated with her relationship to the blood-kin. He analyses

47 Darkness is associated to incest in many myths see Seaford 1990, 76, 83–84.

several tragic examples and concludes that captivity imposed by the father or by the family, a tentative to keep a girl within her blood-kin, symbolises the rejection of marriage, which implies the impossibility for girls to perpetuate life through marriage. At the same time, the girl escaping to the countryside, which symbolises a loss of control, has equally disastrous consequences.[48] Also the opposition between light and darkness play a role in myths related to incest. As a matter of fact, Byblis' myth seems to follow this symbolism. Caunus has to abandon his father land but he starts it anew somewhere else by means of marriage, as explicated in Conon's version. Byblis will not marry, will not be integrated in the οἶκος nor in the πόλις, and will disappear from the civilised world. Thus, the ideas of incest, marriage and access to adulthood through marriage appear over and over again. Separation from the family is widely attested in marriage rituals. Sometimes there is even a symbolic death of the bride.[49] The erotic element and the age of the siblings, who are in some versions twins, also point to a marriage-related background. From a narrative point of view, the separation of Byblis and Caunus from their home is translating the danger of an excessively close kin relationship. Incest must be solved with its opposite element, that is, distance.

Bibliography

Bakhtin, M.M. 1981. Forms of Time and the Chronotope in the Novel. In *The Dialogic Imagination: Four Essays*, ed. M. Holquist, 84–258. Austin.

Biraud, M., D. Voisin and A. Zucker (eds.). 2008. *Parthénios de Nicée. Passions d'amour.* Grenoble.

Bremmer, J. 1987. Oedipus and the Greek Oedipus Complex. In *Interpretations of Greek Mythology*, ed. J. Bremmer, 41–59. London.

Brown, M.K. 2002. *The Narratives of Konon: Text, Translation and Commentary of the Diegeseis.* Munich and Leipzig.

Burkert, W. 1983. *Homo Necans. The Anthropology of Ancient Greek Sacrificial Ritual and Myth.* Berkeley, Los Angeles and London.

Buxton, R. 2009. *Forms of Astonishment. Greek Myths of Metamorphosis.* Oxford.

Calderón Dorda, E. 1998. *Sufrimientos de amor y fragmentos.* Madrid.

Celoria, F. 1992. *The Metamorphoses of Antoninus Liberalis: A Translation with Commentary.* London and New York.

Copley, F.O. 1940. The Suicide-Paraclausithyron: A Study of Ps.-Theocritus Idyll XXIII. *TAPhA* 71: 52–61.

48 Furthermore, a girl's escape to nature would hint at a Dionysiac dimension, for Dionysus is the God that liberates women and brings them together as maenads on the mountainside.

49 Burkert 1986, 62–63; Dowden 1989, 34–40.

Copley, F.O. 1956. *Exclusus Amator: A Study in Latin Love Poetry.* Baltimore.
Cuartero, F. 1982. *Parteni de Nicea, Dissorts d'amor.* Barcelona.
Cummings, M.S. 1996. *Observations on the Development and Code of Pre-elegiac Paraklausithuron* (Ph.D. Diss.). University of Ottawa.
Cyrino, M.S. 2010. *Aphrodite.* London and NewYork.
Delattre, C. 2006. L'ordre généalogique, entre mythographie et doxographie. *Kernos* 19: 145–159.
Delattre, C. 2009. Aitiologia: mythe et procédure étiologique. *Mètis* 7: 285–310.
Del Canto Nieto, R. 2003. *Antonino Liberal, Metamorfosis.* Madrid.
Dowden, K. 1989. *Death and the Maiden. Girls' Initiation Rites in Greek Mythology.* London and New York.
Durbec, Y. 2009. Nicaenetos de Samos: un poéte hellénistique. *ARF* 11: 21–24.
Forbes Irving, P.M.C. 1990. *Metamorphosis in Greek Myths.* Oxford.
Francese, C. 2001. *Parthenius of Nicaea and Roman Poetry.* Frankfurt am Main.
Gallé Cejudo, R. 2013. La prosificación poética en los Amores apasionados de Partenio: el ejemplo de la canción y la maldición. In *Les Études Classiques, Revue de la Societé des Études Classiques de l'Université de Namur*, 247–275.
Giangrande, G. 1982. A Textual Problem in Nicaenetus. *CL* 2: 81–82.
Gow, A.S.F. 1950. *Theocritus. Edited with a Translation and Commentary.* vol. I: Introduction, Text, and Translation. Cambridge.
Grenfell, B.P. and A.S. Hunt. 1919. *The Oxyrhynchus Papyri* 13. Oxford.
Gutzwiller, K. 2017. *A Guide to Hellenistic Literature*, Malden, Oxford and Victoria.
Higbie, C. 2007. Hellenistic Mythographers. In *The Cambridge Companion to Greek Mythology*, ed. R.D. Woodard, 237–254. Buffalo.
Hollis, A.S. 1976. Some Allusions to Earlier Hellenistic Poetry in Nonnus. *CQ* 26: 142–150.
Hornblower, S. 1982. *Mausolus.* New York and Oxford.
Huxley, G. 1970. Nikainetos and Oikous. *GRBS* 11.3: 251–257.
Larson, J. 2001, *Greek Nymphs. Myth, Cult, Lore.* Oxford and New York.
Lightfoot, J.L. 1999. *Parthenius of Nicaea: The Poetical Fragments and the Erotica Pathemata.* Oxford.
Nagle, B.R. 1983. Byblis and Myrrha: Two Incest Narratives in the Metamorphoses. *CJ* 78: 301–315.
Papathomopoulos, M. 1968. *Antoninus Liberalis. Les Métamorphoses.* Paris.
Pellizer, E. 1993. La mitografía. In *Lo spazio letterario della Grecia antica, volume I. La produzione e la circolazione del testo, tomo II. L'ellenismo*, eds. G. Cambiano, L. Canfora and D. Lanza, 283–303. Rome.
Pirenne-Delforge, V. 2010. Flourishing Aphrodite: An Overview. In *Companion to Aphrodite*, eds. A. Smith and S. Pickup, 3–16. Leiden.
Powell, J.U. 1925 (1970). *Collectanea Alexandrina.* Oxford.
Seaford, R. 1990. The Imprisonment of Women in Greek Tragedy. *JHS* 110: 76–90.
Seaford, R. 2012. *Cosmology and the Polis. The Social Construction of Space and Time in the Tragedies of Aeschylus.* Cambridge.
Valverde Sánchez, M. 1989. *El aition en las Argonáuticas de Apolonio de Rodas.* Murcia.
White, H. 1982. Parthenius and the Story of Byblis. *CL* 2: 185–192.
White, H. 2007. Observations on the Text of Theocritus. *Habis* 38: 123–132.

Andreas N. Michalopoulos

Mythological Time and Space in Ovid's Exile Poetry

The decision of the emperor Augustus to banish Ovid to Tomi, on the Black Sea coast, in 8th century AD, marked the poet's life decisively and irrevocably.[1] Ovid struggled to achieve his recall to the capital with the help of his wife[2] and loyal friends who stayed back in Rome. He wrote nine books of elegies (five books entitled the *Tristia* and four entitled the *Epistulae ex Ponto*), in which he depicted his hard life in Tomi and he requested that he may be allowed to move, if not back to Rome, at least to a place closer to the capital, away from the extreme edge of the empire and of the civilized world.

One of the ways Ovid tried to achieve his goal was the crafty and targeted use of myth.[3] The goal of this paper is to examine Ovid's reception of myth and his use of myth as a means of enriching his arguments and of constructing his own exilic *persona*. I will focus on Ovid's utilisation of myth in the third elegy of *Tristia* 1 (1.3).[4] These are the main reasons for choosing this particular elegy:

a) it is one of the first letters of the collection and as such it is programmatic; it sets out some of the most important aspects of Ovid's poetics of exile,

b) there are numerous mythological references in this elegy, which enable us to explore Ovid's use of myth in his exilic poetry.

Tristia 1.3 is preceded by *Tr.* 1.1, in which the poet gives instructions to his new book which is about to travel to Rome for the first time, and by *Tr.* 1.2, in which the poet recounts his journey from Rome to the place of his exile. In *Tr.* 1.3 the poet recalls the night of exile, his last night in Rome, his preparations

1 Throughout his exilic poetry Ovid explicitly states that Augustus was responsible for his exile. See *Tr.* 1.3.5–6, 1.3.85–86, 1.5.61–62, 2.1.7–8, 4.9.11–12, 5.7.7–8, 5.9.11–14, *Pont.* 1.2.59, 1.7.43–48, 2.7.55–56, 3.6.7–10, 3.7.39–40.

2 According to *Tr.* 4.10.69–74, Ovid had three wives in his lifetime. For the identity of Ovid's wife and her portrayal in the exilic elegies see Hinds 1985 and Hinds 1999; Helzle 1989; Öhrman 2008, 151–189; Tissol 2014, 103f. For similarities and differences between Ovid's wife and the elegiac *puella*, see Nagle 1980, 44–46, 51–54; Videau-Delibes 1991, 217–231; O' Gorman 1997, 116; Angulo 2008; Öhrman 2014, 430–431.

3 For a bibliography on Ovid's use of myth in his exilic poetry see Claassen 2001, 11 n. 1. See also Rahn 1958, 115–19; Besslich 1972, 185; Green 1994, xv, xvi; Claassen 2001, 48–56; Claassen 2008, 265–283; Claassen 1999, index s.v. myth.

4 On *Tr.* 1.3 see among others Doblhofer 1980 and Doblhofer 1987, 81–96; Videau-Delibes 1991, 24–49; Rosati 1999; Huskey 2002a, 30–61.

DOI 10.1515/9783110535150-016

and departure. One can clearly see that *Tristia* 1.2 and 1.3 follow the narrative structure of the first two books of the *Aeneid*. Both works begin *in medias res* with descriptions of violent storms at sea, then to be followed by a flashback narrative, in which the protagonist recalls his dramatic departure from his homeland.[5]

Let us have a look at the structure of *Tr.* 1.3. The poem is divided into four sections, framed by an Introduction (1–4) and an Epilogue (101–102):

I. Introduction (1–4)
II. Evening (5–26)
III. Nightfall (27–46)
IV. Pre-dawn hours (47–70)
V. Morning (71–100)
VI. Epilogue (101–102)

Ovid as an Alter Aeneas – *Tr.* 1.3 and *Aen.* 2

The intertextual relationship between *Tristia* 1.3 and the *Aeneid*, mainly Book 2, has long been established. Ovid's use of Verg. *Aen.* 2 is one of the most commented-upon aspects of the poem.[6] Already from the beginning Ovid takes up the role of Aeneas,[7] as he is about to narrate the events of his last night at Rome (*Tr.* 1.3.1–4):

> *Cum subit illius tristissima noctis imago,*
> *quae mihi supremum tempus in urbe fuit,*

5 See Evans 1983, 47–48; Huskey 2002b, 92. On the structure and the organisation of the narrative in *Tristia* 1 see Herrmann 1924, 1–38; Klodt 1996, 273–276; Zimmermann 2005; Tola 2008.
6 See Kenney 1965, 47 n. 1; Nagle 1980, 29; Videau-Delibes 1991, 29–34; Rosati 1999, 788 f.; Huskey 2002b; Zimmermann 2005; Dell'Innocenti Pierini 2007, paragraphs 8–12; Öhrman 2014, 428 with n. 8.
7 For Ovid as Aeneas see among others Luck 1977, 36; Evans 1983, 37, 47 f.; Videau-Delibes 1991, 29–49; Edwards 1996, 121–121; Hardie 2002, 289–290; Huskey 2002b, 88–94. Claassen 1999, 191 sees *Tristia* 1 as a miniature 'elegiac epic' with numerous flashbacks to the fall of Troy. See also Claassen 1988, 166. For the numerous parallels of Ovid's exile poetry with epic see Williams 2002a, 350–353; Gaertner 2005, 144; and Claassen 2009, 174–175. For Ovid as an epic hero in his exilic poetry see Evans 1983, 37, 40, 48–49; Edwards 1996, 121–122; Holzberg 2002, 181–182; Williams 2002b, 236; Galasso 2009, 204.

cum repeto noctem, qua tot mihi cara reliqui,
labitur ex oculis nunc quoque gutta[8] *meis.*

When the saddest memory comes to mind, of that night, my last hour in the city, when I recall that night when I left so much so dear to me, even now tears fall from my eyes.[9]

Ovid clearly picks up the beginning of *Aeneid* 2, where Aeneas begins his account of his own misfortunes in Dido's palace at the request of the Carthaginean queen (Verg. *Aen.* 2.3–13):

Infandum, regina, iubes renovare dolorem,
Troianas ut opes et lamentabile regnum
eruerint Danai, quaeque ipse miserrima vidi
et quorum pars magna fui. quis talia fando
Myrmidonum Dolopumve aut duri miles Ulixi
temperet a lacrimis? et iam nox umida caelo
praecipitat suadentque cadentia sidera somnos.
sed si tantus amor casus cognoscere nostros
et breviter Troiae supremum audire laborem,
quamquam animus meminisse horret luctuque refugit,
incipiam.

O Queen, the sorrow you bid me bring to life again is past all words, the destruction by the Greeks of the wealth of Troy and of the kingdom that will be mourned for ever, and all the horrors I have seen, and in which I played a large part. No man could speak of such things and not weep, none of the Myrmidons of Achilles or the Dolopians of Neoptolemus, not even a follower of Ulixes, a man not prone to pity. Besides, the dewy night is already falling fast from the sky and the setting stars are speaking to us of sleep. But if you have such a great desire to know what we suffered, to hear in brief about the last agony of Troy, although my mind recoiled in anguish when you asked and I shudder to remember, I shall begin.[10]

Just like Aeneas at the night of Troy's fall, Ovid tries to stay at Rome for as long as possible, but is then forced to leave in the midst of wailing and lamentation. Even if some reader fails to realize that Ovid's last night in Rome is intended to echo the fall of Troy and is modelled on Aeneas' experiences[11] – which is nearly

8 Aeneas, too, wheeps at *Aen.* 1.459, 465, 470. Cf. Odysseus' wheeping as he listens to Demodocus' song in Alcinous' palace (Hom. *Od.* 8.521–531).

9 All *Tristia* translations by A. S. Kline, 2003. *Ovid – Poems from Exile* (with slight modifications). http://www.poetryintranslation.com/klineasovidexile.htm, N.p., n.d. Web. 29 Nov. 2015.

10 All *Aeneid* translations by D. West, *Virgil. The Aeneid.* London 1990 (with slight modifications).

11 Ovid's association with Aeneas on the night of Troy's fall is the richer, most meaningful and sustained, however, other mythological associations can be made too. According to Zimmer-

impossible for the average contemporary Roman – Ovid makes sure to spell it out clearly with an impressive disclaimer[12] (*Tr.* 1.3.25–26): *si licet exemplis in parvo grandibus uti,*[13] */ haec facies Troiae, cum caperetur, erat* ("If one might use a great example for a lesser, this was the face of Troy when she was taken").

This disclaimer shows that Ovid is consciously using an *exemplum* to illustrate his situation, and that he is well-aware that he is comparing greater things with smaller; of course, self-irony and self-mockery may be detected here.[14] Ovid models his exilic *persona* on Latin literature's most famous exile,[15] Aeneas, and compares the captured Troy with his home. There are obvious similarities between Ovid and Aeneas in the way that Ovid is shaping his exilic *persona*, but there are also substantial differences:[16] Ovid is about to leave the thriving capital of the world, while Aeneas flees from a captured city; Ovid departs from Italy, which is Aeneas' destination. Ovid leaves behind his wife who is alive and well, while Aeneas' wife, Creusa, is killed on the night of Troy's fall. For Ovid exile equals death, so his departure resembles a funeral.[17] In many aspects, his journey is harder than Aeneas', because he travels alone and his destination is much worse than Aeneas'.[18]

---

mann 2005, 217–218 in *Tr.* 1 Ovid is using three models for his own fate: Aeneas, Odysseus, and Hector (in *Il.* 6). At p. 219, following Froesch 1976, 26, he claims that *Tr.* 1.3.25–26 recall Hector farewell scene with Andromache and Astyanax in *Il.* 6. For Ovid as Hector and Meliboeus see Degl'Innocenti Pierini 2007.

Rosati 1999, 790–796 notes that the farewell scene in *Tr.* 1.3 as a whole picks up the elegiac *topos* of lovers parting at dawn. He also points out a parallel between Ovid and Protesilaus about to depart for Troy (cf. Laodamia's Ovidian letter, *Her.* 13). Furthermore, Claassen 1996, 580 sees Ovid as a 'departed Dido', comparing the wailing of his household with Verg. *Aen.* 4.667–671.

12 There is a similar disclaimer at *Tr.* 1.6.28 (*grandia si parvis adsimilare licet*), where Ovid compares his wife to Livia.

13 The combination *exemplis ... grandibus* frames the phrase *in parvo.*

14 On the irony see Green 1994, 209; Claassen 1996, 580; Amann 2006, 86–93; Öhrman 2014, 429.

15 For Aeneas as an exile see Bruwer 1974; Klodt 1996; Huskey 2002b, 90 n. 9; Gaertner 2007, 16. On mythical characters in general as exemplary exiles see Gaertner 2005 on Ov. *Pont.* 1.3.27–84.

16 Evans 1983, 37; Huskey 2002b, 96–98, 102.

17 See Luck 1977 on *Tr.* 1.3.21–22; Nagle 1980, 23–24; Posch 1983, 156; Claassen 1999, 174 f.; Tola 2004, 126–127. For exile as death see e. g. *Tr.* 1.2.65–66, 1.2.71–72, 1.3.21–24, 1.3.89–98, 1.4.28, 3.3, 5.9.19, *Pont.* 1.8.27, 1.9.17, 4.9.74, 4.16.51. Wistrand 1968, 6–26; Nagle 1980, 22–35; Doblhofer 1987, 166–178; Williams 1994, 8–25; Claassen 1996, 578–585; Claassen 1999, 239–241 with n. 37; Ingleheart 2015. See also Gaertner 2007, 159 with n. 24 for the wordplay *exilium-exitium* in Ennius.

18 Evans 1983, 37; Huskey 2002b, 104 n. 32.

Although scholars have worked intensively on identifying points of contact (and difference) between Ovid and Aeneas in *Tr.* 1.3, they have not sufficiently noted – or even, they completely missed – some other very interesting implications of this relationship. After all, Aeneas is the founder of the *gens Iulia*, the imperial family. Thus, on a first level, Ovid's association with him is a sign of respect towards the emperor. Or *is it?* Is there perhaps a sting in the tail?

Given that Ovid is to be identified with Aeneas on the night of Troy's fall, given that his home is to be identified with Troy, and his family with the Trojans, the ancestors of the Romans, then one is free to think that the one who banished Ovid and destroyed his home, i. e. Augustus, should automatically be identified with Aeneas' enemies, the Greek conquerors, the enemies of the ancestors of the Romans. This comparison is not at all flattering for Augustus; although he proudly draws his origin from the defender of Troy, Aeneas, he is now being put in the role of the enemies of the city. Greek and Roman mythology meet, and so does modern Rome (more precisely, Ovid's home) and ancient Troy, which is actually not very far from the place of Ovid's exile.

Now, since in *Tr.* 1.3 Ovid actually uses the *Aeneid* as literary background, one can look for more undercurrent links which – to the best of my knowledge – have not been picked up by previous scholars. Let us have a look at lines 85–86 (the words of Ovid's wife): *te iubet e patria discedere Caesaris ira, / me pietas: pietas haec mihi Caesar erit* ("Caesar's anger drives you to leave your country, my loyal love orders me. Loyalty will be my Caesar").

Ovid's wife contrasts her own *pietas* with Caesar's *ira*[19] and declares that her *pietas* commands her to follow her husband, whereas Caesar's *ira* forces her husband to leave his homeland. In the *Aeneid* Aeneas is persecuted by Juno's *ira*. Now, in the *Tristia*, Ovid is persecuted by Augustus' *ira*. Again, this is not flattering at all for the emperor. At best, this comparison is humorous; at worst, it is provocative and insulting. In this way Ovid entangles his wife into the game of comparisons and mythological associations and bestows upon her Aeneas' most emblematic quality, his *pietas*. Whereas Augustus, Aeneas' descendant, is possessed by *ira*, Ovid's wife displays *pietas*. Not only that, but she defiantly declares that for her the *pietas* towards her exiled husband plays the part of Caesar. This declaration acquires particular signifance, considering that Augustus had appropriated the *pietas* of his ancestor, Aeneas, and that the Senate had dedicat-

19 For wrath as a prominent feature of Augustus in Ovid's exile poetry (*ira Caesaris*) see Scott 1930; Drucker 1977, 82–171; Syme 1978, 223–225; and McGowan 2009, ch. 6. See also Claassen 1986, section 5.2.1 for a study of the use of the term *ira* and its occurrence in Ovid's exile poetry, and see Claassen 1999, 295 n. 83 for the image of Augustus – the angry god as creation of the poet.

ed in the Curia Iulia a golden shield inscribed with his attributes: *virtus*, *clementia*, *iustitia*, and *pietas* (*Res gestae* 34.2).

But, is Ovid's stance anti-Augustan?[20] Is this open or covert anti-Augustanism? Are his mythological *exempla* perhaps too bold? Before I deal with these questions, let me draw your attention to another *exemplum* in this poem.

The Mettus *Exemplum*

Tr. 1.3.75–76: *sic doluit Mettus tum cum in contraria versos / ultores habuit proditionis equos* ("So Mettus grieved when, punishing his treachery, the horses were driven in different directions").

Mettus Fufetius was an Alban general in the reign of Tullus Hostilius. He was summoned by the Romans to support them against the Fidenae and the Veientines, but he withdrew his troops in mid-battle – from cowardice or treachery – and awaited the outcome of the battle. He was severely punished by the victorious Romans. His punishment is recorded by Livy (1.28.10–11):

> *exinde duabus admotis quadrigis in currus earum distentum inligat Mettium, deinde in diversum iter equi concitati lacerum in utroque curru corpus, qua inhaeserant vinculis membra, portantes.* [11] *avertere omnes ab tanta foeditate spectaculi oculos. primum ultimumque illud supplicium apud Romanos exempli parum memoris legum humanarum fuit; in aliis gloriari licet nulli gentium mitiores placuisse poenas.*

> He then brought up two four-horse chariots, and caused Mettius to be stretched out and made fast to them, after which the horses were whipped up in opposite directions, and bore off in each of the cars fragments of the mangled body, where the limbs held to their fastenings. [11] All eyes were turned away from so dreadful a sight. Such was the first and last punishment among the Romans of a kind that disregards the laws of human-

20 For the light-hearted way in which Ovid often treats Augustus in his exile poetry see Evans 1983, 10–30; Williams 1994, 154–209; Claassen 1999, 219–228. For Ovid's ambivalent discourse about the emperor's *ira* and *clementia* see Bretzigheimer 1991; Tarrant 1995, 73; Gaertner 2005, 9–12; Galasso 2009, 202. Ovid's most defiant statement about Augustus is *Tr.* 3.7.47–52 on the emperor's inability to control the poet's thought, talent and reputation. See Evans 1983, 17–19 and 182 n. 20 for bibliography. For possible interpretations of the poet's bold irreverence towards the emperor see Claassen 2008, 38. More generally, Ovid's 'political' stance towards the Augustan regime is a hotly debated issue. See Kennedy 1992; Galinsky 1996, 261–268; Barchiesi 1997, 1–11; O'Gorman 1997. For bibliography on the subject see Nugent 1990, 241 nn. 7 and 8, 244 n. 12; Habinek 2002, 61.

ity. In other cases we may boast that with no nation have milder punishments found favour.[21]

Some scholars consider the couplet *Tr.* 1.3.75–76 to be spurious, an interpolation.[22] They focus on matters of textual transmission[23] and metre,[24] but mainly claim that Ovid has absolutely no reason to identify or at least compare himself with a traitor, like Mettus, since this would undermine his case. On the other hand, there are those who defend the authenticity of the couplet. Green rightly states that "Arguments for the deletion of this couplet as inappropriate (why introduce the Roman arch-traitor here?) and *therefore* spurious, lack cogency and seriously underrate Ovid's mordant irony".[25] I fully agree with Green; my goal is to contribute a few new arguments in favour of the authenticity of the couplet based on Ovid's use of myth. I also intend to show that this couplet is harmoniously integrated into the elegy and that it perfectly serves Ovid's train of thought and argumentation:

1) First of all, the argument that it makes no sense for Ovid to identify himself with a traitor is weak, because throughout his exilic poems Ovid frequently refers to his *error* and accepts his guilt and culpability.[26]

2) Ovid here does not focus on Mettus' betrayal, but on the fact that he was torn in two by horses running in opposite directions. Right in the previous couplet Ovid declares that he feels torn in two, since a part of him stays behind in Rome and another is forced to leave (*Tr.* 1.3.73 f.): *dividor haud aliter, quam si mea membra relinquam, / et pars abrumpi corpore visa suo est.* "I was torn, as though I had left my limbs behind, and half seemed severed from my body".

3) The Mettus *exemplum* picks up Ovid's association with Aeneas and the *Aeneid.* The harsh punishment of Mettus is depicted on Aeneas' new shield, the work of Vulcan, in Book 8 (8.642–645):

haud procul inde citae Mettum in diversa quadrigae
distulerant (at tu dictis, Albane, maneres!),

21 All Livy translations are by B.O. Foster, 1919. *Livy, History of Rome Vol. I. Books 1–2.* Cambridge, MA.

22 Heinsius deleted the couplet; both Luck 1977 and Hall 1995 agree with the deletion.

23 *Mettus* was the suggestion of Naugerius. The manuscripts read *Priamus.* Hall 1995, on *Tr.* 1.3.75–76 is certain: *nec dubito quin spurium sit hoc distichon: ... nec fuit ille [Ovidius] proditor.*

24 According to Posch 1983, 160 n. 377, the elision between *cum* and *in* (75) is not Ovidian.

25 Green 1994, on *Tr.* 1.3.75–76.

26 On Ovid's diction when referring to his *error* see McGowan 2009, 41–44. Claassen (1987, 32) points out that Ovid discusses his responsibility for this exile less and less as time passes and especially after the death of Augustus.

raptabatque viri mendacis viscera Tullus
per silvam, et sparsi rorabant sanguine vepres.

Close by, four-horse chariots had been driven hard in opposite directions and had torn Mettus in two – Alban, you should have stood by your promises – and Tullus was dragging the deceiver's body through a wood while a dew of blood dripped from the brambles.

Ovid retrospectively inscribes himself on Vergil's epic; Mettus' punishment, which is prophesied on Aeneas' shield, can now be seen as a prefiguration of Ovid's punishment several centuries later, a punishment enforced by Aeneas' own descendant, Augustus.

4) The emphasis of the Mettus *exemplum* is not on the crime he committed, but on the punishment he suffered. Livy (1.28.11, quoted above) points out that Mettus' punishment was extremely harsh and hideous,[27] and that this was the first and last punishment imposed by the Romans which defied the human laws. Henceforth the Romans could boast that their punishments were the mildest among all nations. One should not forget that this is the account and opinion of Livy, a personal friend of Augustus and the leading historian of his time. Since Ovid compares his punishment with Mettus' inhuman punishment, then automatically Augustus, who inflicted this punishment, is to be compared with the king Tullus Hostilius who inflicted Mettus' atrocious punishment. Once again this comparison does not flatter Augustus, who wanted to appear in the eyes of his people as an enlightened leader such as Numa Pompilius and not Tullus Hostilius.[28] Augustus is implicitly accused of transcending the limits of proper Roman behavior.

5) The manner of Mettus' death recalls the death of Sinis (or Pityocamptes 'the pine-bender'); he was the notorious villain who used to tie his victims to two pine trees which he bent down to the ground; he would then let the trees go, tearing his victims apart. Theseus killed him in exactly the same way.[29] But what does Theseus have to do with anything in this poem? Actually, a lot.

---

27 Ogilvie 1965, on Liv. 1.28.10: "The manner of Fufetius' death is unparalleled in Roman criminal history ... not so much for its brutality as for its singularity".

28 Cornell (1997, 119–120) notes that the names of Numa and Tullus Hostilius are historically correct, but otherwise these kings were "little more than contrasting stereotypes".

29 Ovid prefers not to use another well-known example of a man torn by his horses, Hippolytus. I believe this is a proof of his intention to cite a Roman example and associate it with Livy and Augustus. Zimmermann (2005, 220 n. 14) suggests that Ovid's choice of Mettus (a traitor) instead of Hippolytus betrays his subconscious guilt. He also points out at p. 215 n. 8 that Ovid's reference to his separation from his body at lines 73–74 (*dividor haud aliter, quam si mea membra relinquam, / et pars abrumpi corpore visa suo est*) recalls the violent death and dismemberment of Orpheus (Ov. *Met.* 11.50): *membra iacent diversa locis.*

Ovid has already used him as an *exemplum* at lines 1.3.65–66: *quosque ego dilexi fraterno more sodales, / o mihi Thesea pectora iuncta fide!* ("And the friends that I've loved like brothers, O hearts joined to me by Thesean loyalty!") The phrase *Thesea ... fides* picks up the famous male comradeship between Theseus and Peirithous[30] and portrays Theseus as a paragon of fidelity. Ovid's friends still remain loyal to him, despite the hardships he is going through. Since exile equals death, then Ovid can be directly compared to Theseus and Peirithous who descended to the Underworld.

Hence, the Mettus *exemplum* is absolutely pertinent to Ovid's strategy and not at all haphazard; it contributes considerably to the poem's internal unity and cohesion. Lenz[31] had stated that if lines 75f. were to be excised, nothing would be lost. I hope that the discussion above has proven that this is certainly *not* the case.

To sum up:

Ovid's utilisation of myth in his exilic poetry is multifaceted. Through myth Ovid links his elegies to master narratives and other literary texts; myth also helps Ovid form his *persona* in exile, while it even conceals his criticism of Augustus. *Tr.* 1.3 is an excellent case study, because the mythological *exempla* found therein cover a vast span, both spatial and temporal (Greece, Troy, Rome, Theseus, Trojan war, Roman history, Ovid's present). Different periods and places are brought together to provide proper mythological background for Ovid's life and exile. The mythological *exempla* that Ovid uses are in close association to one another and hold the poem together. Various mythological strings run parallel or converge in *Tr.* 1.3 (Aeneas, Theseus and Mettus, to name but a few).

Ovid's real-life memory is joined with his literary memory. He remembers the events of his own life and relates them to literary events. Ovid mainly has in mind Book 2 of the *Aeneid*. By associating himself with Vergil's Aeneas, Augustus is automatically identified with the enemy. The result is extremely ironic at Augustus' expense and it sets the tone for Ovid's exilic poetry. Ovid uses Vergil, the most Augustan poet of all, in order to comfront Augustus. And the means to do this is myth.

30 See e.g. Ov. *Tr.* 1.3.66, 1.5.19, 1.9.31–32, 5.4.26, *Pont.* 2.6.26. Their friendship was proverbial. See Otto 1968 s.v. *Theseus* 347 §1779. Peirithous wanted to abduct Persephone from the Underworld, and Theseus offered to help (Verg. *Aen.* 6.392–397, Apollod. 2.124, *Epit.* 1.23–4). Their mission failed and they were stuck to a rock. Hercules managed to save only Theseus (Apollod. 2.124, *Epit.* 1.23–24, Hyg. *Fab.* 251, 257.1, Hor. *Carm.* 4.7).

31 Lenz 1962, 115.

By projecting himself on renowned mythical characters Ovid adapts well-known myths to his own situation. He communicates with his audience in the common language of myth, using codes most readers are able to understand. He tries to make his personal story common to all by investing it with well-known literary and mythological examples. In this way his life takes on greater dimensions and Ovid becomes a hero of sorts, comparable to great heroes of myth.

Bibliography

Amann, M. 2006. *Komik in den Tristien Ovids.* Basel.

Angulo, E.B. 2008. Un modelo de literatura de amor conyugal: Ovidii exulis Corpus amatorium. *Euphrosyne* 36: 135–148.

Barchiesi, A. 1997. *The Poet and The Prince: Ovid and Augustan Discourse.* Berkeley.

Besslich, S. 1972. Ovids Winter in Tomis. *Gymnasium* 79: 177–191.

Bretzigheimer, G. 1991. Exul ludens: zur Rolle von relegans und relegatus in Ovids Tristien. *Gymnasium* 98: 39–76.

Bruwer, S.M. 1974. *The Theme of Exile in Virgil's Aeneid* (Ph.D. Diss.). Stellenbosch.

Claassen, J.-M. 1986. *Poeta, exsul, vates: A Stylistic and Literary Analysis of Ovid's Tristia and Epistulae ex Ponto* (Ph.D. Diss.). Stellenbosch.

Claassen, J.-M. 1987. Error and the Imperial Household: An Angry God and the Exiled Ovid's Fate. *AClass* 30: 31–47.

Claassen, J-M. 1988. Ovid's Poems from Exile: The Creation of a Myth and the Triumph of Poetry. *A&A* 34: 158–169.

Claassen, J.-M. 1996. Exile, Death and Immortality: Voices from the Grave. *Latomus* 55: 571–590.

Claassen, J.-M. 1999. *Displaced Persons: The Literature of Exile from Cicero to Boethius.* London.

Claassen, J.-M. 2001. The Singular Myth: Ovid's Use of Myth in his Exilic Poetry. *Hermathena* 170: 11–64.

Claassen, J.-M. 2008. *Ovid Revisited. The Poet in Exile.* London.

Claassen, J.-M. 2009. *Tristia.* In *A Companion to Ovid*, ed. P.E. Knox, 170–183. Chichester.

Cornell, T.J. 1997. *The Beginnings of Rome: Italy and Rome from the Bronze Age to the Punic Wars (c. 1000–264 BC).* London.

Degl'Innocenti Pierini, R. 2007. Quantum mutatus ab illo.... *Dictynna* 4 (http://dictynna.revues.org/153).

Doblhofer, E. 1980. Ovids Abschied von Rom – Versuch einer Modellinterpretation von *Trist.* 1,3. *AU* 23.1: 81–97.

Doblhofer, E. 1987. *Exil und Emigration: zum Erlebnis der Heimatferne in der Römischen Literatur.* Darmstadt.

Drucker, M. 1977. *Der verbannte Dichter und der Kaiser-Gott: Studien zu Ovids späten Elegien* (Ph.D. Diss.). Heidelberg.

Edwards, C. 1996. *Writing Rome: Textual Approaches to the City.* Cambridge.

Evans, H.B. 1983. *Publica Carmina: Ovid's Books from Exile.* Lincoln and London.

Froesch, H. 1976. *Ovid als Dichter des Exils.* Bonn.
Gaertner, J.F. (ed.) 2005. *Ovid Epistulae ex Ponto, Book 1. Edited with Introduction, Translation, and Commentary.* Oxford.
Gaertner, J.F. (ed.) 2007. *Writing exile. The Discourse of Displacement in Greco-Roman Antiquity and Beyond.* Leiden and Boston.
Galasso, L. 2009. *Epistulae ex Ponto.* In *A Companion to Ovid*, ed. P.E. Knox, 194–206. Chichester.
Galinsky, G.K. 1996. *Augustan Culture: An Interpretive Introduction.* Princeton.
Green, P. 1994. *The Poems of Exile. Tristia and the Black Sea Letters, With a New Foreword.* Berkeley.
Habinek, T. 2002. Ovid and Empire. In *The Cambridge Companion to Ovid*, ed. P. Hardie, 46–61. Cambridge.
Hall, J.B. (ed.) 1995. *P. Ovidi Nasonis Tristia.* Stuttgart and Leipzig.
Hardie, P.R. 2002. *Ovid's Poetics of Illusion.* Cambridge.
Helzle, M. 1989. Mr and Mrs Ovid, *G&R* 36: 183–193.
Herrmann, K. 1924. *De Ovidii Tristium libris V* (Ph.D. Diss.). Leipzig.
Hinds, S. 1985. Booking the Return Trip: Ovid and *Tristia* 1. *PCPhS* 11: 13–32.
Hinds, S. 1999. First Among Women: Ovid, *Tristia* 1.6 and the Traditions of "Exemplary" Catalogue. In *Amor: Roma. Love and Latin Literature*, eds. S.M. Braund and R. Mayer, 123–142. Cambridge.
Holzberg, N. 2002. *Ovid: The Poet and His Work.* Ithaca.
Huskey, S.J. 2002a. *Ovid's Tristia I and III: An Intertextual Katabasis* (Diss.). Iowa.
Huskey, S.J. 2002b. Ovid and the Fall of Troy in *Tristia* 1.3. *Vergilius* 48: 88–104.
Ingleheart, J. (ed.) 2010. *A Commentary on Ovid, Tristia, Book 2.* Oxford.
Ingleheart, J. 2015. *Exegi monumentum:* Exile, Death, Immortality and Monumentality in Ovid, *Tristia* 3.3. *CQ* 65: 286–300.
Kennedy, D. 1992. 'Augustan' and 'Anti-Augustan': Reflections on Terms of Reference in Roman Poetry and Propaganda in the Age of Augustus. In *Roman Poetry and Propaganda in the Age of Augustus*, ed. A. Powell, 26–58. London.
Kenney, E.J. 1965. The Poetry of Ovid's Exile. *PCPhS* 191: 37–49.
Klodt, C. 1996. Verkehrte Welt: Ovid, *Trist.* 1.4. *Philologus* 140: 257–276.
Lenz, F. 1962. Ovid *Tristia* 1, 3, 75 f. *Maia* 14: 109–116.
Luck, G. 1977. *P. Ovidius Naso Tristia.* Heidelberg.
McGowan, M.M. 2009. *Ovid in Exile: Power and Poetic Redress in the Tristia and Epistulae ex Ponto.* Leiden.
Nagle, B.R. 1980. *The Poetics of Exile. Program and Polemic in the Tristia and Epistulae ex Ponto of Ovid.* Brussels.
Nugent, S.G. 1990. *Tristia* 2: Ovid and Augustus. In *Between Republic and Empire: Interpretations of Augustus and His Principate*, eds. K.A. Raaflaub and M. Toher, 239–257. Berkeley.
Ogilvie, R.M. 1965. *A Commentary on Livy Books 1–5.* Oxford.
O'Gorman, E. 1997. Love and the Family: Augustus and Ovidian Elegy. *Arethusa* 30: 103–123.
Öhrman, M. 2008. *Varying Virtue. Mythological Paragons of Wifely Virtues in Roman Elegy.* Lund.

Öhrman, M. 2014. Fake Farewells: The Elegiac Cast of Ov. *Trist.* 1.3. In *Latinet i tiden. Festskrift till Prof. Hans Aili*, eds. E. Andersson, E. Kihlman and M. Plaza, 427–437. Stockholm.
Otto, A. 1968. *Sprichwörter und sprichwörtlichen Redensarten der Römer.* Nachträge. Eingeleitet und mit einem Register hrsg. von Reinhard Häussler. Hildesheim.
Posch, S. 1983. *Tristia I: Interpretationen.* Innsbruck.
Rahn, H. 1958. Ovids elegische Epistel. *A&A* 7: 105–120.
Rosati, G. 1999. L'addio dell'esule morituro (*Trist.* 1.3): Ovidio come Protesilao. In *Ovid. Werk und Wirkung. Festgabe für Michael von Albrecht zum 65. Geburtstag*, vol. 2, ed. W. Schubert, 787–796. Frankfurt am Main.
Scott, K. 1930. Emperor Worship in Ovid. *TAPA* 61: 43–69.
Syme, R. 1978. *History in Ovid.* Oxford.
Tarrant, R.J. 1995. Ovid and the Failure of Rhetoric. In *Ethics and Rhetoric. Classical Essays Presented to Donald Russell on his Seventy-Fifth Birthday*, eds. D. Innes, H. Hine and C. Pelling, 63–74. Oxford.
Tissol, G. 2014. *Ovid: Epistulae ex Ponto Book I.* Cambridge.
Tola, E. 2004. *La métamorphose poétique chez Ovide: Tristes et Pontiques. Le poème inépuisable.* London.
Tola, E. 2008. Chronological Segmentation in Ovid's *Tristia:* The Implicit Narrative of Elegy. In *Latin Elegy and Narratology. Fragments of a Story*, eds. G. Liveley and P.B. Salzman-Mitchell, 51–67. Columbus.
Videau-Delibes, A. 1991. *Les Tristes d'Ovide et l'élégie Romaine. Une poétique de la rupture.* Paris.
Williams, G.D. 1994. *Banished Voices: Readings in Ovid's Exile Poetry.* Cambridge.
Williams, G.D. 2002a. Ovid's Exilic Poetry: Worlds Apart. In *Brill's Companion to Ovid*, ed. B.W. Boyd, 337–381. Leiden.
Williams, G.D. 2002b. Ovid's Exile Poetry: *Tristia*, *Epistulae ex Ponto*, and *Ibis*. In *The Cambridge Companion to Ovid*, ed. P.R. Hardie, 233–245. Cambridge.
Wistrand, E. 1968. *Sallust on Judicial Murders in Rome. A Philological and Historical Study.* Gothenburg.
Zimmermann, B. 2005. Ovids Abschied von Rom. Zur Struktur des I. Buchs der *Tristien.* *Eikasmos* 16: 211–221.

Part IV **Shifting Chronotopes**

James Andrews

Kairos: The Appropriate Time, Place and Degree in Protagoras' Myth of Origins

ἦν γάρ ποτε χρόνος ὅτε θεοὶ μὲν ἦσαν, θνητὰ δὲ γένη οὐκ ἦν ...

The myth that the Platonic Protagoras says will explain Athenian beliefs about political virtue begins with gods and time. Clearly, this will be a creation story. Other such accounts were prepared to dispense with the gods of myth and allow time itself to serve in some sense as the "father of all".[1] More conventional, Protagoras will make the gods our creators. But while *chronos* remains throughout this myth nothing more than a receptacle of action, Protagoras will elevate the element of *timing* in the actions of the divine agents. It is this aspect of time, not its slow, creative, evolutionary passage but its incitement to action, that the present study will emphasize.

Time that incites agents to action is *kairos*. But there is more to *kairos* than time. Thus, even though the meaning most commonly assigned to it is, as LSJ puts it, "*exact* or *critical time, season, opportunity*", we must resist limiting its meaning to the temporal dimension. Rather, we must understand *kairos*, with Malcolm Heath, as "the time *or place* at which, *or degree* in which, something is appropriate" (italics mine).[2] All four terms of this definition – time, place, degree, and appropriateness – are essential to *kairos*. Moreover, the standard by which we know that a time, place, or degree is a *kairos* may at one time be its "appropriateness" (τὸ πρέπον), but at other times its "necessity" (τὸ δέον), and other times still its "proper measure" (τὸ μέτριον). If we focus on *kairos*

1 Pind. *Ol.* 2.17: χρόνος ὁ πάντων πατήρ. Commenting on Moschion 6.18 Nauck/Snell (ὁ τίκτων πάντα καὶ τρέφων χρόνος), Xanthakis-Karamanos (1981, 412–413) writes: "already in archaic literary sources, in Orphic cosmogonies and Pherecydes of Syros ... time is declared to be the first principle of the creation of the Universe; and in the fifth century it is widely regarded as a prime creative power". Xanthakis-Karamanos goes on to cite (in addition to the Pindar and Moschion passages) Bacch. 7.1; Soph. *OC* 618; Eur. *Supp.* 787–788, *Heracl.* 900. We may add DK 88B18 (Crit. *Sisyphus*). As for the *Ol.* 2 passage, Pindar is speaking of the inability even of time, which brings everything into being, to *undo* the results of past actions. For narrative χρόνος (ἦν χρόνος, ὅτ', etc.), see below, n. 17.

2 Heath 1989, 30. See Pfister 1938, 137: the word "den begrenzten, bestimmten Abschnitt, die bestimmte Stelle, daher auch die richtige Stelle im Raum oder in der Zeit bedeutet, und da das Wesentliche an einem bestimmten Abschnitt seine feste Begrenzung und Ausdehnung ist, so bedeutet καιρός auch diese Begrenzung, das bestimmte richtige Maß". Regarding etymology see Pfister 1938, 137–138.

DOI 10.1515/9783110535150-017

in this broad sense – without regard for the term itself, which appears nowhere in Protagoras' myth – we will find that the myth is a dramatic account of mastery in the management of place and degree as well as time.[3]

The *Protagoras* begins with Socrates agreeing to introduce Hippocrates to Protagoras and to inquire on Hippocrates' behalf what the young man may expect if he becomes the student of the sophist. After a good deal of questioning by Socrates, Protagoras finally declares that Hippocrates will gain mastery in *euboulia*, sound decision-making. And when Socrates asks whether he has correctly understood this as a claim involving *arete* and *techne*, Protagoras agrees that it is so: Hippocrates may expect to achieve technical mastery in political excellence. But as soon as Protagoras has agreed that his profession of *euboulia* is tantamount to the claim that he teaches entechnic *arete*, Socrates objects: as he sees it, arete simply cannot be acquired through instruction. He gives three reasons for holding this view, all of them based on his observation of his fellow Athenians. Protagoras lets Socrates raise his objections, and then undertakes, in his Great Speech, to refute all three of them, each one after the other. However, since his myth, which is our present concern, is designed to address only the first of Socrates' objections, we need mention only this one.

Socrates observes that, when the Athenians meet in their citizen assembly, they readily call upon experts for their expert advice on such matters as construction projects and ship building. If on those occasions anyone other than an expert attempts to give advice, they flatly refuse to give him a hearing. However,

> when some matter of state policy comes up for consideration, anyone can get up and give his opinion, be he carpenter, smith or cobbler, merchant or ship-owner, rich or poor, noble or low-born, and no one objects to them as they did to those I mentioned just now, that they are trying to give advice about something which they never learnt, nor ever had any instruction in. So it's clear that they don't regard that as something that can be taught.[4]

And since the Athenians are universally recognized for their great wisdom, Socrates takes it as compelling proof that *arete* is not teachable (319b3 – d7).

The essence of Protagoras' response is that Socrates has falsely assumed that expertise is in all cases the possession of a select few. In fact, when it

---

3 A good place to begin examination of kairos is the diachronic inventory and analysis in Wilson 1980 and 1981. Other studies: Race 1981; Pfister 1938; Levi 1924; Rostagni 1922; Pohlenz 1933; Craik 1998, 209 – 213; Steidle 1952, 270 – 274 et passim; Guillamaud 1988; Tordesillas 1992 (esp. 82 – 84); Wareh 2012, 13 – 75 passim; Poulakos 1983, 59 – 62; and the various contributions and bibliography in Sipora and Baumlin 2002.

4 Unless otherwise noted, all translations are taken from Taylor 1991.

comes to matters of state policy, expertise is shared by all citizens. That the Athenians invite any and all citizens to advise on such matters shows that they correctly understand, as Socrates does not, that all citizens share in *politike arete*, and that this *arete* is a *techne*. Of course, the paradox of a *techne* in which no one is not an expert obviously demands explanation, and Protagoras is ready to provide one.

The explanation comes in the form of a myth, and begins with the emergence from earth and fire of the various creatures that were to inhabit this world.[5] Such was the physiological nature of all the other species that their survival was assured. But the human species that was to emerge into the light of day lacked the necessary physiological attributes, and would have perished had it not been for the intervention of Prometheus, who endowed us with a native talent for inventing life-supporting crafts: agriculture, the manufacture of clothing and homes, and so on. Yet even then our survival was not assured, for our species still lacked the skill of living together in the polis, without which we were incapable of such collective action as would keep us safe from wild beasts.[6] Now it was Zeus who intervened, bringing into the world the two attributes necessary to maintain life in a polis-community: *aidos* and *dike*, respect for the feelings and the rights of others.[7] And whereas the *techne* that Prometheus had given resulted in a division of labor – some people farming, others manufacturing shoes, clothing, or homes, and so on – Zeus saw to it that this one *techne*, the art of living in the polis, would be distributed to *all* human beings. Moreover, he laid down a law, according to which "he who cannot share in respect and justice is to be killed as plague of the city" (322d).[8] So Protagoras' answer to Socra-

5 Manuwald 2003, 40 and 2013, 167 sees "good reasons for ascribing the basic ideas of the myth to the historical Protagoras", and many scholars are similarly inclined: Guthrie 1971, 63 – 64; Kerferd 1981, 125; Balaban 1999, 151; Zilioli 2007, 98 as well as Beresford 2013. According to the latter (139), the myth conveys "a rationalist and naturalist account of the origin of animals and the early development of human beings" (regarding the problem of the "divine apparatus" in Protagoras' myth, see below, n. 8). Many have linked this theory of origins specifically to the *Περὶ τῆς ἐν ἀρχῇ καταστάσεως* ascribed to the historical Protagoras (D-K 80 A1). But Dodds 1973, 9 n. 4 (following Diels' note at D-K 80B8b) reasonably surmises that "this title has been merely inferred from Plato's myth".

6 But see Coby 1987, 56 on the first emergence of communal life.

7 Dodds 1973, 10. Zilioli (2007, 100 – 102) reconciles this universal dispensation with "the doctrine of ethical relativism endorsed by the sophist in the *Theaetetus*, on the basis of which each community decides on which ethical values to ground the peaceful cohabitation of its members".

8 According to Beresford (2013, 143), "Protagoras is setting out, behind a veil of myth, a Presocratic theory about the non-divine, natural origins of life, humanity and morality". His divine

tes' question how *politike arete* could possibly be a *techne* shared by all is that Zeus decreed that *everyone* must be an expert in this field. Protagoras will go on to argue that as a result of this divine decree, all citizens devote themselves to the study of "justice and the rest of the excellence of a citizen" (323a6–7, b2), ever and always teaching and learning from one another. For only if everyone shares in dike and *aidos* can cities exist.[9]

Protagoras has used a myth to prove that all citizens share in the practice of entechnic *arete*. Why a myth? He suggested at the outset that he was equally prepared to give a rational explanation of his view that citizen excellence is a learned *techne* and indeed invited his audience to choose whichever explanation, the mythic or the analytic, they preferred. And that Protagoras did not proceed analytically is due first to Socrates, who remained silent when asked for his preference, and to the rest, who wanted Protagoras himself to choose.[10] Protago-

apparatus is intended "to pre-empt the charge of atheism". It is there, says Beresford, "to conceal, or at least soften, the godlessness of the underlying theory". Havelock 1957, 87 and 91–94, who likewise is interested in the myth as testimony to an authentically Protagorean rationalist account of the origins of life and human society, sees the addition of the gods as simply Plato's way of "keep(ing scientific anthropology) under the control of metaphysics". Morgan (2000, 136 n. 4) objects to Havelock's failure to appreciate that "the codification of societal assumptions in mythological form is a common sophistic practice". According to Coby (1987, 56) "the very deities responsible for human inventiveness are themselves products thereof ... Thus, Protagoras already intimates the extent of his impiety: Zeus and Prometheus are human inventions". Less strained is Manuwald (1999, 176): "Protagoras will sicher mehr eine Aussage über die Sonderstellung des Menschen im Vergleich zu anderen Lebewesen machen als eine Aussage über die Götter oder die Realgrundlage des (als empirisches Phänomen natürlich nicht zu leugnenden) Kultes" (see Manuwald 1999, 192). There are still simpler ways of dealing with the divine apparatus, e. g. Kerferd (1981, 168): "the fact that it is a myth deprives it of any possible conflict with Protagoras' agnosticism" and Brisson (1975, 9 n. 3): "Protagoras utilise ... la mythologie populaire sans forcément en soutenir la validité". See Manuwald 2013, 167. Then too (though some scholars resist this) there is the possibility that the gods are simply an effective element of the mythic "charm offensive" (χαριέστερον εἶναι μῦθον ὑμῖν λέγειν). However, from the perspective of the present paper, presenting a mythic account of entechnic *arete* is occasioned by Protagoras' desire to dramatize decision-making in the face of kairos at the dawn of time, and it is this that requires divine agents.

9 Manuwald (2013, 164–165) notes the frequent recurrence of this thesis in the subsequent sections of the speech (323a2–3; 323b7–c2; 323c4; 324d7–e2; 326e7–327a2) and suggests that this is "the anthropological-political equivalent of [Protagoras'] epistemological principles". See Manuwald 1999, 175.

10 Morgan (2000, 138): "By posing the question of audience preference as he does [stressing the aptness of mythic explanation for this younger audience and for his elderly authority], Protagoras manipulates his listeners into allowing him his choice of approach, and makes it seem as though the two approaches are equivalent and easily distinguishable. It will later become evident that the choice of myth is indispensable" (see Lampert 2010, 50). One may go further: inras' response: "Well, I think that it will be more enjoyable to tell you a story" (320c6–7).

In availing himself of a mythic explanation, Protagoras says that he will address Socrates and the rest as an older man speaking to his juniors" (320c2: ὡς πρεσβύτερος νεωτέροις). We may interpret this remark in two ways. On the one hand, Protagoras is, as he said a bit earlier, "a good age now, ... old enough to be the father of any of you" (317c), so that he truly is "an older man speaking to his juniors", Socrates included; and in proceeding by way of myth, he is speaking paternalistically to grown men. Given the patronizing and reprimanding tone with which he sometimes addresses Socrates,[11] it is indeed reasonable to interpret his choice of myth as a form of condescension. But some among the assemblage – most notably, the novice Hippocrates – really are *quite* young, not far removed from those school years in which myth normally plays a large role in a child's education. This is a point Protagoras himself will make later in his speech (325e2–326a4):

> when (children) have learned their letters ... , (their teachers) set before them at their desks the works of good poets to read, and make them learn them by heart; they contain a lot of exhortation, and many passages praising and eulogizing good men of the past, so that the child will be fired with enthusiasm to imitate them, and filled with the desire to become a man like that.

Of course, it is one thing to read Homer's hallowed stories of Achilles and Diomedes as a child and another to hear Protagoras telling his own peculiar version of the myth about Zeus and Prometheus, and indeed telling it to young men who, like Hippocrates, are eager to proceed to a loftier and more sophistic form of education. Nonetheless, it is worth our while to consider whether Protagoras' myth, in addition to laying to rest Socrates' doubts about entechnic political virtue, aims also to sow in the young Hippocrates a desire to imitate the actors of the story.[12] To see how this might be so, let us examine the myth in greater detail.

---

tending in any event to deliver both a *mythos* and a *logos*, Protagoras will be able to claim after the fact that his speech has satisfied his audience regardless of their preference for mythos or logos.

11 He speaks to Socrates as one who is helpless and confused (324d2, 324e1–2, 326e3) and as failing to realize what a spoiled existence he is living (327e1: τρυφᾷς).

12 Morgan (2000, 133) stresses the sophistic use of myth as protreptic pedagogy, noting how Protagoras' myth and Hippias' *Trojan Dialogue* (Pl. *Hp. Mai.* 286a–b) share this aim: "Both sophists convey conventional wisdom and the means to take advantage of it. In the *Protagoras* this knowledge is characterised as 'Promethean' forethought". See also Manuwald 1999, 171 ("in gewisser Weise kann der Protagorasmythos auch [wie Prodikos' Geschichte von Herakles am Schei-

Once upon a time the gods shaped living creatures from earth and fire and the various compounds of these two elements. However, it was left to Prometheus and Epimetheus to apply the finishing touch to this creation. They were given the task of outfitting each species with its own distinctive power or strength (320d5–6: κοσμῆσαί τε καὶ νεῖμαι δυνάμεις ἑκάστοις ὡς πρέπει).[13] We may assume that the foresightful Prometheus was to distribute the attributes and powers, and Epimetheus "the examiner" would inspect the distribution once it was completed.

As they prepared to execute their assignment, Epimetheus persuaded Prometheus to exchange roles with him: it was Epimetheus who, by distributing the various means of survival, would exercise foresight on behalf of the creatures, and Prometheus who would inspect the results, ἐπισκέψασθαι.[14] At first, Epimetheus acquitted himself admirably well: as the prototype of each species was physically different, each assignment Epimetheus approached was a new situation, calling for a new decision. He gave to some strength or size or the weapons of claws and horns, while those denied strength were given speed. Those denied size or weapons were given winged flight or subterranean refuge. And so Epimetheus equipped each species with the means of survival *appropriate* to its needs (320d5–6: ἑκάστοις ὡς πρέπει), "mak[ing] sure that no species should be wiped out" (321a1). Survival of the threat of mutual wholesale destruction was thus assured. But the preservation of the species required another distribution as well, one designed to guard them against the seasons and environment (321a3–4: αἱ ὧραι). And so Epimetheus provided those exposed to the cold with thick hair, and those exposed to the heat tough skin. Some were shod in hooves, others in thick, bloodless skin. As regards sustenance, he suited some species to the pasturelands and others to the fruits of woodlands, while to still others he gave roots for sustenance and to others still, the power to prey on other species. Predators, however, "he made less prolific, [whereas] to those on whom they preyed he gave a large increase, as a means of preserving the species" (321b5–6). A powerful species thus might prey on a weaker, but again, the species were preserved from widespread destruction (321a3: ἀλληλοφθορίαι).

dewege] als ein Protreptikos zum Gut-Sein verstanden werden"); Knudsen 2012; Manuwald 2013, 167.

13 Not the survival of any one creature, of course, since every living creature, then and now, was mortal (θνητόν), but rather the survival of each species as a whole. See Guthrie (1957, 86): "nature's devices for preservation seem to operate only at species level and to ignore the individual".

14 320d7. Cf. 321c3–4 (Προμηθεὺς ἐπισκεψόμενος) and Adam 1893, 109. See Coby (1987, 53): "Prometheus, in a lapse of foresight, agreed to his brother's proposal".

It is worth noting that, just as Epimetheus neatly balances powers "one against another" (321a1: ἐπανισῶν), so does Protagoras with the phrases and clauses used to narrate the distribution:

νέμων δὲ
τοῖς μὲν ἰσχὺν ἄνευ τάχους προσῆπτεν
τοὺς δ' ἀσθενεστέρους τάχει ἐκόσμει
τοὺς δὲ ὥπλιζε
τοῖς δ' ... διδοὺς
ἃ μὲν ... ἤμπισχεν
ἃ δὲ ηὖξε
ταῦτα δὲ ἐμηχανᾶτο
ἐπειδὴ δὲ ... ἐπήρκεσε
πρὸς τὰς ἐκ Διὸς ὥρας ... εὐμάρειαν ἐμηχανᾶτο
ἀμφιεννὺς αὐτὰ
ἱκανοῖς μὲν ...
δυνατοῖς δὲ ...
καὶ εἰς εὐνὰς ἰοῦσιν ὅπως ὑπάρχοι ... στρωμνὴ
καὶ ὑποδῶν
τὰ μὲν ...
τὰ δὲ ...
τοὐντεῦθεν ... ἐξεπόριζεν
τοῖς μὲν ...
ἄλλοις δὲ ...
τοῖς δὲ ...
ἔστι δ' οἷς ἔδωκεν
καὶ τοῖς μὲν ... προσῆψε
τοῖς δ' ... πορίζων.[15]

Each successive allocation parades the details in a long sequence of μέν ... δέ ... constructions before finally concluding, in the same way he distributed all the other things, balancing one against another (321a1). The whole is composed, as H. Hansen observes, in a "naïve style",[16] one well-suited to a story that begins "once upon a time." But while we may agree with scholars that Protagoras' myth

15 Metrical difficulties arise only in the first foot of the first line and the missing second-foot diaereses. Each line must be assumed to be catalectic. See further n. 16.
16 Hansen, "Greek 701: Greek Rhetoric and Prose Style" (http://greek701.ws.gc.cuny.edu/essay-on-style/#plato1). To be sure, the passage is not without its sophistication: with Hansen, we may note chiasmus, variatio, and anaphora. Moreover, we may say of Protagoras' opening isocolon that it is a striking instance of what Halliwell (2012, 269) identifies in his discussion of the *Helen* as the Gorgianic "near-metrical element of rhythmic patterning". Still, the overall impression remains one of simplicity. For a discussion of the "strung-together" style of the myth (λέξις εἰρομένη), see Norden 1913, 368–70.

is specifically a fable,[17] and that the style is appropriate to fable, the prolonged μέν ... δέ ... itemization of the allocations is at the same time a conspicuous retardation of narrative time to story time.[18] Moreover, the quick and decisive intervention that immediately follows – Prometheus' theft – brilliantly highlights just how leisured and dilatory this preceding narrative is.

It was certainly available to Protagoras simply to say that Epimetheus failed to provide human kind with a proper power of survival, and thereby failed at his *overall* task of "equip[ping] *each* kind with the powers it required" (320d5–6, italics added). But in fact Epimetheus never really embraced this overall assignment. This, says O. Balaban, is because Epimetheus "indulge(s) in the activity for its own sake, not for the sake of something else [and] want(s) to extend [time] as much as possible ...".[19] Indeed, we may say that, whereas Prometheus and Zeus are concerned with *kairos*, Epimetheus is a heedless consumer of *chronos*, the flow of time, lacking a final critical moment. But that is not all.

According to A. Beresford, Epimetheus' distribution of the powers of survival is an allegory corresponding to what was in the sophist's day the "standard naturalist theory of origins". According to this theory,[20] "mindless and purely natu-

17 For the "Märchenstil" signaled by the opening words (ἦν γάρ ποτε χρόνος ὅτε ...), see Nestle 1931, 92, who offers parallels; cf. Denyer 2008, 100. Kurke (2011, 284) remarks: "Protagoras' μῦθος must be understood as 'fable' not 'myth' (as it is still conventionally read by many scholars of ancient philosophy) ...". See Manuwald 1999, 171 (to be added to Kurke's excellent bibliographical note 62), for whom Protagoras' representation of the μῦθος as an alternative to λόγος suggests that this myth is quite distinct from the Socratic use of myth elsewhere in Plato. Manuwald concludes: "die mythische Form bereits dem historischen Protagoras gehört". See also Manuwald 2003 (an investigation of the relevance of this *sophistic* myth for Platonic myth-making, and in particular for the Platonic Jenseits-Mythen).

18 When Socrates later complains that Protagoras makes his answers too long, Protagoras responds that it is for him to judge what length of speech is called for from one time to the next (334d9: ὅσα δεῖ). For further comment by Socrates on Protagoras' mastery in macrological and brachylogical discourse, see 329b1–5. According to Philostratus (DK80 A2), Plato uses the μύθῳ μακρῷ to highlight Protagoras' tendency to speak μακρολογώτερον τοῦ συμμέτρου. I must defer discussion of Protagorean macrologia to another occasion.

19 Balaban 1999, 161–162. Balaban (1999, 164) reads the myth in terms of "the three basic values that constitute the motivation for human activity", assigning pleasure (consumption) to Epimetheus, utility (efficiency, production) to Prometheus, and morality ("positive repression") to Zeus.

20 Although "various later sources give us a reliable picture of rationalist, Presocratic theories about the origin of life and of human society" (Beresford 2013, 140), pinning down the particulars of theory and author is quite difficult. Nonetheless, Beresford (2013, 140 n. 6) agrees with Guthrie that the late sources are drawing chiefly on Democritus. See Cole 1990, 51 who stresses that even "the most naturalistic possible interpretation of the contents of the myth [fall far short

ral forces" gave rise over time to innumerable life-forms, most of which were not viable. Beresford continues:

> Those that did survive were lucky winners of nature's lottery: amid the myriad failed experiments, a few animals happened to emerge with structures and features that enabled them to persist... In the allegorical version of the same theory Protagoras needs to find a god who can represent this absence of divine providence and thoughtful design. But how could any god stand for the absence of gods? He chooses Epimetheus, whose name expressly signifies Lack of forethought: the careless and thoughtless god who is "not intelligent at all" (321b7: οὐ πάνυ τι σοφὸς ὤν) and who never notices a problem until it is too late ...[21]

Two observations are in order. First, if it is true that this standard naturalist theory of origins underlies Protagoras' account, then the protraction and repetitiveness of the narrative of Epimetheus' distribution invites us to imagine the long evolutionary process that this theory presupposes.[22] Second, Beresford is certainly correct that Epimetheus does indeed, in the last analysis, grossly bungle: the task was "to equip each kind with the powers it required", and yet there in the end is man standing helpless in the face of the elements and without defense against his predators. But one may object: the blunder comes only at the end of a series of viable "adaptations" – the assignment to one species after another of a suitable power of survival. In achieving at least this much, Epimetheus laid the groundwork for his brother's completion of the task. Thus, in the same way that we speak of "trial and error" in the long evolutionary processes of the "standard naturalist theory of origins", so too Epimetheus' momentary failure illustrates the "trial-and-error" ultimately leading to a "successful" evolutionary outcome.

---

of] the more careful and detailed naturalistic reconstruction of [technological and social] history" which his study associates with Democritus. See n. 31 below.

21 Beresford 2013, 144–145.

22 What makes zoogony so slow and lengthy is its "trial-and-error" process. See Beresford (2013, 145 n. 15): "Darwin proposes that biological evolution proceeds mindlessly, thoughtlessly, stumbling upon good 'design' (i.e., successful adaptations) by making blind, unguided modifications and suffering the consequences, usually bad, occasionally good. This basic idea, that some sufficiently large number of blind trials will inevitably generate at least some lucky successes – eliminating the need for a conscious designer – is thus common to both ancient and modem biological naturalism, regardless of the considerable differences in the mechanisms of generation that they propose, and it is a central philosophical insight of both".

Let us pause to admire, *in terms of kairos*, what the dilatory, "chronic" Epimetheus did get right,[23] before the consequences of his carelessness became manifest. His "compensatory distribution"[24] of powers – fangs, claws, hooves, fur, hide, flight, concealment, swiftness of foot, diet, strength, size, and prolificity – was in each instance perfectly well suited to environment and the threat of predation. That is one way in which his distribution was fitting and appropriate (ὡς πρέπει). But there is another: his distribution of powers resulted in no species being more fit for survival than the next (again, apart from the human species). All species now had this fitness for survival *to the same degree*, and this too makes his action πρέπον. But of course Epimetheus did not have long to bask in the glory of his achievement, for he soon realized, too late, that he had exhausted the entire supply of powers before providing for the last of the species, the human species. Thus, he had failed to see to it that *each and every* species shared in the distribution of powers (*ἑκάστοις* ὡς πρέπει), and *to the same degree*.

The critical day, the one on which it was fated that the human race would enter the world, arrived (321c6: ἤδη δὲ καὶ ἡ εἱμαρμένη ἡμέρα παρῆν). Conducting a last-minute inspection of Epimetheus' work, Prometheus found the human species naked and hoofless, without covering of fur or hide, and altogether lacking claws and fangs. Eager to rescue humanity from the consequences of his brother's blunder, Prometheus slipped into the workshop that Hephaestus shared with Athena, taking from it the fiery *techne* of the forge and all the accompanying skills necessary if human beings were to find their own means of sustaining themselves. Thus, whereas the distribution to other creatures drew on the aforementioned pool of material resources and physical attributes, the human race was given a power of an altogether different order. Soon, human beings were inventing (presumably with the aid of forged tools)[25] an entire array of means for sustaining themselves: houses, clothing, shoes, bedding, and food production. And they also invented articulate sounds for naming such objects. Moreover, understanding that this fiery entechnic skill, given to them at their creation, had come from the gods, they saw it as an element of divinity in their na-

---

23 Beresford (2013, 145) offers a different perspective on Epimetheus' successful adaptations: "Of course, in the story this bungling, thoughtless god produces beautifully designed animals. But that is precisely the paradox behind the allegory: that even mindless, natural processes, even a thoughtless creator, can generate intricate and complex adaptations".

24 Calame 2012, 131.

25 The empyric nature of the *techne* (322d1–2: τὴν ἔντεχνον σοφίαν σὺν πυρί; cf. 321e12: ἔμπυρον τέχνην) suggests as much, and at the same time gives greater prominence to Hephaestus and his forge. Thus, the demiourgic *technai* (321d5) are banausic in origin.

ture.[26] Theirs was an *Olympian* allotment, through which they enjoyed a special relationship with Hephaestus, Athena, and the rest of the gods. Indeed, human beings claimed kinship with the gods, whom they honored with various religious practices, including the construction of altars and sacred images.

The human race was now sheltered against the elements – and *perhaps*, through its invention of religion, also the wrath of the gods, whose Olympus had been violated by Prometheus' theft. Yet there remained the threat posed by wild beasts. To meet this danger, human beings sought strength in numbers. They started building cities as fortified bases for waging collective war against the brutes. They understood that this collective enterprise required the polis, but they failed to realize that living within a polis itself required a techne of an altogether different nature from the "craftiness" that they had brought with them from birth into the world. Lacking this additional *politike techne*, they were constantly inflicting harm on each other, until they finally were forced to scatter to their individual abodes, once again exposing themselves to the depredations of the wild beasts.

Unable to survive outside the city but then again, unable to survive within it, the human race would have perished but for Zeus, who instructed Hermes to place justice and respect[27] at their disposal, so that there might arise that collectivization of personal lives, bound up in ties of friendship, that we know as lawful polis society.[28] Hermes was prompt to comply, though not before asking whether this gift, *politike techne*, should be distributed in the same fashion as the demiourgic *technai* had come to be distributed. Was Zeus' gift to be distributed in a similar fashion, that is, only to specialists? Or were justice and respect to be distributed to all human beings alike? "Let all share in them", said Zeus, "for cities could not come into being, if only a few shared in them as in other crafts". But Zeus did not, perhaps could not, remove our natural tendency to harm one another.[29] That this natural tendency persisted even after Zeus' gift is evident from Zeus' final instruction to Hermes: "lay down on my authority a

---

26 Adam's deletion of 322a4 τὴν τοῦ θεοῦ συγγένειαν (1905, 112) is surely mistaken. See Guthrie 1957, 141 n. 10 and Manuwald 1999, 176, 192.

27 Some translators render *aidos* "reverence". This, however, carries religious overtones which, in light of the fact that the human race had long since begun worshipping the gods, are inappropriate. Taylor translates 'conscience'.

28 322c3: πόλεων κόσμοι τε καὶ δεσμοὶ φιλίας συναγωγοί. See Pl. *Plt*. 311b9 – c1, where the *politike techne* that brings lives together in friendship is identified with the kingly art: ὁμονοίᾳ καὶ φιλίᾳ κοινὸν συναγαγοῦσα αὐτῶν τὸν βίον ἡ βασιλικὴ τέχνη … . See also Pl. *Leg*. 793b4 (speaking of ἄγραφα νόμιμα / πάτριοι νόμοι): δεσμοὶ γὰρ οὗτοι πάσης εἰσὶν πολιτείας.

29 In what sense did this gift altered human nature? See below, n. 32.

law that he who cannot share in reverence and justice is to be killed as a carrier of a disease to the city" (322d1–5).

This is a story of two interventions, each of which rescues the human species at a moment of crisis. The first occurs when "the appointed day was already at hand, on which man too had to come out of the earth to the light of day" (321c6–7). At that critical moment, Epimetheus, for all his success when assigning individual powers, one after the other, realized that he had failed to achieve τὸ πρέπον.[30] This was the first attempt to master kairos, and it ended in error. Prometheus immediately made a second trial. Intervening in the most timely way, Prometheus once and for all[31] rescued the human species from its exposure to the elements and its want of daily sustenance. But it remained for Zeus to resolve the second crisis, that of protecting the human species from its predators and from its own self-destruction. As the human population was slowly being extinguished, Zeus intervened in a timely fashion, instructing Hermes to give humans *politike techne*, the art of the polis-community (322b5).[32]

Zeus' power to rescue a species, at a moment's notice and without danger to himself, was never in doubt. The same is not true of Prometheus' intervention. In the Prometheus story, as the young Hippocrates would have known it from his schooldays and as we know it from Hesiod (*Theog.* 521–569), Zeus was no friend to the human race, and Prometheus' intervention on our behalf was fraught with danger to himself, and indeed resulted in the most horrific punishment. A sim-

30 This is not to say that Epimetheus had any earlier concern for an overall final outcome, but only that he realized, too late, that he ought to have been guided by such a goal. We may, however, concede this much (pace Balaban), that, relative to each species, each in isolation from the next, he did display a sense of purpose.

31 Whereas other accounts of the origins of human culture posit stages in the development of the arts, each a human response to a new crisis, Protagoras imagines the theft of Prometheus affording *all* the necessary means for managing life (Manuwald 1999, 179). Thus, the only crisis that matters (apart from the one solved by almighty Zeus' intervention) is the one successfully met by Prometheus.

32 If we are to judge from 322d4–5 (τὸν μὴ δυνάμενον αἰδοῦς καὶ δίκης μετέχειν) and 323c5 (αὐτὴν οὐ φύσει ἡγοῦνται εἶναι), Zeus' gift to early man did not alter human nature but resulted in an inherited predisposition to act with respect and justice. It is I believe in this sense that "Protagoras sees morality ... tied to human nature" (Beresford 2013, 150, who rightly speaks of Protagoras holding the view that "the finished virtues (are) a product of instruction and training acting upon *natural predispositions*" (151, italics mine)). Less nuanced and therefore (in my opinion) incorrect is Reale 2004, 139: "secondo il mito, la virtù è dono divino; il che equivale a dire che essa è dono che l'uomo riceve per natura". Reale consequently regards 322d4–5 and 323c5–8 as evidence of self-contradiction in Protagoras' argument (2004, 140–141). On "second nature" in Aristotle, see McDowell 1998, 184–185, who speaks of "the *acquisition* of a second nature, involving the moulding of motivational and evaluative propensities" (italics added).

ilarly gruesome outcome was possible in Protagoras' retelling of the myth. The studious Hippocrates, then, must ask himself why Prometheus, in this sophistic retelling of the myth, evades his crucifixion, and how it is that a kindly, philanthropic Zeus becomes our second benefactor. Reflecting on the myth, he will find the answer in the titan's practice of the art of making the right decision – right, that is, in terms of what is fitting and necessary relative to time, place, and degree.

According to Hesiod, Prometheus stole fire from Zeus, who was withholding it from man (*Theog.* 563: οὐκ ἐδίδου). In consequence of this transgression against the Olympian, Prometheus was subjected to a gruesome punishment: chained to a column (perhaps even impaled on it),[33] he suffered the torment of the eagle's daily feeding on his liver. But this is not the story of Promethean crime and punishment as Protagoras tells it (321c7–322a2).

> Prometheus was at his wits' end to find a means of preservation for mankind, so he stole from Hephaestus and Athena their technical skill along with the use of fire ... That is how man acquired his practical skill, but he did not yet have skill in running a city; Zeus kept watch over that. Prometheus had no time to penetrate the citadel of Zeus – moreover the guards of Zeus were terrible – but he made his way by stealth into the workshop which Athena and Hephaestus shared for the practice of their arts, and stole Hephaestus' art of working with fire, and the other art which Athena possesses, and gave them to men. And as a result man was well provided with resources for his life, but afterwards, so it is said, thanks to Epimetheus, Prometheus was charged with theft.[34]

Protagoras has Prometheus steal fire together with the associated technical skills.[35] But he tells us that Prometheus contemplated stealing *politike techne* as well, and explains why he proceeded with the one theft but not both. In order to steal *politike techne*, he would have had to breach the Olympian citadel, eluding its formidable guards.[36] Furthermore, he was under constraint of time: "the appointed day was at hand, on which man too had to come out of the earth to the light of day". Prometheus, beset with *aporia* (321c7–8), clearly had little time to spare, and certainly not enough to assail the well-guarded palace on the citadel either *before* or *instead of* the easier and quicker task of invad-

33 On the meaning of *Theog.* 522 (μέσον διὰ κίον' ἐλάσσας), see West 1966, 313; Gantz 1993, 155 and n. 6.

34 Here I substitute Lombardo's translation of κλοπῆς δίκη μετῆλθεν for Taylor's "paid the penalty for theft".

35 For the deviations from Hesiod found in the *Prometheus Bound*, see Calame 2012, 135–136; Gantz 1993, 159.

36 Perhaps we are intended to think of the figures of Kratos and Bia, as they are presented in the *PV.*

ing the workshop of Athena and Hephaestus (this, as the commentators note, presumably lay in the workers' quarters *beneath* the mighty citadel).[37] Thus, the myth's most remarkable divergence from the account in the *Theogony* is that Prometheus transgressed the honour *not* of Zeus but of Hephaestus and Athena. As a result, he was never in danger of suffering the terrible punishment narrated in the *Theogony.* Instead, the lesser gods who were the offended party brought a private prosecution for theft (322a2: κλοπῆς δίκη μετῆλθεν).[38] Aware as he is of the Hesiodic version of the story (and perhaps also the *Prometheus Bound*), Hippocrates must find this detail about petty theft and a private lawsuit oddly anticlimactic.[39] And in light of this, he must admire the thievery of Prometheus as a model of smart decision-making.

Protagoras teaches *euboulia*, and in order to show Hippocrates, his prospective student, what this skill entails, he offers Prometheus as its mythic paradigm.[40] Thus, in a dire moment, [41] when first Epimetheus and then Prometheus himself felt helpless to rescue a situation caused by Epimetheus' failure to achieve τὸ πρέπον in the overall task, Prometheus hit upon a remedy. His first thought was to secure the twin *technai* of craftsmanship and of politic (*demiourgike, politike*). But then, after assessing both the temporal constraints (321d6: οὐκέτι ἐνεχώρει) and the effectiveness of striking at one target, an unprotected workshop, rather than the other, Zeus' lofty and heavily-guarded citadel, Prometheus limited his sights to the first, for it was (to use a term familiar from

37 Manuwald 1999, 191. See Adam 1893, 111. Calame 2012, 133 observes: "the mention of the collaboration of these two divinities helps implicitly to centre the story on Athens". We may go further, and identify τὸ τῆς Ἀθηνᾶς καὶ Ἡφαίστου οἴκημα τὸ κοινόν with the Athenian Hephaisteion and Zeus' citadel with the Acropolis. For the former was certainly shared by the two deities: the cult statuary, completed by 416/415 (*IG*^{3} 472) included both Athena and Hephaistos. And if we wish to think of the workshop imagined by Protagoras as located in the workers' quarters, lying humbly beneath the royal citadel, we may recall that the hill on which the Hephaisteion stands, the Kolonos Agoraios, gave its name to the hired workmen who assembled there (Harpocration s.v. κολωνέτας). And of course, it certainly lay beneath the regal Acropolis. I wish to express my gratitude to Professor Marion Meyer of the University of Vienna for helping me see and appreciate these delightful archaeological facets of Protagoras' myth.

38 Gantz 1993, 159.

39 And perhaps humorous too: according to Dem. *In Tim.* 105, the court could levy, in addition to material compensation for theft, five days and nights in the stocks – a day on the beach compared to the column and vulture. See Cohen 1983, 62–68 (who regards *In Tim.* as "the main evidence" for the *dike* klopes). And the hint that Athene and Hephaistos have recourse to Athenian law further adds to the humor.

40 See Morgan on sophistic "socially generic 'advice to young men'" and Manuwald on the myth as "protreptikos to virtue" (above n. 12).

41 321c2–8: ἠπόρει ... ἀποροῦντι ... ἀπορίᾳ σχόμενος.

Homer) καίριον, the right place to strike.[42] Moreover, he moderated the degree of his offense, achieving in this respect τὸ μέτριον, thereby exposing himself merely to a private prosecution rather than capital punishment. And yet, as a result of humanity's piety and worship, consequences of Prometheus' petty theft, the *politike techne* that Prometheus could not have stolen from Zeus' awesome palace without extreme peril is freely given by the Olympian to worshipful humanity. It is Zeus himself who gives his pious human worshippers the art of living in harmony in the polis, without which the species could never have survived. If, then, we take the myth as our guide, successful decision-making in difficult circumstances – we may call it *euboulia*, the very thing that Protagoras says Hippocrates is keen to learn – is nothing other than *eukairia*, correctly knowing "the time or place at which, or degree in which, something is appropriate" and then acting on that knowledge. And if we focus our attention on merely the temporal aspect, *euboulia* and *eukairia* derive the maximum benefit from one-time interventions rather than thoughtless Epimethean, "chronic" iteration.[43]

It is not within the scope of the present paper to explore the importance of *kairos* for the dialogue as a whole. We may nonetheless note how we are invited from the very outset of the dialogue to view matters in these terms. There, an unnamed acquaintance, accompanied by an unspecified number of companions, chances upon Socrates in an unspecified public space. The acquaintance asks Socrates where he has been. But before he can answer, the acquaintance does so for him: "No, don't tell me. It's pretty obvious that you've been hunting the ripe and ready Alcibiades" (309a1–2: τὴν Ἀλκιβιάδου ὥραν).[44] Again, before Socrates can confirm or deny any such predatory activity, the acquaintance admonishes him, saying that, while Alcibiades certainly is handsome (*kalos*), he is no longer at that select moment of youth when Socrates' pederastic attention was not inappropriate: he is now a man, a beard just emerging, and Socrates must restrain himself (309a1–5) from such poorly timed, inappropriate behavior.

---

42 At *Il.* 8.84, an arrow shot by Paris strikes Menelaus' lead horse squarely in the forelock, which the poet considers as μάλιστα δὲ καίριον. Cf. *Il.* 8.326. Says Wilson 1980, 180: the adjective here "is used to mark the lethal or critical point for the body to receive a wound". Pfister (1937, 139) notes that one of the ways that we see Latin tempus serving a semantic role similar to *kairos* is in the use of tempora to denote "the fatal place of the body".

43 We may note the remark of Diog. Laert. (9.53) that Protagoras was "the first to expound on the function of *kairos*" (πρῶτος ... καιροῦ δύναμιν ἐξέθετο).

44 The translation, which captures both the crassness of the acquaintance and most importantly the meaning of ὥραν, is taken from Lombardo and Bell 1992. Taylor translates, "... chasing around after that handsome young fellow Alcibiades".

Socrates replies to this charge of shameful, untimely erotic pursuit with a quote from Homer, to the effect that the emergence of a beard is precisely that moment when youth (ἥβη) is endowed with its greatest charm. Not surprisingly, the elegance of this response falls leaden at the feet of the vulgar acquaintance, who persists: "so what's up? Were you just with him? And how is the young man disposed towards you?" (309b3 – 4).

Though he does not use the term, the acquaintance is faulting Socrates for erotic akairia – that is, for an egregious failure to see how terribly untimely he is in his erotic pursuits. Socrates, however, manages, with surprising grace and wit, to change the subject. Protagoras is in town, and he, Socrates, has spent the day with him in the home of the sophist's host, Callias, basking in the superior beauty of his wisdom. He then tells how this meeting with Protagoras came about.

We could go on to speak of the ensuing flurry of pre-dawn and quite untimely activities of Socrates' young friend Hippocrates, and then go on to discuss much else in this dialogue that concerns *kairos*. But the degree to which that story would be told is great indeed, and must be left for another more fit time and place. In the words of the poet, καιρὸς δ' ἐπὶ πᾶσιν ἄριστος.[45]

Bibliography

Adam, J. and A.M. Adam. 1905 (1893¹). *Platonis Protagoras. With Introduction, Notes and Appendices*. Cambridge.
Balaban, O. 1987. The Myth of Protagoras and Plato's Theory of Measurement. *HPQ* 4: 371 – 384.
Balaban, O. 1999. *Plato and Protagoras: Truth and Relativism in Ancient Greek Philosophy*. Lanham, MD.
Beresford, A. 2013. Fangs, Feathers, and Fairness: Protagoras on the Origins of Right and Wrong. In Van Ophuijsen, van Raalte and Stork 2013, 139 – 162.
Brisson, L. 1975. Le mythe de Protagoras: Essai d'analyse structurale. *QUCC* 30: 7 – 37.
Calame, C. 2012. The Pragmatics of 'Myth' in Plato's Dialogues: The Story of Prometheus in the *Protagoras*. In Collobert, Destrée and Gonzalez 2012, 127 – 143.
Coby, P. 1987. *Socrates and the Sophistic Enlightenment*. London and Toronto.
Cohen, D. 1983. *Theft in Athenian Law*. Munich.
Cole, T. 1990. *Democritus and The Sources of Greek Anthropology*. Atlanta, GA.
Collobert, C., P. Destrée and F.J. Gonzalez (eds.) 2012. *Plato and Myth: Studies on the Use and Status of Platonic Myths*. Leiden.
Craik, E.M. 1998. *Hippocrates Places in Man*. Oxford.
Denyer, N. 2008. *Plato Protagoras*. Cambridge.

45 Hes. *Op.* 694.

Gantz, T. 1993. *Early Greek Myth: A Guide to Literary and Artistic Sources.* Baltimore.
Guillamaud, P. 1988. L'essence du kairos. *REA* 90: 359–371.
Guthrie, W.K.C., 1957. *In The Beginning. Some Greek Views on The Origins of Life and The Early State of Man.* Ithaca.
Halliwell, S. 2012. *Between Ecstasy and Truth: Interpretations of Greek Poetics from Homer to Longinus.* Oxford.
Hansen, H. Greek 701: Greek Rhetoric and Prose Style. http://greek701.ws.gc.cuny.edu/essay-on-style/#plato1
Havelock, E.A. *The Liberal Temper in Greek Politics.* New Haven.
Heath, M. 1989. *Unity in Greek Poetics.* Oxford.
Kerferd, G.B. 1981. *The Sophistic Movement.* Cambridge.
Knudsen, R.A. 2012. Poetic Speakers, Sophistic Words. *AJP* 133: 31–60.
Kurke, L. 2011. *Aesopic Conversations.* Princeton.
Lampert, L. 2010. *How Philosophy Became Socratic. A Study of Plato's Protagoras, Charmides, and Republic.* Chicago.
Levi, D. 1924. Il concetto kairos e la philosofia di platone. *Rendiconti della Reale Accademia dei Lincei Classe di Scienze Morali Storiche e Filologiche* 33: 93–188.
Lombardo, S. and K. Bell. 1992. *Plato Protagoras.* Indianapolis.
Manuwald, B. 1999. *Platon Protagoras.* Göttingen.
Manuwald, B. 2003. Der Mythos im Protagoras und die Platonische Mythopoiie. In *Plato's Protagoras. Proceedings of the Third Symposium Platonicum Pragense*, eds. A. Havlicek and F. Karfik, 39–59. Prague.
Manuwald, B. 2013. Protagoras' Myth in Plato's *Protagoras:* Fiction or Testimony? In Van Ophuijsen, Van Raalte and Stork 2013, 162–177.
McDowell, J. 1998. *Mind, Value, and Reality.* Cambridge, MA.
Morgan, K. 2000. *Myth and Philosophy from the Presocratics to Plato.* Cambridge.
Nestle, W. 1978. *Protagoras Platon.* Stuttgart (Leipzig and Berlin 1931[1]).
Norden, E. 1956. *Agnostos Theos.* Stuttgart (Leipzig and Berlin 1913[1]).
Pfister, F. 1938. Kairos und Symmetrie. In *Würzburger Festgabe: Heinrich Bulle dargebracht zum 70. Geburtstag am 11. Dez. 1937*, ed. R. Herbig (Würzburger Studien zur Altertumswissenschaft 13), 131–150. Stuttgart.
Pohlenz, M. 1933. ΤΟ ΠΡΕΠΟΝ. *Nachrichten von der Gesellschaft der Wissenschaften zu Göttingen.* Fachgruppe 1, nr. 16: 53–92.
Poulakos, J. 1983. Toward a Sophistic Definition of Rhetoric. *Ph&Rh* 16: 35–48.
Race, W.H. 1981. The Word Καιρός in Greek Drama. *TAPA* 111: 197–213.
Reale, G. 2004. Il mito in Platone con particolare riguardo al mito nel Protagora. In *Il Protagora di Platone: struttura e problematiche*, ed. G. Casertano, 128–144. Naples.
Rostagni, A. 1922. Un nuovo capitolo nella storia della retorica e della sofistica. *SIFC* 2: 148–201 (transl. by J. Sipora, A New Chapter in the History of Rhetoric and Sophistry. In Sipora and Baumlin 2002, 77–88).
Schofield, M. 1986. Euboulia in the Iliad. *CQ* 36: 6–31.
Sipora, P. and J.S. Baumlin (eds.) 2002. *Rhetoric and Kairos.* Albany, NY.
Steidle, W. 1952. Redekunst und Bildung bei Isokrates. *Hermes* 80: 257–296.
Taylor, C.C.W. 1991. *Plato Protagoras.* Oxford.

Tordesillas, A. 1992. Kairos dialectique, kairos rhétorique. Le projet platonicien de rhétorique philosophique perpetuelle. In *Understanding the Phaedrus: Proceedings of the I Symposium Platonicum*, ed. L. Rossetti, 77–92. St. Augustin.
Van Dijk, G.-J. 1997. *AINOI, ΛΟΓΟΙ, MYΘOI. Fables in Archaic, Classical, and Hellenistic Greek Literature. With a Study of the Theory and Terminology of the Genre.* Leiden, New York and Cologne.
Van Ophuijsen, J.M., M. van Raalte, P. Stork (eds.) 2013. *Protagoras of Abdera: The Man, His Measure.* Leiden.
Van Riel, G. 2012. Religion and Morality. Elements of Plato's Anthropology in the Myth of Prometheus (Protagoras, 320d-322d). In Collobert, Destrée and Gonzalez 2012, 145–164.
Wareh, T. 2012. *The Theory and Practice of Life: Isocrates and the Philosophers.* Cambridge, MA.
West, M.L. 1966. *Hesiod Theogony.* Oxford.
Wilson, J.R. 1980. Kairos as Due Measure. *Glotta* 58: 177–204.
Wilson, J.R. 1981. Kairos as Profit. *CQ* 31: 418–420.
Xanthakis-Karamanos, G. 1981. Remarks on Moschion's Account of Progress. *CQ* 31: 410–417.
Zhelezcheva, T. and J.S. Baumlin 2002. A Bibliography on *Kairos* and Related Concepts. In Sipora and Baumlin 2002, 237–245.
Zilioli, U. 2007. *Protagoras and the Challenge of Relativism.* Hampshire, England and Burlington, VT.

Christos A. Zafiropoulos

From Here to Eternity: *Mythologein* in Plato's *Phaedo*

The starting point of this paper is a passage at the very beginning of Plato's *Phaedo*, one that remains rather overlooked or little commented by scholars.[1] This is surprising enough, given the volume of bibliography on Plato and on the *Phaedo* in particular, and also given our awareness of Plato's tactics as regards the introductions of his dialogues, besides their dramatic qualities, namely to embed them in the ideological content of the main text.[2]

Socrates has just had his feet released from their prolonged fettering for a whole month (due to the sacred law that forbade the shedding of blood during Apollo's sacred festival, the Delia). He rubs one leg and as blood re-circulates properly in it (a detail that shall be counterbalanced in the concluding scene by the progressive numbing of Socrates' feet) he feels relief and then he addresses the following words to his gathered friends and disciples, actually his very first ones in their last conversation:[3]

> What a peculiar thing it seems to be, my friends, this thing that people call 'pleasure'. What a surprising natural relation it has to its apparent opposite, pain. I mean that the two of them refuse to come to a person at the same time, yet if someone chases one and catches it he is pretty much forced always to catch the other one too, as if they were two things but joined by a single head. And I do believe, he said, that if Aesop had reflected on them (εἰ ἐνενόησεν αὐτὰ Αἴσωπος), he would have composed a fable (μῦθον): that they were [once] at war, and that god wanted to reconcile them, and that, finding himself unable to do so, he joined their heads together, the result being that [thenceforth] if one of them comes to somebody the other too will later follow in its train. That is precisely what seems to be happening to me too. Because pain was in my leg from the fetter, pleasure seems to have come in its train. (60b3–c7)

1 I would like to thank the Organizing Committee of the *Time and Space in Greek Myth and Religion* Conference and especially Menelaos Christopoulos, Effimia Karakantza and Athina Papachrysostomou for their invitation and for all their toil in putting together the conference and the present volume. I also owe special thanks to the anonymous evaluator for her / his suited and most helpful comments.

2 On this 'programmatic' aspect of Plato's dialogues see Capuccino 2014; Zafiropoulos 2015, 20–21, 32–33, 40, 73.

3 The Greek text is Duke, Hicken, Nicoll, Robinson and Strachan 1995. The translations of the passages follow A. Long's translation in Sedley and Long 2010, 45–46. Italics in the texts and additions in square brackets are mine.

DOI 10.1515/9783110535150-018

Cebes the Pythagorean interrupts him and reports the worries of the poet Euenus from Paros as regards "those poems you've been composing, your versifications of *Aesop's tales* (ἐντείνας τοὺς τοῦ Αἰσώπου λόγους) and the proem to Apollo ..., what on earth your idea was in composing them when you came here, given that you had never composed poetry before" (60c9 – d4). Socrates denies [ironically?] being competitive to poets and he reveals that he did compose these verses in response to his experiencing a voice in a repeated dream, one that urged him always with the same utterance "Socrates, it said, compose music and work at it" (ὦ Σώκρατες, ἔφη, μουσικὴν ποίει καὶ ἐργάζου) (60e6 – 7). Socrates then notes with respect to that utterance:

> In the past I used to suppose that it was encouraging me and cheering me on to do what I was doing, like those who cheer runners. I took the dream to be cheering me on in the same way to do just what I was doing, composing music, on the grounds that philosophy is the greatest music, and that that was what I was doing. But now since the trial was over and the god's festival was holding up my death, I thought that just in case the dream might after all be instructing me to compose music as commonly understood, I should not disobey it but should start composing [that particular kind of music]. For it seemed safer not to depart before I had honoured my sacred obligation by composing poems in obedience to the dream. So that is how I came to start by making a composition dedicated to the god whose festival was currently being held. But, after I had attended to the god, I reflected that the poet, if he is to be a poet, should compose stories, not arguments (ποιεῖν μύθους ἀλλ' οὐ λόγους). I myself was not a story-teller (καὶ αὐτὸς οὐκ ἦ μυθολογικός), so I took the stories I had ready to hand and knew, those of Aesop (μύθους ... τοὺς Αἰσώπου), and made compositions out of the ones that first came to mind. So, Cebes, tell all this to Euenus, give him my best wishes and tell him, if he is in his right mind, to come after me as soon as possible. I leave, it seems, today: so the Athenians command. (60e7 – 61c1)

Socrates' fable *à-la* Aesop justifies its labelling. It shares both the structure and the application of Greek fables, especially the 'aetiological' ones. An initial state of crisis between two antagonists is mediated by the intervention of a third agent – in aetiological fables a divine one – who provides a permanent solution as well as an explanation for the ensuing and current state of affairs.[4] It also matches most of the characteristics of Plato's own *mythoi* in his dialogues, i.e. it is a hortatory, monological narrative with an entertaining effect, narrated by an older speaker to a younger audience, it is placed at the beginning of a philosophical conversation, and it recites events that are set in the distant past, yet they are unverifiable and in an unspecified time and space.[5] Furthermore, it calls to

4 On the structure of Greek fables see Betegh 2009, 84 – 85; Holzberg 1993, 95 – 102; Nøjgaard 1964, 74 – 82, 141 – 170; Zafiropoulos 2001, 7 – 8.
5 See Most 2012, 16 – 18.

mind Aristophanes' aetiology on the *androgynon* in the *Symposium* (189c – 193d), which is included by many in Greek fable material.[6] In fact, compared to Aristophanes' story, the *Phaedo* fable presents quite the reverse outcome, for here god unites two seemingly opposites into a single whole, instead of cutting an undifferentiated whole in two pieces.

Socrates' fable on pain and pleasure has both a *promythium*, an introductory summary of its message, and an *epimythium*, a concluding one.[7] It is a brief, plain and unadorned narrative, one that is easy to memorize, and it is used as an illustration of Socrates' state, hence it is in line with the fable's main application in antiquity.[8] As regards its relation to the philosophical content of the dialogue, it introduces the theme of the erroneous beliefs of the multitude ("what people call pleasure") and offers a first glimpse at the argument from opposites that Socrates shall soon bring forward (69e – 72e); therefore, it is embedded in the philosophical content of the dialogue.[9] On the other hand, this fable clearly states an authorship that stands outside what was customary for fables (it is neither anonymous nor Aesopic) and accordingly it stresses its fictitiousness, an interesting detail that marks a move from 'traditional', as in anonymous and/ or unverifiable, to philosophical *mythologein*.

A puzzle in our passage that calls for further investigation is the conspicuous confusion of terminology for 'fable'. Apparently, Cebes' λόγος for 'fable' is in accordance with the classical period terminology, whereas Socrates' μῦθος is not. In particular, to summarize bibliography on the issue, *ainos*, the term for exhortative animal allegory in verse during the Archaic times, as in Hesiod and Archilochus, was replaced in the Classical period by *logos*, at a time when the genre was also connected with Aesop as its *euretes*, its founder, whereas *mythos* gradually became the exclusive term for 'fable' from the Hellenistic period onwards.[10] Nevertheless, occasionally the two terms can be used alternately, as in our pas-

6 Zafiropoulos 2015, 25, 111. Contra: Van Dijk 1997, 670 – 671.

7 Nøjgaard 1964, 359 – 380.

8 See Van Dijk 1997, 40 – 42, 72 – 75; Zafiropoulos 2015, 122 – 124. The narrative and persuasive qualities of fables made Aristotle recommend their use as *exempla*, παραδείγματα of the ideas and the messages that the orator would wish to infuse to his audience, *Rh.* 1393a28 – 31, 1393b8 – 1394a18.

9 Betegh 2009, 78.

10 See Nøjgaard 1964, 122 – 128; Van Dijk 1993, 171 – 174; Van Dijk 1997, 79 – 88, 105 – 107; Zafiropoulos 2001, 2 – 3; Zafiropoulos 2015, 118 – 119. Terminology for fable in Greek literature prior to Plato: a) *ainos:* Hes. *Op.* 202 – 212; Archil. fr. 174 – 181, 185 – 187 West. b) *logos:* Hdt. 1.141, 2.134 (Aesop as a λογοποιός); Ar. *Av.* 651 – 653, *Pax* 127 – 134, *Vesp.* 1258 – 1260, 1399 – 1405; Antiphon Soph. 87 B 54 DK. c) *mythos:* Aesch. fr. 139 Radt.

sage.[11] In such cases, scholars propose that *logos* refers to prose, whereas *mythos* to verse and more fictive compositions. Accordingly, it has been suggested that an evaluative differentiation between the two terms is hinted at here, with *logos* bearing a higher status than *mythos*, the latter suggesting fictitiousness.[12] On the other hand, the only term for 'fable' that is used explicitly in Aristophanic comedy is *logos* and a passage from the *Wasps* points to the inappropriate, vulgar, and comic quality that these particular Aesopic *logoi* carry.[13] Consequently, terminology for 'fable' in the surviving sources before Plato secures no safe ground for evaluative criteria on its content.[14]

As regards our passage in particular, a first point is that Plato's Socrates (who knows his Aristophanes pretty well) terms fables *mythoi* without ascribing any lesser status to them.[15] Actually, Plato was quite familiar with fable material. As a matter of fact, his dialogues provide one of our main sources on fable in the Classical times, second to Aristophanes.[16] Apart from his own fable on pleasure and pain, Socrates also recounts the aetiological fable on the cicadas in the *Phaedrus* (259b-d) and the anecdote on Thales and the Thracian woman in the *Theaetetus* (174a; included in later fable collections as the fable of the astronomer). Moreover, in the *first Alcibiades* (considered to be spurious by many scholars) Socrates narrates Aesop's *mythos* of the fox and the lion, whereas in the *Republic* Adeimantos alludes to the Archilochean fable of the fox and the ape (365c4–6) and in the *Laws* the Athenian Stranger probably alludes to the fable of the wolves and the dogs (906d–e). Many scholars also include in fable material Aristophanes' aforementioned aetiology in the *Symposium* and

11 Cf. two passages from Aristophanes, *Pax* 129–132 and *Vesp.* 1174–1182.

12 Van Dijk 1997, 85, 90.

13 Ar. *Vesp.* 1174–1182. This passage also testifies to the aforementioned differentiation of the two terms. Philocleon's response to his son's call to recount the kind of stories, *logoi*, seemingly grave, that it is fit to narrated in the presence of eminent men, as in a symposium, is to barrage him with myth and fable material, which Bdelycleon, in disgust, terms *mythoi*. The passage also testifies to confusion or indifference for terminology as regards popular use of *mythos* and *logos*. See Fowler 2011, 63.

14 Van Dijk 1997, 90. Fable combines blame language with illustrative didacticism on proper behaviour and practical ethics.

15 Aristotle, who connected *mythos* with fiction and falsehood, applies both terms for 'fable'. Λόγος: *Rh.* 1393a28–31, 1393b8–22. Μῦθος: *Hist. an.* 618a18–20; *Mete.* 356b13–15.

16 On fables in the classical period see Van Dijk 1997 and Zafiropoulos 2015, 118–119.

Protagoras' famous *mythos* on the origins of mankind in the dialogue that bears his name (320c–322d).[17]

The conflation of fable terminology in the introduction of the *Phaedo* seems also to point to an association of *mythos* with poetry (actually with the versification of prose fables – *enteinas*) as well as with fiction and of *logos* with (prose) philosophical argumentative discourse and practice.[18] This not only seems to verify the aforementioned semantics of fable terminology, but it also calls to mind the notorious juxtaposition of *mythos* to *logos*, an interpretative pattern that since the work of eminent scholars such as W. Nestle, B. Snell and F. Cornford, to name but a few, remains a major issue in the study of Greek thought.[19] Plato's work holds a pivotal place in the assumed progress from myth-based thought to reasoning, specified in the contrast between the fictive, symbolic, narrative, disordered, poetic world of *mythos* and the true, rational, argumentative, systematic, philosophical world of *logos*. Plato is seen as a major factor in the demerit of the authority that *mythos* and its variants held from the epic onwards and from his and Aristotle's work onwards the prevalence of *logos* was secured by the scrutiny of argumentative discourse.[20]

In the last decades, however, the idealism and the teleology of such interpretative patterns have been noted and criticized and scholars now opt for the interaction, if not the blending of *mythos* with *logos* in the Classical period.[21] There are many such readings now in Platonic studies, readings according to which despite the marked difference in status appointed to the two terms in many passages (e.g. *Cra.* 408c, *Resp.* 522a, *Ti.* 26c, e), where *logos* is something incontrovertible whereas *mythos* is incapable of proof, still in many other passages there is "no such clear-cut opposition between myth and argumentative discourse".[22] In addition, despite Plato's criticism on myth as used by non-philosophers, Plato's Socrates turns out in most of the dialogues turn to be most *mythologikos*. The dialogues abound in metaphors, allegories, fictive accounts, and of course in myths, including some of the most memorable and influential

---

17 Xenophon's Socrates recounts the *logos* of the dog (*Mem.* 2.7.13–14). For further discussion on fables and allusions to fables in Plato see Van Dijk 1997, 317–321, 324–336, 667–671 and Zafiropoulos 2015, 111–112.

18 Brisson 1998, 45–46.

19 For an outline see Most 1999. The contrast, of course, is not limited to philosophers. See, for example, Thuc. 1.22.4.

20 Brisson 2004, 19–25; Buxton 1999b, 1–4; Fowler 2011; Most 1999.

21 See, for example, Buxton 1999b, 5–13; Calame 1999; Most 1999.

22 Calame 1999, 125 n. 9.

ones from Greek literature as a whole.[23] And this seems to be most true for the *Phaedo*, which seems to be embedded in myth, as it opens and closes in mythological imagery, language and context; it starts with the *aition* for the Delia (58a-b) and Socrates' reported *mythopoiein*, and concludes in Socrates' long account on the afterlife of the soul and the visionary cartography of the Earth and the heavens.[24] Therefore, he contradicts his introductory assertion of not being *mythologikos* himself, which can then be read as another irony of his.[25] In fact, Plato's Socrates fashions and presents his argumentation both in *mythos* and in *logos*.[26]

It may also be significant that compound words with the two terms – such as μυθολογία, μυθολόγημα, μυθολογεῖν, μυθολογητέον – frequent Plato's dialogues, which actually provide the *terminus post quem* for most of them.[27] As for our passage, it provides our first evidence of the surprisingly rare in antiquity term μυθολογικός, and a hapax until the second century AD, when it resurfaces in Pollux's *Onomasticon*.[28] The numbers, the frequency and the uniqueness of these compounds in the dialogues, together with Plato's connection of *mythologein* with philosophical activity, allow us to suggest that whether he was the originator of those terms or not, it is possible that he was the first to use them and to explore their semantics to such extent.

Plato had a special predilection for mythologizing and he deemed it as a most appropriate and effective tool to assist argumentative discourse, regardless of its fictitiousness and against our modern assumptions.[29] Quite the opposite,

---

23 For example, from the account on Atlantis in the *Critias* to the phantasmagoria of the charioteer in the *Phaedrus*, up to the great eschatological myths on the afterlife of the soul and its judgment in the otherworld in the *Gorgias*, the *Republic* and the *Phaedo*. The latter are also 'eschatological' in their narrative function, for they serve as concluding illustrations (and as indirect verifications) of the dialogues' core argument, see Clay 2007, 215.

24 Betegh 2009, 77; Clay 2007, 215; Halliwell 2000; Most 2012, 13–14.

25 The phrasing at 61b5 recalls in form – and perhaps in irony – Socrates' οὐ γάρ εἰμι ποιητικός in the *Republic* (393d8), a parallel that Professor Theodoros Stephanopoulos brought to my attention.

26 Contra Betegh 2009, 83–84.

27 See Brisson 1998, 40 and appendices I-II; Most 2012, 13. Cf. Xen. *Symp.* 8.28. This led Glenn Most to suggest that "they were most likely coined" by Plato. See Most 2012, 13. However, we should perhaps be more reserved with respect to Plato's agency, given the Homeric μυθολογεύειν (*Od.* 12.450, 453 – yet with a rather different meaning, i.e. "to tell word for word") and especially Aeschylus' διεμυθολόγησεν ("to utter", *PV* 889) and Isocrates' μυθολογεῖν, "to tell a legend or a mythic tale" (2.49; 6.24). Cf. Sappho fr. 18.4. See Calame 1999, 127–128.

28 See Rowe 1999.

29 On Plato's acceptance of fiction, even on the grounds of a useful kind of falsehood, a noble lie, γενναῖον ψεῦδος, see Gill 1993, esp. 52–55. Apparently, a century after Plato, another Peloponnesian and a fellow-citizen of Echecrates who features at the introduction of the *Phaedo*,

impressed by the power of myth to persuade via its unique blending of pleasure with exhortation, of emotional involvement with persuasion, and above all because of its illustrative qualities, Plato saw myth to be a most suitable vehicle for discourse that can help the philosopher and his disciples attain knowledge on issues that test the limits of reasoning. That is, to perceive the imperceptible and to articulate what is difficult to express by argumentative discourse proper. As such, in the dialogues myth is often an integral part of the philosopher's argumentation in search of true knowledge. Accordingly, (its) fictitiousness turns into a beneficial means to the greatest end attainable by the human intellect.[30]

Accordingly, there is a tendency in the dialogues to leave the semantic boundaries between *logos* and *mythos* vague or even to blur them in order to better serve persuasiveness, which also characterizes our introductory passage from the *Phaedo*. Hence, Socrates terms an eschatological account, for example, a *logos* in the *Gorgias* (523a1–2), but a *mythos* in the *Phaedo* (101b1–2).[31] It should not surprise us that Plato's Socrates may chastise a *mythos* (*Grg.* 527a5–6) on the otherworld and the afterlife of the soul as an "old woman's" tale and still unfold in great detail his own powerful and impressive eschatological myths on the punishments that await the vicious many. Strong imagery, fictive tales, allegories, likely accounts, εἰκότες μῦθοι are more than welcomed to assist the philosopher's task to educate and enlighten his audience. And nowhere else in the dialogues are the philosopher's *mythoi* and *logoi* thus collocated and interwoven, nowhere else is their interplay as dense and explicit as in the *Phaedo*.[32] This justifies the following closer look at *mythos* and *logos* in this particular dialogue.[33]

Timon of Phlius, jokingly yet perhaps quite revealingly for our discussion on Plato's *mythologein* punned on his name as suggestive of his talent in 'fabricating', in *plattein*. Clay 2007, 214.

30 Hence, the philosopher's *mythoi* and his myth-like *logoi* are exonerated for their adoption of fabrications, similar to those of the poets, for theirs are examined and approved by the philosopher. Their μυθολογία is deemed to be "falsehood in words" only, not "falsehood in the soul" or true falsehood (*Resp.* 382c–d). Therefore, it is a harmless and at the same time advantageous account that can implant the proper dispositions in the souls of its hearers. It functions like the notorious "noble lie" (414b–c), the fabricated μῦθος (415c2) that the Guards have to narrate (μυθολογοῦντες, 415a3) to the beguiled – for their own good – citizens of the ideal state, namely a useful for their education mixture of truth with falsehood. See Gill 1993, esp. 46–52, 56–57.

31 Or, as in the *Timaeus*, a narrative that is obviously a myth is termed a true *logos* (the story of Atlantis, 26e4–5), whereas what is clearly a philosopher's account on the ideal state (in the *Republic*) is termed a *mythos* (26c9), see Murray 1999, 260. On this deliberate terminological confusion see also Brisson 1998, 45–46, appendices I-II; Brisson 2004, 19–25; Clay 2007, 212–214; Edelstein 1949, 465–468; Fowler 2011, 50; Halliwell 2000; Lloyd 1989, 181–183; Morgan 2004, 155–159; Murray 1999.

32 On myth in Plato see, among many, Betegh 2009; Clay 2007, 212; Edelstein 1949; Gill 1993, 39–41; Morgan 2004, esp. 162–164; Murray 1999; Tarrant 1990; Yossi 1996, 29–51.

Socrates' first *mythos*, his fable on pain and pleasure, visualizes and provides an *aition* for a major and recurrent theme of the dialogue, namely his own and his disciples' empirical finding that man's life is characterized by a mixture of pleasure with pain, yet always favouring the latter. Actually, this realization and the sentiments that it generates, which becloud and threaten reasoning, set from the very start the pervading tone in Phaedrus' account. And they constitute a threat to their last conversation with Socrates that Phaedrus and all the other disciples bring along with them into his cell. As Phaedrus states, they all shared and bore "a quite peculiar experience, an unusual *mixture blended together from both the pleasure and the pain*" at the realization that this would be their last conversation with the master (59a3–7).[34] In the rest of the dialogue and until his own end, Socrates shall keep confronting this problem, the moral shortcomings and the false beliefs that stem from materiality and are shared by the many, as well as their detrimental effect to philosophical reasoning and conversation. First, right after his reply to Euenus, he agrees to explain his non-conformist statement that the philosopher and anyone interested in philosophy should welcome death (61c–d). To do so, he has to argue on the afterlife of the soul, on what is perhaps a most inaccessible topic, given the limitations of human experience and understanding. By undertaking such a task Socrates – as often in the dialogues – stands out above the many, a heroic and exemplary exception to man's attitude at the face of death. Instead of letting himself be consumed by similar disturbing thoughts and feelings, he courageously chooses to expand on the seemingly unbounded (the posthumous wanderings of man's imperishable and eternal part) from his limited space and time, in his small cell while time counts down to his execution. And the method that he chooses for such contemplation contributes to Socrates' *atopia* on his last day: as he points out, he, and anyone in his situation, he says, should resort to a combination of philosophical scrutiny with fictitious narrative (διασκοπεῖν τε καὶ μυθολογεῖν, 61e1–2).[35]

What ensues is an exchange of arguments, of *logoi* (e.g. 62b3, 63a2, 8), between Socrates and two Pythagoreans, Simmias and Cebes, on Pythagorean doctrines (i.e. of the body as a prison for the soul – quite appropriate given Socrates' own situation – and on gods guarding our souls). Facing a deadlock in their con-

33 See also Betegh 2009, 77–84; Clay 2007, 215; Morgan 2004, 192–200.

34 Ὡς ἐν φιλοσοφίᾳ ἡμῶν ὄντων ὥσπερ εἰώθεμεν – καὶ γὰρ οἱ λόγοι τοιοῦτοί τινες ἦσαν – ἀλλ' ἀτεχνῶς ἄτοπόν τί μοι πάθος παρῆν καί τις ἀήθης κρᾶσις ἀπό τε τῆς ἡδονῆς συγκεκραμένη ὁμοῦ καὶ ἀπὸ τῆς λύπης, ἐνθυμουμένῳ ὅτι αὐτίκα ἐκεῖνος ἔμελλε τελευτᾶν.

35 Morgan 2004, 194–195. On Socrates' *atopia* in the dialogues and particularly in the *Phaedo*, see Zafiropoulos 2015, 45–46, 184–188.

versation, Socrates undertakes to deliver a new defense of his thesis, an *apologia* as he terms it (69d7, cf. 63b2), in a second 'trial' of him, this time a philosophical one in which he seeks to persuade his own disciples.[36] The *logoi* that he then delivers on the posthumous fate of the soul are rich in imagery (e. g. 66b – 67b), that is, they combine scrutiny with mythologizing, as promised. And until the end of the conversation *logoi* with mythologizing traits shall constitute a good part of Socrates' argumentation.

Socrates first realizes that instead of *logoi* proper his argumentation is in need of further argumentation to answer Cebes' fear – and its disheartening effect on those present – that after death the soul is simply scattered away and vanishes, just like smoke (70a). For this he will resort to a new round of the previous strange mixture, of a thorough (δια-) mythological and discursive examination (διαμυθολογῶμεν ... διασκοπεῖσθαι, 70b6 – c3), a combination of proof with soothing and reassuring words, of πίστις with παραμυθία (70b2 – 3), on the immortality of the soul and on the eternal blissfulness that awaits the philosopher's soul. So he delivers an old *logos* (λόγος παλαιός, 70c5 – 6), a combination of contemplation with speculation grounded on tradition and hearsay, namely the "argument from opposites", from which it is deduced that the soul existed before birth in heavens where it wondered at the divine spectacle of the Forms. And there it shall return after death. Calls for further argumentation on the Forms and the soul's recollection of them are answered by Socrates, but Cebes' distrust of *logoi* (ἀπιστεῖν τοῖς λόγοις, 77a9) causes a new deadlock to the conversation. He is fine with the part about the existence of the soul before birth, but the other half, the posthumous existence of the soul, needs further argumentation, he notes (77c).[37]

It is at this point that Socrates makes another comment that helps us decipher his declared need for use of *mythologein*. He explains his interlocutors' repeated mistrust to his *logoi* (and perhaps of all the disciples who have been listening in silence) in the light of psychological processes. Perhaps, he says, in the soul of each of us lies a childish fear at the thought of death (that on leaving the body our soul might be blown and scattered away by the wind). And the kind of words that seem suitable for sending away such bogeymen of the soul are the charms, the ἐπῳδοί, that were applied in antiquity alongside medical treatment to cure children (77d5 – e2). In like manner, the philosopher must seek recourse

---

36 He argues that the philosopher's soul must remain untroubled by the engagements caused by the body and oppose pleasures, pains, and the false beliefs which impede its progress to true knowledge. Instead, the philosopher must look upon the virtues to fortify his soul and to secure its dwelling among the divine in its afterlife.

37 See Morgan 2004, 194 – 196.

to a similar remedy in order to complement the therapeutics of men's souls and shed away the bogeyman that feeds on the failings from our materiality.[38] Socrates' disciples – his 'children' and practically the ones he chose to spend his last day with instead with his own biological sons – are more susceptible than him to such fears and seemingly have been overcome by it. Their reactions illustrate Socrates' diagnosis for men and accordingly advocate a similar therapy by the philosopher. Hence, this passage helps explain the call for and the (limited at the moment) recourse to blending imagery and fiction with inquiry and reasoning. More importantly, it prescribes as therapeutic means the account that will follow soon after.

As a result, Socrates then combines argumentation (on the need to purify one's soul) with accounts of strong imagery and dramatic qualities (like the one on the soul being nailed to the body by passions and its struggles to escape, 80d-81d). However, contrary to Socrates' expectations, their condition deteriorated. In the end, *everyone* but him felt unpleasant, they fell in utter disappointment and distrust as regards their own ability – and the very possibility of philosophy in general – to attain truth and knowledge at all (88c). This development brought forth the philosopher's ultimate fears, worse than his own physical death: first that his argument so far might die; second that his disciples might suffer from the disease of *misologia*, hatred of arguments (89b–d), which would threaten them with the death of philosophy inside them. So, to cure them Socrates resorts to a long mythologizing account, his swan-song and his prophesizing θεῖος λόγος that he delivers before his death in order to be applied in such therapeutics of their own souls and of those of other men (cf. 83a, 84e–85b, 85d1–4). So he embarks on a second sail to true, yet hard to perceive knowledge, his great and filled with imagery account on the afterlife of the soul.[39]

This is why at this particular point (110b1–4), at the other end of the dialogue, thereupon forming a ring composition with our introductory Aesopic passage, Socrates repeats emphatically (twice in two adjacent phrases) the term *mythos*. He does so to describe the account that follows right after (and to contradict once again, yet now decidedly, his introductory denial of being *mytholo-*

38 Philosophical discourse conceived as a kind of incantation to be sung for the therapy of the soul, but also as a task to be practiced on a daily and lifelong basis, these are well known themes in Plato's dialogues, especially in the *Charmides*. For passages and bibliography, see Zafiropoulos 2015, 183–184 n. 33.

39 It starts with the description of the true shape of the earth and of its regions, and proceeds to the wanderings and the sightings of the soul in the heavens, then its judgment and finally its life in the otherworld (110b–114c).

gikos).[40] And the myth that he now composes is not an imitation of the myths of fabulists. So it is by means of extended mythologizing controlled by the philosopher that he seeks to widen the limits of the conversation and to provide his present audience (but for future hearers and readers too) with a treatment for the disease of hatred of argumentation, that of *mythologia* against *misologia*.

A final evidence of the special importance that *mythologein* holds in the argumentation and the philosophical therapeutics of the *Phaedo* as a whole is provided at the end of Socrates' conversation with his disciples. Perhaps, as he points out,

> it does not befit a man of intelligence to insist that these things are as I have described them. However, since the soul turns out to be immortal, I think that for someone who believes this to be so it is both fitting and worth the risk – for fair is the risk – to insist that either what I have said or something like it is true concerning our souls and their dwelling places. One must, so to speak, chant such things to oneself, which is why I myself have been drawing out my myth for a long time.[41]

That is, when discourse reaches an impasse, especially with respect to issues that transcend the limits of human experience (as well as the limitations of the reasoning abilities of the many, as death and the beyond), *mythologein* is valuable as the philosopher's last resort due to its illustrative potential, his last discursive armament in stressful and inimical to reasoning conditions, despite his awareness of its possible deficit in truth status. In such cases, belief and faith in the claims to truth of philosophical myths turn to be worthwhile counterparts to knowledge. Thus, philosophy meets medico-religious therapeutics and Socrates prescribes his *mythologein* for daily recital.[42]

From this perspective, Socrates attempts not only to save *logos* from *misologia* in the *Phaedo*, but also to show a way to philosophers to restore the status of *mythos* by means of philosophically administered *mythologein*.[43] This said, one

---

40 Εἰ γὰρ δὴ καὶ μῦθον λέγειν καλόν, ἄξιον ἀκοῦσαι, ὦ Σιμμία, οἷα τυγχάνει τὰ ἐπὶ τῆς γῆς ὑπὸ τῷ οὐρανῷ ὄντα. Ἀλλὰ μήν, ἔφη ὁ Σιμμίας, ὦ Σώκρατες, ἡμεῖς γε τούτου τοῦ μύθου ἡδέως ἂν ἀκούσαιμεν.

41 114d1–7: τὸ μὲν οὖν ταῦτα διισχυρίσασθαι οὕτως ἔχειν ὡς ἐγὼ διελήλυθα, οὐ πρέπει νοῦν ἔχοντι ἀνδρί· ὅτι μέντοι ἢ ταῦτ' ἐστὶν ἢ τοιαῦτ' ἄττα περὶ τὰς ψυχὰς ἡμῶν καὶ τὰς οἰκήσεις, ἐπείπερ ἀθάνατόν γε ἡ ψυχὴ φαίνεται οὖσα, τοῦτο καὶ πρέπειν μοι δοκεῖ καὶ ἄξιον κινδυνεῦσαι οἰομένῳ οὕτως ἔχειν – καλὸς γὰρ ὁ κίνδυνος – καὶ χρὴ τὰ τοιαῦτα ὥσπερ ἐπᾴδειν ἑαυτῷ, διὸ δὴ ἔγωγε καὶ πάλαι μηκύνω τὸν μῦθον.

42 See also Gill 1993, 60; Morgan 2004, 199–200; Murray 1999, 256.

43 Actually, Plato saw myth as something that had to be saved, using a metaphor, τὸν μῦθον σῴζειν, on the heroic and therapeutic attributes of dialectics. See *Leg.* 645b1–2; *Resp.* 621b8–c1; *Tht.* 164d8–10. See also Brisson 1998, 60–61 (following a note by E.R. Dodds). On the corporal-

might object that at the end of the day (and of the dialogue) myth seems to die out: despite Socrates' lengthy and impressively illustrative eschatological narrative, as soon as the conversation is over and the time for him to drink the *pharmakon* is announced, all the disciples burst into tears and desperation. So, seemingly his *mythologein* has failed. Yet this is not necessarily the case. We must keep in mind that Socrates endowed his disciples – and philosophers to follow – with a *lifetime* therapeutic task: in order to be effective, philosophical *mythologein*-as-incantation has to be sung on a daily basis in order and, accordingly, one (and us, future readers of the *Phaedo*) must not haste into conclusions when judging from possible momentary failings (as with the disciples in this case).

To sum up, Socrates' introductory Aesopic *mythos* provided a first, yet defective approach to materiality and the upheavals it sets upon the soul, for it was anchored upon his own, temporary somatic condition, thus it was limited to an empirical statement. However, his *logoi* on the matter and on the immortality of the soul that followed, they reached a deadlock and faced skepticism, as they tried to argue and persuade on the imperceptible. Even worse, they threatened to spread disbelief among the interlocutors upon the effectiveness and even the worthiness of philosophizing at all. At this crucial point Socrates introduces philosophical *mythologein* as a discursive tool that may help him visualize issues that go beyond the limits of human perception. His mingling of *logos* with *mythos*, characteristic of him in many dialogues, is particularly evident throughout the *Phaedo*, thus refuting his introductory declaration and portraying him as the most *mythologikos*.[44]

Bibliography

Betegh, G. 2009. Tale, Theology and Teleology in the *Phaedo*. In *Plato's Myths*, ed. C. Partenie, 77–100. Cambridge.

Brisson, L. 1998. *Plato The Myth Maker.* Chicago and London (transl. G. Naddaf).

Brisson, L. 2004. *How Philosophers Saved Myths. Allegorical Interpretation and Classical Mythology.* Chicago and London (transl. C. Tihanyi).

Buxton, R. (ed.) 1999a. *From Myth to Reason? Studies in the Development of Greek Thought.* Oxford.

Buxton, R. 1999b. Introduction. In Buxton 1999a, 1–21.

ity of myth in Plato, its description as having a body or as being 'headless', see esp. *Phdr.* 264c2–5 and *Grg.* 505c10–d3; *Leg.* 752a2–4; *Ti.* 69a6–b2. See also Yossi 1996, 45.

44 See also Brisson 1998, 90,116–117; Brisson 2004, 19; Clay 2007, 214; Edelstein 1949, 468–475; Morgan 2004, 160–161; Murray 1999, 257–262.

Calame, C. 1999. The Rhetoric of Muthos and Logos: Forms of Figurative Discourse. In Buxton 1999a, 119–143.

Capuccino, C. 2014. *ΑΡΧΗ ΛΟΓΟΥ. Sui proemi platonici e il loro significato filosofico.* Florence.

Clay, D. 2007. Plato Philomythos. In *The Cambridge Companion to Greek Mythology*, ed. R.D. Woodard, 210–236. Cambridge.

Duke, E.A., W.F. Hicken, W.S.M. Nicoll, D.B. Robinson and J.C.G. Strachan. 1995. *Platonis Opera.* vol. 1. Oxford.

Edelstein, L. 1949. The Function of the Myth in Plato's Philosophy. *JHI* 10: 463–481.

Fowler, R.L. 2011. Mythos and Logos. *JHS* 131: 45–66.

Gill, C. 1993. Plato on Falsehood – Not Fiction. In *Lies and Fiction in the Ancient World*, eds. C. Gill and T.P. Wiseman, 36–87. Exeter.

Halliwell, S. 2000. The Subjection of Muthos to Logos: Plato's Citations of the Poets. *CQ* 50: 94–112.

Holzberg, N. 1993. *Die Antike Fabel. Eine Einführung.* Darmstadt.

Lloyd, G.E.R. 1989. *The Revolutions of Wisdom.* Berkeley, Los Angeles and Oxford.

Morgan, K.A. 2004. *Myth and Philosophy from the Presocratics to Plato.* Cambridge.

Most, G.W. 1999. From Logos to Mythos. In Buxton 1999a, 25–50.

Most, G.W. 2012. Plato's Exoteric Myths. In *Plato and Myth. Studies on the Use and Status of Platonic Myths*, eds. C. Collobert, P. Destrée and Fr.J. Gonzalez, 13–24. Leiden and Boston.

Murray, P. 1999. What is a Muthos for Plato? In Buxton 1999a, 251–262.

Nøjgaard, M. 1964. *La fable antique.* vol. 1. Copenhagen.

Rowe, C. 1999. Myth, History, and Dialectic in Plato's *Republic* and *Timaeus-Critias.* In Buxton 1999a, 263–278.

Sedley, D. and A. Long. 2010. *Meno and Phaedo.* Cambridge.

Tarrant, H.A.S. 1990. Myth as a Tool of Persuasion in Plato. *Antichthon* 24: 19–31.

Van Dijk, G.-J. 1993. Theory and Terminology of the Greek Fable. *Reinardus* 6: 171–183.

Van Dijk, G.-J. 1997. *ΑΙΝΟΙ, ΛΟΓΟΙ, ΜΥΘΟΙ. Fables in Archaic, Classical, and Hellenistic Greek Literature. With a Study of the Theory and Terminology of the Genre.* Leiden, New York and Cologne.

Yossi, M. 1996. *Μύθος και λόγος στον Σοφοκλή.* Athens.

Zafiropoulos, C.A. 2001. *Ethics in Aesop's Fables: The Augustana Collection.* Leiden, Boston and Cologne.

Zafiropoulos, C.A. 2015. *Socrates and Aesop. A Comparative Study of Plato's Phaedo.* Sankt Augustin.

Myrto Garani

Ovid's Temple(s) of Vesta (*Fasti* 6.249–460)

In the sixth and last book of his *Fasti*, in which Ovid delves into the origins of the Roman festivals that take place in the month of June, the poet devotes more than two hundred lines in order to account for the shape of the temple of Vesta in the Forum Romanum, the etymology of the Goddess' name, as well as the origins of the Vestalia, i.e. the people's festival that was held in her honour in the 9th of June (*Fast.* 6.249–460). In this paper, I explore Ovid' answer to what Carole Newlands describes as "Augustus' rewriting of Rome's past and shaping of its present through control of the city's monuments and calendar";[1] as I will argue, by his elegiac treatment, Ovid questions and destabilizes the recent Augustan integration of Vesta, the ancient guarantor of Roman safety, into the Roman space and the calendar.

Ovid opens his account with a hymnic invocation to Vesta to grant him favour in his endeavor to spell out the origins of her cult (6.249–256):[2]

> *Vesta, fave! tibi nunc operata resolvimus ora,*
> *ad tua si nobis sacra venire licet.*
> *in prece totus eram: caelestia numina sensi,*
> *laetaque purpurea luce refulsit humus.*
> *non equidem vidi (valeant mendacia vatum)*
> *te, dea, nec fueras aspicienda viro;*
> *sed quae nescieram quorumque errore tenebar,*
> *cognita sunt nullo praecipiente mihi.*

> O Vesta, grant me your favour! In your service now I open my lips, if it is lawful for me to come to your sacred rites. I was wrapt up in prayer; I felt the heavenly deity, and the glad ground gleamed with a purple light. Not indeed that I saw you, O goddess (far from me be the lies of poets!), nor was it meet that a man should look upon you; but my ignorance was enlightened and my errors corrected without the help of an instructor.

The poet appears to be concerned about what is lawful for him to reveal and creates the expectation to the reader that the Goddess herself will enlighten him. He confesses that he has not seen her with his own eyes, since this would be impos-

1 Newlands 2002, 226.

2 The text and translations of Ovid's *Fasti* are by Frazer 1931, revised by Goold 1996 (slightly adapted).

DOI 10.1515/9783110535150-019

sible for a man; he rather claims to have been inspired by her *numina* (6.251), even though he emphatically renounces the lies of the poets (*mendacia vatum*, 6.253). Still he justifies his putting on the vatic mantle on account of his need to rely on a divine mentor, who would help him to retrieve remote poetic memories regarding the foundation of the temple by Numa, in Rome's fortieth year (6.257–264). In other words, in the beginning of his account about Vestalia, Ovid introduces his basic method for the reconstruction of the past, i.e. the oscillation between truth and memory, so as to obliterate their alleged contrast.[3]

Before we proceed with the core of Ovid's account of the Vestalia festival, we should recall that this is not the first appearance of Vesta within the Ovidian work. The Goddess is omnipresent even if fragmented within the Ovidian calendar over the preceding half of the year, with the month of February being the only exception. To quote Herbert-Brown: "[Vesta's image is] strategically fragmented and augmented across March, April, May and June".[4] But is there a decipherable strategy behind the Ovidian portrayal of Vesta?

In the opening book of his work, which is devoted to January, more precisely on 11th of January, within the framework of the festival of Carmentalia, Carmentis, Evander's mother, utters a prophecy about Rome's forthcoming worldwide power (1.527–530):

iam pius Aeneas sacra et, sacra altera, patrem
adferet: Iliacos accipe, Vesta, deos!
tempus erit, cum vos orbemque tuebitur idem,
et fient ipso sacra colente deo

Anon pious Aeneas shall hither bring his sacred burden, and, burden no whit less sacred, his own sire; Vesta, admit the gods of Ilium! The time will come when the same hand shall guard you and the world, and when a god shall in his own person hold the sacred rites.

Carmentis makes a double *proleptic projection* into two periods of the Roman history: she gives a prophetic order to Vesta to welcome the Trojan Penates into Rome[5] and then a God as her high priest; she thus associates Vesta with the defense of Rome and predicts the glory of Augustus, who by the time of the *Fasti* was the chief priest of Rome (i.e. pontifex maximus). Carmentis' utterance reaches correspondingly two topographical ends, first the temple of Vesta in the Roman Forum which allegedly was founded in the regal period by Numa and

3 Williams 1991, 184 notes that "the information of this inspired *vates* of supposedly privileged insight is substantially derivative".

4 Herbert-Brown 2009, 136.

5 Vesta had her origins in Ilium. Cf. Verg. *Aen.* 2.296, 2.567, 5.744, 9.259.

then Vesta's shrine which was established up on the Palatine hill by Augustus, soon after he became Pontifex Maximus in 12 BC. I will revisit below the importance of this particular religious and political decision of Augustus to establish a new shrine in honour of Vesta, instead of moving himself down to the Forum in the Regia, adjoining the Temple of Vesta and the house of the Vestals as it would be otherwise expected. For now let us bear in mind that, by introducing both the mythical as well as the Augustan Vesta, Carmentis encompasses *the whole temporal and spatial span* eventually granted to the Roman Goddess.

Vesta is already present in Ovid's narration of the *Romulean period*; within the framework of March, the month which was devoted to Mars, the god of war and father of the mythical founder of Rome, Ovid describes that, while Rhea Silvia, the Vestal, was being raped by Mars, Vesta's images had their eyes covered with their virgin hands (*Silvia fit mater; Vestae simulacra feruntur / virgineas oculis opposuisse manus*; "Silvia became a mother. The images of Vesta are said to have covered their eyes with their virgin hands", 3.45–46). Vesta is also invoked by Romulus to stand by him along with Jupiter and Mars in connection with the celebration of the foundation of the city on the 21st of April and the festival of Parilia (4.827–832):[6]

vox fuit haec regis: "condenti, Iuppiter, urbem
et genitor Mavors Vestaque mater, ades;
quosque pium est adhibere deos, advertite cuncti.
auspicibus vobis hoc mihi surgat opus.
Longa sit huic aetas domitaeque potentia terrae,
sitque sub hac oriens occiduusque dies."

The king spoke thus: 'O Jupiter, and Father Mavors, and Mother Vesta, stand by me as I found the city! O take heed, all you gods whom piety bids summon! Under your auspices may this my fabric rise! May it enjoy long life and dominion over a conquered world! May East and West be subject unto it!"

Apart from these brief references to the *pre-Augustan* Vesta, scholars agree that, in the course of the first five books of his *Fasti* Ovid's main *temporal focus*, as regards the figure of Vesta, is the *Augustan era*. In fact, Ovid seems to create what Gareth Williams calls "an Augustan climax".[7] On the 6th of March Ovid commemorates the fact that on the death of Lepidus Augustus was elected *Pontifex Maximus* (3.415–428):

6 The Vestals participate in the Parilia festival, during which they burn on their altar a calf and blood from the tail of the victorious horse in the October race (Ov. *Fast.* 4.721–734). For more on Vestals and their virginity, see Beard 1980 and 1995; Wildfang 2006.
7 Williams 1991, 196.

Sextus ubi oceano clivosum scandit Olympum
Phoebus et alatis aethera carpit equis,
quisquis ades castaeque colis penetralia Vestae,
gratare, Iliacis turaque pone focis.
Caesaris innumeris, quos maluit ille mereri,
accessit titulis pontificalis honor.
ignibus aeternis aeterni numina praesunt
Caesaris: imperii pignora iuncta vides.
di veteris Troiae, dignissima praeda ferenti,
qua gravis Aeneas tutus ab hoste fuit,
ortus ab Aenea tangit cognata sacerdos
numina: cognatum, Vesta, tuere caput!
quos sancta fovet ille manu, bene vivitis, ignes:
vivite inexstincti, flammaque duxque, precor.

When the sixth sun climbs up Olympus' steep from ocean, and through the ether takes his way on his winged steeds, all you, whoever you are, who worship at the shrine of the chaste Vesta, wish the goddess joy and offer incense on the Ilian hearth. To Caesar's countless titles, which he has preferred to earn, was added the honour of the pontificate. Over the eternal fire the divinity of Caesar, no less eternal, you preside: the pledges of empire you see side by side. You gods of ancient Troy, you worthiest prize to him who bore you, you whose weight did save Aeneas from the foe, a priest of the line of Aeneas handles your kindred divinities; Vesta, do you guard his kindred head! Nursed by his sacred hand, you fires live well. O live undying, flame and leader both, I pray.

This was the day that Vesta, the goddess of the hearth acquired for the first time a male priest.[8] What is even more important, Ovid claims that it was then that Augustus became a *cognatus* of Vesta and hence calls upon the Goddess to guard Augustus' kindred head (*cognatum, Vesta, tuere caput!* 3.426), since the latter is meant to be the one who will take care of the pledges of the empire (*imperii pignora*, 3.422). As Herbert-Brown argues, the '*cognatio* theme' was an Ovidian creation: "alone by re-creating and conveying the new mythology behind that cult, Ovid shows how the ancient institution of Vesta's worship in the Forum was *subordinated* to the new Pontifex Maximus and the Palatine".[9]

---

8 Herbert-Brown 1994, 71–72. As Herbert-Brown 1994, 69 also notes, this was the first time that in practice the emperor got involved in the cult more intimately and this change is reflected by the topographical link between Vesta's new home in the residence of the new Pontifex Maximus on the Palatine: "The idea of Vesta having a male priest did not exist before Augustus became Pontifex Maximus ... A primary relationship of a sacred nature between Augustus and the Goddess of the Roman hearth has been created". See also Bömer 1987.

9 Herbert-Brown 1994, 80. As a descendant of Aeneas, Augustus descends from Jupiter through Venus and so is *cognatus* 'collateral kinsman' of Vesta. Cf. Ov. *Met.* 15.864–865, *Fast.* 3.425–426. See also Fraschetti 1988, 956–957; Fantham 1998, 275 ad 949–951.

On the 28th of April, just a few weeks after Augustus' election as *pontifex maximus*, Ovid refers to Augustus' consecration of Vesta's shrine on the Palatine (4.949–954):

aufer, Vesta, diem: cognati Vesta recepta est
limine; sic iusti constituere patres.
Phoebus habet partem: Vestae pars altera cessit:
quod superest illis, tertius ipse tenet.
state Palatinae laurus, praetextaque quercu
stet domus: aeternos tres habet una deos.

O Vesta, take your day! Vesta has been received in the home of her kinsman: so have the Fathers righteously decreed. Phoebus owns part of the house; another part has been given to Vesta; what remains is occupied by Caesar himself. Long live the laurels of the Palatine! Long live the house wreathed with the oaken boughs! A single house holds three eternal gods.

Vesta becomes then one of the three deities on the Palatine, along with Apollo and the emperor himself; accordingly, instead of the Forum, it is the emperor's palace that is now established as the sacred center of the city. We cannot be sure whether Augustus *transferred* Vesta's cult from the Forum to his palace or just duplicated Vesta's hearth, although according to the historians the latter case is more plausible.[10] Dio Cassius reports that, since the Pontifex Maximus had to live in a public residence, Augustus made a portion of his own house public property".[11] In other words, Augustus transformed his *private space* into *public*; the blending of those two spaces, which were so far symbolically separate made possible the fusion of the private cults with those of the state. Barchiesi rightly stresses the fact that "the impact of this initiative can hardly be overestimated,

10 For the discussion about the problem of what was exactly dedicated to Vesta on Palatine see Herbert-Brown 1994, 74–81, who concludes that the cult of Vesta was reproduced on, not transferred to the Palatine. On the various temples of Vesta see Hill 1989, 23–24, 31–32; LTUR vol. V s.v. Vesta (1999). For the problem of the existence of Vesta's image on the Palatine see also Barchiesi 1997, 206–207: "According to the diction of the official calendars Augustus had welcomed the arrival of a *signum* of the goddess in his house. What relationship is there between this 'portrait' and the mysterious 'vacuum' in the traditional '*aedes Vestae*' in the Roman Forum?".
11 Dio Cass. 54.27.3 καὶ οὔτε ἐκεῖνα ἔτ' ἐκυρώθη οὔτ' οἰκίαν τινὰ δημοσίαν ἔλαβεν, ἀλλὰ μέρος τι τῆς ἑαυτοῦ, ὅτι τὸν ἀρχιέρεων ἐν κοινῷ πάντως οἰκεῖν ἐχρῆν, ἐδημοσίωσεν. τὴν μέντοι τοῦ βασιλέως τῶν ἱερῶν ταῖς ἀειπαρθένοις ἔδωκεν, ἐπειδὴ ὁμότοιχος ταῖς οἰκήσεσιν αὐτῶν ἦν. "That measure therefore, now failed of passage, and he also received no official residence; but inasmuch as it was absolutely necessary that the high priest should live in a public residence, he made a part of his own house public property. The house of the rex sacrificulus, however, he gave to the Vestal Virgins, because it was separated merely by a wall from their apartments". Dio's translation is by E. Cary (Loeb vol. VI, 1927).

because, for the first time in Rome's history, it brings about a close integration between state cults and 'private' cults, a fusion that is naturally only made possible by the great polyvalence of the self-image constructed by Augustus, which is in continual oscillation between formerly separate symbolic spaces."[12]

What is accordingly the impact of Vesta's spatial transportation upon the identity of the Goddess herself? To quote Feeney, "it was an extraordinary transformation for *Vesta publica populi Romani Quiritum*, the guarantor of the city's identity and continuity, whose whole raison d'être consisted in remaining *fixed* in her *sedes*".[13] This decisive imperial deed of fiddling with one of Vesta's intrinsic traits, i.e. her stability, is also mirrored in the Ovidian text. While the consecration of the Palatine shrine coincides with the first day of the Floralia, the five-days licentious and bawdy fertility festival in honour of Flora, the Goddess of the blossoming flowers, Ovid gives an eccentric command to Vesta to seize the day (*aufer, Vesta, diem* / "O Vesta, take thy day!", *Fast.* 4.949). Despite the fact that Vesta could just have shared the day with Flora, the poet ostracizes the latter's story to be narrated in the next book (5.183–378) and reserves the *textual space* only for Vesta.

So far, we have seen that in association with Augustus' priesthood and the foundation of the Palatine shrine, Vesta appears to be the *passive receiver* of the poet's orders; let us now turn to the passages in which the presence of Vesta is associated with the *Augustan concept of revenge*; as we will soon grasp, in these cases the relationship between the Goddess and the poet is capsized: it is Vesta who now appears to be the authoritative agent.

In his narration of the events associated with the Ides of March (15th of March), Ovid devotes the major part of his textual space to the story of another plebeian festival, that in honour of Anna Perenna, the *numen* of the recurring year; a festival similar to the Floralia in rowdiness and public reveling. Vesta, however, interrupts Ovid's narration and urges the poet to deflect his focus and delve into his memory regarding the events of Caesar's assassination; in doing so, he initiates a *temporal retrogression* (*Fast.* 3.697–702):

> *Praeteriturus eram gladios in principe fixos,*
> *cum sic a castis Vesta locuta focis:*
> *"ne dubita meminisse: meus fuit ille sacerdos,*

---

12 Barchiesi 1997, 204–205. See also Fantham 1998, 275 who remarks that "This public status of his domestic precinct and blending of his private and public identities made it possible to represent his private shrine of Vesta and the Penates as public cult, although the public cult continued in the forum temple".

13 Feeney 1991, 215.

sacrilegae telis me petiere manus
ipsa virum rapui simulacraque nuda reliqui:
quae cecidit ferro, Caesaris umbra fuit."

I was about to pass by in silence the swords that stabbed the prince, when Vesta spoke thus from her chaste hearth: 'Doubt not to recall them: he was my priest, it was at me these sacrilegious hands struck with the steel. I myself carried the man away, and left naught but his wraith behind; what fell by the sword was Caesar's shade.

In a way similar to what Vesta earlier did with regard to Flora, she vindicates once again extra textual space, in the case in question that of Anna Perenna, and the corresponding space reserved for the latter within the Roman calendar. At the same time, Vesta usurps the *space of action* which was granted to Venus in Ovid's *Metamorphoses* (15.843–851), since the goddess of the hearth replaces Venus as the divinity who rescues the body of Julius Caesar and thus becomes the agent of his apotheosis. By this double expansion at the expense of both Anna Perenna and Venus, the Ovidian Vesta succeeds in integrating Julius Caesar *anachronistically* into the contemporary time-span of her influence.

With her own place established within the Ovidian text, the Roman calendar and the Roman space, Vesta reappears in Book 5 to celebrate the foundation of the temple of Mars Ultor in the 12th of May (5.573–577):

"si mihi bellandi pater est Vestaeque sacerdos
auctor, et ulcisci numen utrumque paro,
Mars, ades et satia scelerato sanguine ferrum,
stetque favor causa pro meliore tuus.
templa feres et, me victore, vocaberis Ultor."

"If my father, Vesta's priest, is my warrant for waging war, and I do now prepare to avenge both his divinity and hers, come, Mars, and glut the sword with knavish blood, and grant your favour to the better cause. You shall receive a temple, and shall be called Avenger, when victory is mine".

Although the urban space of Rome was not a traditional place for Mars,[14] it is striking that it is Augustus who invites him within the urban space of Rome and activates his role as avenger, authorizing thus the presence of war within the elegiac framework.[15]

14 Barchiesi 2002, 5. See also Newlands 1995, 92–93; Herbert-Brown 2009, 135.
15 Herbert-Brown 2009, 136: "In the 4 Julian passages considered above, we see that Ovid invokes and enhances Vesta as the authoritative agent who introduces three new Julian gods into the Roman calendar".

Let us now turn back to Ovid's account of Vestalia in book 6, in which the reader is faced with two critical reversals.

In the first instance, Ovid shifts the focus from Vesta's Palatine shrine to a detailed description of the temple of *Vesta publica populi Romani Quiritium* which was situated in the Roman Forum (6.265–282):

forma tamen templi, quae nunc manet, ante fuisse
dicitur, et formae causa probanda subest.
Vesta eadem est et terra: subest vigil ignis utrique:
significant sedem terra focusque suam.
terra pilae similis, nullo fulcimine nixa,
aëre subiecto tam grave pendet onus.
ipsa volubilitas libratum sustinet orbem,
quique premat partes, angulus omnis abest,
cumque sit in media rerum regione locata,
et tangat nullum plusve minusve latus,
ni convexa foret, parti vicinior esset,
nec medium terram mundus haberet onus.
arte Syracosia suspensus in aëre clause
stat globus, immensi parva figura poli,
et quantum a summis, tantum secessit ab imis
terra; quod ut fiat, forma rotunda facit.
par facies templi; nullus procurrit in illo
angulus; a pluvio vindicat imbre tholus.

Yet the shape of the temple, as it now exists, is said to have been its shape of old, and it is based on a sound reason. Vesta is the same as the Earth; under both of them is a perpetual fire; the earth and the hearth are symbols of the home. The earth is like a ball, resisting on no prop; so great a weight hangs on the air beneath it. Its own power of rotation keeps its orb balanced; it has no angle which could press on any part; and since it is placed in the middle of the world and touches no side more or less, if it were not convex, it would be nearer to some part than to another, and the universe would not have the earth as its central weight. There stands a globe hung by Syracusan art in closed air, a small image of the vast vault of heaven, and the earth is equally distant from the top and bottom. That is brought about by its round shape. The form of the temple is similar: there is no projecting angle in it; a dome protects it from the showers of rain.

Although the original *aedes* which was built by Numa had to be restored after its destruction by fire in 41 BC and so its thatched roof was replaced with a bronze or copper one (*Fast.* 6.261), what remained *unchanged* ever since was the temple's *circular tholos shape*, as it is attested by coins.[16] In order to account for

16 For the representation of the temple on coins see Hill 1989, 23–24; LTUR s.v. Templum Vestae.

this atypical building-shape, Ovid follows Verrius Flaccus' scientific explanation and equates Vesta's temple with the *earth*, as both of them possess an eternal fire (*Fast.* 6.267; cf. Festus p. 320 Lindsay). Ovid overlooks the difference between the circular temple and a sphere, so as to impose the *central place* of the temple within the Roman space, by making use of what De Jong calls "the symbolic function of space".[17] As Gee has discussed, this idea is associated with the Stoic idea of the rotatory motion of the earth within the *geocentric* universe, itself maintained in equipoise by centripetal force.[18] In order to further exemplify this *centrality*, the temple is also compared with the sphere of Archimedes, the 3rd century Syracusean mathematician; this sphere which was brought to Rome by M. Claudius Marcellus after his conquest of Syracuse was a miniature globe with movable parts representing the vast heavens (*Fast.* 6.277–278).[19] Ovid thus makes a double *spatial projection*, comparing the temple both with heavens and then with its astronomical miniature, so as to emphasize that Vesta's hearth orients human space by being located at the *centre* of the cosmos, this overlapping with the Roman Forum.[20] Gee comments that "Ovid's positioning of the Sphere of Archimedes inside the Vestalia, a Roman festival most important to the Julians after 12 BC, superimposes the presence of the ruler on the globe in an analogous function."[21] As Gareth Williams puts it: "If Vesta is the earth (267) then Augustus is its military master, and if the earth lies at the centre of the universe, then by implication Vesta's Roman hearth is the focal point of the whole universe".[22] Ovid then adds *multiple etymologies* about the name of Vesta (6.299–304):

stat vi terra sua: vi stando Vesta vocatur,
causaque par Grai nominis esse potest.
at focus a flammis et quod fovet omnia, dictus;
qui tamen in primis aedibus ante fuit.

17 De Jong 2012, xiv "when space becomes semantically charged and acquires an additional significance on top of its purely scene-setting function".
18 Gee 2000, 94–107.
19 The sphere of Archimedes occurs only three times in literature before Ovid (Cic. *Rep.* 1.21–22, *Tusc.* 1.63, *Nat. D.* 2.88).
20 For the identification of Vesta with the earth and the conception of her temple as a symbolic representation of the globe see Dion. Hal. *Ant. Rom.* 2.66.3; Festus s.v. *Rutundam aedem* (p. 320 Lindsay) and Servius *auct.* on *Aen.* 2.296. In Plut. *Num.* 11, Vesta is identified not with the earth but with the whole universe. Cicero derives it from the Greek Hestia (*Nat. D.* 2.67). See also Varro (Aul. Gell. XIV 7.7) who says that it was not a temple but a round building, merely a copy of the round hut of the early kings, with walls of osiers and a primitive thatched roof.
21 Gee 2000, 123.
22 Williams 1991, 196.

> *hinc quoque vestibulum dici reor; inde precando*
> *praefamur Vestam, quae loca prima tenet.*
>
> The earth stands by its own power; Vesta is so called from standing by power (*vi stando*); and the reason of her Greek name may be similar. But the hearth (*focus*) is so named from the flames, and because it fosters (*fovet*) all things; yet formerly it stood in the first room of the house. Hence, too, I am of opinion that the vestibule took its name; it is from there that in praying we begin by addressing Vesta, who occupies the first place.

Among other etymologies, he derives Vesta's name from *vi stare* (*Fast.* 6.299–304), stating that Vesta as earth should remain immobile.[23] It should not go unobserved that Ovid does not refer to the Varronian etymology of Vesta as the one who is vested in flowers (Varro *apud* Augustine *De Civ. D.* 7.24 fr. 268 Cardauns: *tellurem, inquit (Varro), putant esse ... Vestam, quod vestiatur herbis* "they think *Tellus* is ... Vesta because she is vested in flowers"), this being the etymology to which he has himself already hinted at the end of his Book 4 (*Fast.* 4.945: *mille venit variis florum dea nexa coronis*, "Vesta comes plaited with varied garlands of a thousand flowers").

Once Ovid leads the reader from Vesta's Palatine temple down the hill, back to the Forum, he subdues the Goddess as well to a process of *spatial transportation*, which – as I would like to argue – subsequently results in the *disruption* of the semantics of the newly organized Augustan space.

In order to stress further the *centrality* of Vesta's original temple within the Roman space, Ovid underlines also Vesta's *spatial primacy.* The poet points to the fact that, since Vesta corresponds to the Greek Hestia, i.e. the hearth, she is the same as *sacred fire* (6.301); hence, he derives the etymology of *focus* from *flamma* and *foveo.* Given the fact that the hearth stood in the first room of the house (*in primis aedibus*, 6.302), the poet almost arbitrarily associates her name with that of the entrance of the house, the so-called *vestibulum.*[24] With the purpose of underscoring her *primacy*, Ovid goes on to argue that, since she is located first in houses -with the word *vestibulum* (6.303) meaning also the beginning or opening part of a speech-, she comes first also in prayers (6.304). In doing so, the poet obfuscates the fact that this suggested liturgical primacy goes counter to the

---

23 Vernant 1974. Gee (2000, 116) notes that the Greek name Hestia was said by scholars in antiquity to come from ἑστάναι (ἵστημι, Ovid's *stando*). Other ancient etymologies derive Hestia from ἕζομαι (*Homeric Hymn* 5.30). Ovid applies *vi stat* to the earth, while in the exegetical tradition it is usually attached to the hearth (presumably the flame, which "stands up by itself".
24 Gell. 16.5.6–12, Macrob. 6.8.19–20 (from *ve-stabulato* 'standing for a long time').

Roman practice, according to which in prayers Vesta comes last (cf. Cic. *Nat. D.* 2.67).[25]

Scholars have been puzzled by the double identity of the Ovidian Vesta as both earth and fire, up to the point to consider Ovid's narration to be a conundrum.[26] Whatever the case may be, by means of his multiple aetiological explanations, Ovid firmly defines Vesta's *focal place* within the Roman space.

The *centrality* and *primacy* of people's Vesta is stressed also in textual terms. Ovid does not draw up the boundaries of his narration about the Vestalia within the narrative text devoted to the 9th of June. In June the 6th (6.227–234), he talks about the period of cleansing her temple. And in June 15th again (6.713–714), he informs us that the debris of her temple was swept into the Tiber and carried away to the sea. By being extended backwards and forwards in both calendrical and textual terms, the Ovidian narration strives to enlarge once again the space – textual as well as literal – that Vesta occupies within the framework of the Roman calendar.

Last but not least, the contemporary reader should also call to mind the fact that from the 7th to the 15th of June on the occasion of the festival, the entrance of the temple, which was by definition limited only to the Vestals, opened to the public for the only time of the year, but still only for married women. Contrary to what Augustus did in association with the Palatine shrine, Vesta's private space in the Forum is only temporarily and partially transformed into public. Despite the fact that the emperor wishes to control boundaries within Roman space, he has to compromise with the fact that in the case of Vesta's shrine he can only act from the outside.

We have just explained that in the Vestalia passage, Ovid presents Vesta as *immobile*. In the beginning of Ovid's narration, the reader is faced with yet another reversal: up to Book 6, during both the Romulean and the Augustan era the Goddess has figured in her *anthropomorphic disguise*; the poet, however, sheds now serious doubts upon what he has narrated so far; it is time for both himself and the reader to get disillusioned: Vesta is *aniconic* (6.295–298):

esse diu stultus Vestae simulacra putavi,
mox didici curvo nulla subesse tholo:
ignis inexstinctus templo celatur in illo,
effigiem nullam Vesta nec ignis habet.

25 Barchiesi 1997, 206; Gee 2000, 117–118.

26 Gee 2000, 117. Servius auct. *ad Aen.* 2.296 may resolve the difficulty, since one is said to contain the other.

> Long did I foolishly think that there were images of Vesta: afterwards I learned that there are none under her curved dome. An undying fire is hidden in that temple; but there is no effigy of Vesta nor of the fire.

Ovid explains that as symbol of the Roman hearth, she is a living flame (*vivam flammam*, 6.291) and as such, she is incorporeal and has no cult statue (*effigiem nullam*, 6.298).

Let us pause for a moment to reiterate Ovid's double reversal: at odds with the Augustan Goddess with which the reader has so far become familiar and plausibly also at odds with the Roman practice, since there was no religious prohibition regarding the statues of Vesta (cf. Cic. *De Or.* 3.2.10), Ovid states that the original Vesta, to be found in the Forum, is immobile and without a cult image. Ovid's account of the Vestalia is thus revealed to be particularly significant, regarding his stance towards the Augustan religious reformation; in fact I would go so far as to claim that in this context Ovid creates an Augustan *anti-climax:* if Vesta is immobile, Ovid seems to suggest, then her relocation by Augustus is vain or – even worse – deceptive.

In the remaining verses of his account, however, Ovid's subversive stance does not appear to be crystallized. In the three aetiological stories that follow, the poet strikingly invents the *temporal trajectory* from myth into history, with Vesta granted again her anthropomorphic face. Due to the restrictions of space, I will not go over the details of the stories; I will rather underscore only the points which are somehow relevant to my argument.

In order to account for the fact that Vesta became the patroness of bakers and millers, Ovid narrates the story of her attempted rape by Priapus, which was interrupted by the braying of an ass (6.319–348).[27] Within the framework of this story with its farcical tone, Vesta's nationalistic identity is undermined; her virginity, which has been just associated with the Goddess' nature as fire, is also threatened.[28] That is why, as Newlands argues, by means of this story Ovid further destabilizes the poem's aetiological quest.[29]

In the next story (6.349–394), the one about the so-called Baker Jupiter [Jupiter Pistor] Ovid makes a gradual transition *from mythical into historical time*, by

---

27 Fantham 1983, Williams 1991. Littlewood 2006, 100 ad 6.311 points out that Vesta's image, in her human form, between the two Lares, with her companion donkey, traditionally adorned the *lararia* of Pompeian bakers, since she was the patroness of bakers and donkeys. For the duplication of the story in Book 1 and Book 6 see Newlands 1995, 124–145; Littlewood 2006, 103–105.
28 Newlands 1995, 131 also remarks that "Livia modeled herself upon Vesta in official cult as a mother figure of exemplary chastity". Cf. Ov. *Pont.* 4.13.29.
29 Newands 1995, 129.

proposing a version which is at variance with the historical tradition:[30] during the siege of the Capitol by the Gauls in 396 BC gods made a council, so as to decide how to assist the Romans. In the presence of Venus and Mars, Jupiter suggests that the Romans should bake bread and throw it over to the enemies, in order to trick them regarding their lack of supplies. Vesta becomes then the "head chef", in Newlands' expression,[31] being the supervisor of both the grinding and the baking. Vesta eventually saves the Capitol from the Gauls and thus gets closely associated with the *defense of Rome*. Newlands stresses the fact that despite the narrative transfer from Priapus' mythical story to the historical context, once again Ovid sticks to the comic character of his narrative "making fun of the project of aetiology".[32]

Vesta's role as Rome's defender is revisited in Ovid's last aetiological story; while Ovid first draws attention to her own role as guardian of the Palladium, this role is soon transferred to Metellus, who figures as her high priest, since he is the one who despite the prohibition enters the shrine and rescues the statue when Vesta's shrine caught fire (6.417–436). Whereas Vesta and Minerva are both associated with what Newlands calls "concept of inviolate chastity", the passage abounds in sexual ambiguity, with latent hints at Metellus' metaphorical rape of the Goddess;[33] scholars stress the fact that in a way similar to what Ovid did in connection with Julius Caesar's priesthood (cf. 3.699, 5.573), the idea of the Pontifex Maximus as Vesta's priest is –to quote once again Herbert-Brown – a "retrojection of an essentially Augustan concept into Republican times: to convey a sense of continuity and tradition to legitimate and camouflage the new meaning of the office of Pontifex Maximus held by Augustus".[34]

---

30 According to the historians' account (Liv. 5.48.8–9, Diod. Sic. 14.116, Val. Max. 7.4.3, Plut. *Cam.* 22.9), Brennus' Gauls ended the siege by receiving a ransom of a thousand pounds of gold. See Williams 1991, 185–186; Littlewood 2006, 122–123 ad 6.393–394.

31 Newlands 1995, 133.

32 Newlands 1995, 135.

33 Newlands 1995, 137–138. For the concept of inviolate chastity see especially the expressions of the temple's amorous burning (*arsit*, 6.438) and of Minerva's rescued image as being raped (*rapta*, 6.453). Newlands 1995, 138 points out that "Metellus' literal penetration into Vesta's innermost sanctuary and removal of her sacred objects is presented as a metaphorical act of sexual defloration".

34 Herbert-Brown 1994, 71. Newlands 1995, 137: "The Vesta / Priapus myth in Book 6 acts as a counterpart to the myth of Vesta and Metellus, and the two myths frame the comic aetion of Baker Jupiter. Both stories concern the goddess as victim of rape, one in actual but mythical terms, the other in metaphorical but historical terms ... 139: Vesta's latest *pontifex maximus*, Augustus, takes over Metellus' dominant role by guaranteeing the protection of her sacred flame and the sexual purity of her cult".

By way of these three aetiological stories, the poet appears to follow anew the *spatial trajectory* from the Forum up to the Palatine. In order to do so, he has first to break the boundaries between past and present and immerse into *the past*; in this way, he ends up by questioning the patriotic identity of Vesta, since this contradicts with her physical vulnerability.

Just before bringing his account of the Vestalia to an end, the poet creates a cycle which looks back to the introductory verses of the passage: while he reminds the reader of the fact that the doom of unchaste Vestals was to get buried alive as a punishment, he repeats his claim that Vesta and the earth are the same element (6.459–460 *sic incesta perit, quia quam violavit, in illam / conditur: est Tellus Vestaque numen idem.* "Because she is put away in the earth which she contaminated, since Earth and Vesta are one and the same deity"). In doing so, the poet seems to be decisively tipping the balance towards the (Republican) Vesta and hence destabilizing Vesta's place within Roman space, precisely in accordance with the recent establishment of the goddess by Augustus on the Palatine.

By means of a typical Ovidian reversal, in the very end of the Vestalia passage the poet reserves four verses to refer to the *spatial trajectory* from the Forum up to the Palatine. Vesta looks back at Carmentis in Book 1 and delivers herself a prophecy about the retrieval of the lost standards and Augustus' revenge (6.465–468):

Crassus ad Euphraten aquilas natumque suosque
perdidit et leto est ultimus ipse datus.
"Parthe, quid exsultas?" dixit dea "signa remittes,
quique necem Crassi vindicet, ultor erit".

Crassus lost the eagles, his son, and his soldiers at the Euphrates, and perished last of all himself. "Why exult, you Parthian?" said the goddess; "you shall send back the standards, and there will be an avenger who shall exact punishment for the slaughter of Crassus".

Herbert-Brown considers this reference to be the confirmation of the "conceptual annexation of the old cult by the new".[35] And I would be willing to accept this claim unconditionally, if it were not for the serious cracks that Ovid's preceding account has caused upon Vesta's nationalistic identity.

To sum up, within a poem that strives to organize the concept of time into an ordered whole for didactic purposes, when it comes to Vesta, Ovid employs a number of proleptic and retrospect references that result in her temporal fragmentation over the Ovidian months as well as her temporal projection into the

35 Herbert-Brown 1994, 79.

Roman past. Whereas Ovid subjects Vesta's temple in the forum to the process of – what Barchiesi calls – *verbalization of the monument*,[36] by underscoring its immobility and its lack of cult image, the Goddess herself constantly oscillates between her ancient dwelling and the one recently founded on the Palatine hill. To put it differently, on the basis that old and new, i.e. the popular and the Augustan Vesta, were also so separate topographically, Ovid keeps throughout his account the interplay between the two facets of the Goddess, ultimately undermining her imperial credentials.[37]

Bibliography

Barchiesi, A. 1997. *The Poet and the Prince: Ovid and Augustan Discourse.* Berkeley (transl. of *Il poeta e il principe. Ovidio e il discorso augusteo.* Rome 1994).

Barchiesi, A. 2002. Martial Arts: *Mars Ultor* in the Forum Augustum. In *Ovid's Fasti: Historical Readings at Its Bimillennium*, ed. G. Herbert-Brown, 1–22. Oxford.

Beard, M. 1980. The Sexual Status of the Vestal Virgins. *JRS* 70: 12–27.

Beard, M. 1995. Re-reading (Vestal) Virginity. In *Women in Antiquity: New Assessments*, eds. R. Hawley and B. Levick, 166–177. London.

Bömer, F. 1958. *Die Fasten, Band II, Commentar.* Heidelberg.

Bömer, F. 1987. Wie ist Augustus mit Vesta verwandt? Zu Ov. *Fast.* III 425 f. und IV 949 f. *Gymnasium* 94: 525–528.

Cardauns, B.M. 1976. *Terentius Varro Antiquitates Rerum Divinarum Teil I: Die Fragmente.* Wiesbaden.

De Jong, I.J.F. (ed.) 2012. *Space in Ancient Greek Literature. Studies in Ancient Greek Narrative,* vol. 3 (Mnemosyne, Suppl. 339). Leiden and Boston.

Fantham, E. 1983. Sexual Comedy in Ovid's *Fasti:* Sources and Motivation. *HSPh* 87: 185–216.

Fantham, E. 1998. *Ovid, Fasti Book IV.* Cambridge.

Feeney, D. 1991. *The Gods in Epic: Poets and Critics of the Classical Tradition.* Oxford.

Fowler, W.W. 1916. *The Roman Festivals of the Period of the Republic.* London.

Fraschetti, A. 1999. Augusto e Vesta sul Palatino. *ARG* 1.2: 174–183.

36 Barchiesi 2002, 21

37 Newlands 1995, 132: "In his treatment of Vestalia, Ovid offers us three basically incompatible views of Vesta: the rational view that sees her as a symbol of fire or earth (267–282, 289–294, 460); the popular, Republican view that sees her as a goddess of the people and, in particular, the lowest social classes such as bakers (311–318); and at the other end of the social scale, the Augustan view that sees her as protectress of the imperial family and their empire (455–468). Ovid makes no attempt to reconcile these different perspectives." Newlands 1995, 139: "The narrator presents us with the popular and the Augustan Vesta, treating one with *urbanitas* and the other with *gravitas*".

Frazer, Sir J.G. 1996. Ovid's *Fasti* with an English translation by Sir James George Frazer, revised by G.P. Goold (first published 1931, reprinted with corrections 1996; Loeb Classical Library 253).

Gee, E. 2000. *Ovid, Aratus and Augustus. Astronomy in Ovid's Fasti.* Cambridge.

Green, C.M.C. 2002. Varro's Three Theologies and their Influence on the Fasti. In *Ovid's Fasti: Historical Readings at its Bimillenium*, ed. G. Herbert-Brown, 71–99. Oxford.

Green, S.J. 2004. *Ovid, Fasti 1. A Commentary* (Mnemosyne, Suppl. 251). Leiden.

Herbert-Brown, G. 1994. *Ovid and The Fasti: A Historical Study.* Oxford.

Herbert-Brown, G. 2009. *Fasti:* The Poet, the Prince and the Plebs. In *A Companion to Ovid*, ed. P.E. Knox, 120–139. Oxford.

Hill, P.V. 1989. *The Monuments of Ancient Rome as Coin Types.* London.

Littlewood, R.J. 2006. *A Commentary on Ovid's Fasti, Book 6.* Oxford.

Newlands, C.E. 1995. *Playing with Time: Ovid and The Fasti.* Ithaca, NY.

Newlands, C.E. 2002. Contesting Time and Space: *Fasti* 6.637–48. In *Ovid's Fasti: Historical Readings at Its Bimillenium*, ed. G. Herbert-Brown, 225–250. Oxford.

Scullard, H.H. 1981. *Festivals and Ceremonies of the Roman Republic.* London.

Wildfang, R.L. 2006. *Rome's Vestal Virgins: Vestal Priestesses in the Late Republic and Early Empire.* New York.

Williams, G.D. 1991. Vocal Variations and Narrative Complexity in Ovid's *Vestalia: Fasti* 6.249–468. *Ramus* 20: 183–204.

Vernant, J.-P. 1974. Hestia-Hermès. Sur l'expression religieuse de l'espace et du mouvement chez les Grecs. In *Mythe et pensée chez les Grecs. Étude de psychologie historique*, 2 vols., ed. J.-P. Vernant, 155–201. Paris.

Sophia Papaioannou

Carmenta in the *Fasti*: A Tale of Two Feasts

Julius Caesar, Augustus and the Politics of Time-management

In the last three decades, the interpretation of Ovid's erudite treatment of the Roman calendar in the *Fasti* is conditioned by the interlocking of poetics and politics.[1] It is commonly agreed that Ovid with admirable sophistication 'rewrote' the calendar and, in doing so, produced an inventive criticism of the reorganisation of traditional religion that spearheaded the ideological revival of *Romanitas* under Augustus. Augustus' restoration of Roman religion was founded on the conviction that the Roman *pietas* of old times had to come back; that the civil wars were a parenthesis now closed for ever, and that the civic experience in post-Actium Rome, built on the tradition of the Roman republican *mores*, would prove this continuity.[2] Nonetheless, the ideology of traditional religion was impossible to revive, given that its experience was irreversibly gone; a legendary and romanticised memory rather than a lost reality among the Romans of the Augustan age. As a result, the projected revival was essentially a thoughtful reconstruction through which several key reforms tied to the alleged restitution of the lost age of republican innocence would become readily acceptable. Given the interconnection of religion and calendar at Rome, the revival of the for-

1 The bibliography on the peculiarly political and uniquely intelligent poetics of the *Fasti* is large, and the last thirty years have generated a number of important and widely-read monographs, including Miller 1991; Mackie 1992; Herbert-Brown 1994 and 2002; Newlands 1995; Barchiesi 1997; Wiseman 1998, chapters 2, 3, 4, 7, and 2008, chapters 6, 7, 13; Pasco-Pranger 2006; La Bua 2010. The bibliographies at the end of each of these books, longer by the year, attest firmly to the unabated interest the study of the *Fasti* continues to generate.

2 Aptly phrased by Scheid (2005, 192): "[Augustus'] cautious innovations [in the administration of Roman religion] were successful because the Romans, and most of all the Roman elite, were open to innovation but nevertheless remained very conservative". Scheid 2005 is essential for defining and understanding Augustus' religious policy. Other notable recent studies on the matter include Galinsky 2011; Wallace-Hadrill 2005, on Augustus' treatment of religion as a domain of special knowledge to be manipulated; Galinsky 1996, 288–331; Price 1996.

DOI 10.1515/9783110535150-020

mer inevitably signalled the rewriting of the latter, a process evidently not just cultural but profoundly political.[3]

Until the days of Julius Caesar, the organisation of the calendar (*fasti*), meaning not just the reckoning of time but more importantly the administration of public life – since the *fasti* designated the days for public business, meetings of the courts, political business, and religious celebrations – , was in the hands of priests and magistrates, all of whom were members of the aristocracy. Julius Caesar in his major reorganisation of the calendar took a first step that would progressively lead to the removal of the calendar from the control of the aristocracy, by appointing expert professionals (mathematicians and astronomers) to undertake a precise calculation of real time and use this calculation as the basis for the restructuring of festal and civic time.[4] Contrary to the republican calendar of the pontiffs that was flexible (e. g. feasts were celebrated on different dates every year, and their celebration was determined by the pontiffs and was disclosed on short notice, while the pontiffs often abused their authority over the intercalation), Caesar's calendar 'fixed' time once and for all, by appointing specific dates through the year for every aspect of civic and religious life at Rome.[5]

For Caesar, this fixidity translated into a subtle but clear statement of his uncontested control within the Roman state. Augustus took the cultural potential of the concept of fixed time a major step further, and capitalised on the political significance that was invested, firstly, in the annual, predictable recurrence of the same feast on the same day, and, secondly, in the binary of exact repetition and continuity, which theoretically could extend as far back as the beginnings of the *Respublica*. A fixed calendar, in other words, created cyclical time and en-

3 Rüpke 1995 is fundamental on the role of the calendars in the administration of Roman policy, during the Republic and in the aftermath of Caesar's calendrical reforms; and more recently Feeney 2007.

4 Wallace-Hadrill 2005, 74: "In slipping from the nobility, Roman time becomes the property of all Romans"; or so Augustus, later on, wished Romans to believe. See also Galinsky 2012, 81–83.

5 On the political motivation behind the philosophy and the structure of the Julian calendar see most recently Stern 2013, 205–227; the reference study on the Julian calendar now is Feeney 2007, where earlier important bibliography is recorded and discussed. It should be noted that the extant calendars only record the *feriae stativae*, those holidays that were celebrated on the same day every year, not the *feriae conceptivae*, that is, the holidays whose day of celebration was not fixed but announced every year by the priests or the magistrates; the latter included the *feriae imperativae*, special public holidays held at the discretion of the leading magistrates or, later, the emperor; about the celebration and the frequency of these moving holidays in the Late Republican and the Augustan period, we know very little; on the public Roman holidays or *feriae publicae* and their various categories, see Lipka 2011, 31–39, with bibliography in the notes.

couraged the canonisation of repetition (a-chrony), inasmuch as it bridged past and present by rewriting the former on the model of the latter (and therefore transforming dia-chrony into syn-chrony). At the same time, fixidity was combined with flexibility. As Mary Beard has argued, despite its canonical precision the calendar was an ideological document through which the Roman national identity was getting formed anew under Augustus. In this sense it was meant to reflect the flexibility embodied into the ever-evolving Augustan regime.[6] Augustus added to the calendar several feasts in celebration of his own accomplishments or in honour of the members of his family, added new meanings to traditional festivals, and rearranged the celebration of several festivals as to fall on dates of notable events and anniversaries for the imperial household.[7] The insertion of the new feasts and the rescheduling of festivals was a methodical process, for it sought to identify or forge ideological and thematic links with the feasts already in the calendar, especially those of special significance for the foundation of Rome.[8]

Ovid's *Fasti* stands as the informed reply to Augustus' reconstruction of the Roman calendar. As he stresses his mission not just to record but to 'research' into the aetiology of the various Roman festivals, Ovid seemingly sides with the Augustan effort to trace and bring back to life, and make them part of the now fixed by annual repetition Roman calendar, long-forgotten cults whose precise ritual and origin had to be devised a new. At the same time, he exposes the imperial practice of devising festal aetiologies that would link established feasts to members of the Augustan household. After the paradigm set by the regime, Ovid organises and presents the material by following certain rules that produce schemes and elaborate patterns of thematic correspondence within books / months and between books / months. And like Augustus, the poet advances his own rewriting of the calendar, by choosing, for instance, untraditional dates to commemorate feasts (he celebrates the Ara Pacis, for instance, not on the day it was consecrated, 4 July 13 BCE, but on the day it was dedicated, 30 January 9 BCE); or by defying the observance of a pattern for the discussion

6 Beard 1987.

7 Newlands 1995, 12: "With Augustus' central insertion into Roman time, Roman national identity became bound up with veneration of the imperial family and respect for the values that family chose to promote".

8 On Augustus' manipulation of the Roman *calendar* to honour himself and his family see Wallace-Hadrill 1987; Herbert-Brown 1994, 23; Zanker 1990, 114; Galinsky 2012, 80–81; Galinsky points out that Augustus' manipulating the calendar by adding festivals in honour of himself and the members of his family, was cleverly combined by the opening up of the *fasti*, the religious calendar, to the people in several cities of Italy.

of festivals that last several days, opting to discuss some of them on the first day, while others he treats on their last day (e.g. the *Parentalia*, which he discusses not on February 13 but on February 21).[9] In essence, Ovid antagonises the regime: in his antiquarian-like zeal, with which he searches for the origins of the Roman festivals, and in the considerable reasoning, regularly enforced due to lack of sufficient evidence, that this process involves, the poet exposes manipulation as a leading factor in the process of religious revival. The manipulated origins of Roman festivals in the *Fasti* underscore precisely the intelligent repairing of fragmented evidence at the foundation of Augustus' moral and religious revival, and comment on it, often with a good deal of outspokenness involved as befits Ovid's customary wit.

The double festival of the *Carmentalia*, peculiarly celebrated on two different but not sequential days, represents a prime case study for Augustus' manipulation of the interrelation of time and politics at Rome, and exemplifies a masterful treatment of antiquarianism, comparable to the Augustan quest for the desired origins of ancestral religion. Ovid's revision of the tradition behind the particular festival is to be approached as his critical response or rather intelligent reaction to the cultural and political power of time-management. It exposes the haphazardness of religious aetiology and of rewriting time around the integration of the imperial household in the *fasti* calendar, and instructs to look always for the political at the core of the religious. At the centre of Ovid's treatment of the *Carmentalia* is the institutionalisation of morality and family life in the aftermath of the *leges Iuliae*. Not to be neglected, an ever-present concern to be accounted for in the study of every Ovidian composition is the realisation that the poet's ingeniousness emanates from the energy-instilling anxiety that accompanies his ongoing emulation of Vergil's artistry.

Carmentalia, a Festival of Fragmented Time

Carmentis (or Carmenta), according to the version of the Roman calendar reported in the *Fasti*, is celebrated on two different feast days (*Carmentalia*)[10] of the

9 Nagle 1995, 6–7. Concerning the *Parentalia*, Augustus' decision to celebrate the feast on the last day may be explained by the fact that only the last day of the *Parentalia*, the *Feralia*, were *feriae publicae*; see Lipka 2009, 46, who cites the detailed study of Radke 1963, 318–325.

10 On the *Carmentalia*, see Aust, *RE* 3 (1899) 1594, s.v. *Carmentis/Carmentalia*; Wissowa, *Roschers Mythologische Lexicon I* (1986) 851–854, s.v. *Carmenta*; Wissowa 1912², 220–221; Graf, *Neue Pauly* 2 (1997) 991 s.v. *Carmentis*.

same month, separated by an interval of three days (January 11 and 15).[11] Each feast day honours a different capacity of the goddess. The first day, which in the *Fasti* receives extensive treatment, honours Carmentis' role as prophetess. The second day was devoted to the goddesses' capacity as helper of women in childbirth. In Ovid's account, however, the originary tales for Carmentis' two feast days had practically nothing in common. The first entry for the feast on January 11, reported at *Fast.* 1.461–586, is a celebration of Evander's mother Carmentis, who according to tradition is the spiritual founder of Pallanteum, the leading force behind the Arcadian settlement in Latium. This story is related briefly in *Aeneid* 8.333–341, by Evander to Aeneas. In his treatment, Ovid emphasises Carmentis' prophetic skill; he does so deliberately, in order to toy with the etymological association of *Carmentis* with *carmen* as both prophecy and poem. The details in Ovid's account of Carmentis' arrival to Latium, further, and her pivotal role in the foundation of Pallanteum clearly elicit comparison with Vergil's brevity on the topic: Ovid revisits the *Aeneid* and restores Carmentis to her deserved role in the epic foundation of Rome, the role that Vergil failed to give her.

The second entry for the feast of Carmentis on January 15 is much more intriguing, because it is connected to a false etymology, evidently devised by Ovid. This inventive entry purports to show that Ovid essentially embraces Vergil's practice of advertising his poetic talent by appropriating and rewriting a little-known story. This entry, set at *Fasti* 1.617–636, associates etymologically Carmentis and the *carpenta*, the covered two-wheeled carriages. According to tradition, prior to Ovid known exclusively from Livy, the Roman matrons were given the right to ride in carriages in 396 BCE, when they contributed funds for the construction of a temple to Apollo following the capture of the Veii (5.25.7–10). This privilege was recanted 180 years later, with the *lex Oppia* of 215 BCE, but strong and prolonged protest on the women's part led the Romans to restore this right to the matrons in 195 BCE. Ovid's second aetiology for the *Carmentalia* is inspired by the events of 195 BCE. Even so, apart from the fact that he completely distorts Livy's tradition, his proposed etymology of the *carpenta* and Carmentis is tenuous at best.

My discussion of Ovid's treatment of the *Carmentalia* focuses on the way in which the selection of the particular fictitious para-etymological aetiology behind the second entry for the *Carmentalia* is engineered to entwine with the

---

11 Of the Roman festivals, four, those marked in capital letters on the *fasti* and therefore assumed to be the oldest festivals, were celebrated more than once every year: the *Agonalia* are marked four times each year, the *Lemuria* are marked three times and on three non-consecutive days, while the *Carmentalia, Consualia, Lucaria, Vinalia* and *Tubliustrum* are noted on two days each; see Orlin 2010.

first one, which reproduces material known to everybody. I argue that Ovid invented the second *Carmentalia* aetiology aware that it would raise eyebrows and challenge his informed audience's specialised knowledge of Roman antiquarian material. The obscure (most likely constructed) tie to the *carpenta* is juxtaposed to the better-known tradition of Carmentis which in turn has been considerably developed in the *Fasti* narrative. The thematic discordance of the two accounts is further underlined by their being celebrated on two different dates that are not sequential. The temporal distance between the first and the second day of the festival stresses the split between the two versions and suggests that each be considered independently. And yet, the breach of the same feast on two non-sequential days cannot be realised, since tradition dictates that the celebration of the *Carmentalia has to* occur on both days in order to be complete. Thus the double celebration of the *Carmentalia* becomes a bridge that unifies the festivals around these days which bear close association with Augustus' and his family. Also, by analogy to the pattern he adopted for the *Carmentalia* (that is, the detailed account of the *aetion* behind the feast), Ovid treats the events of January 13 and 16, which celebrate the mythological origins for certain honours newly awarded to the *domus Augusta*, all on the first day.[12] The discordance in the two aetiologies for the *Carmentalia* effectively projects them as complementary pieces of political philosophy and poetic experimentation, even though they seem to have little in common: both expose the subtle manipulation, the former in the rewriting of the Roman past as planned by Augustan ideology, the second in the objective behind Augustus' social reforms, specifically the *leges de maritandis ordinibus*, for the revival of the senatorial elite.

Carmentis in Vergil vs. Ovid[13]

I shall begin with an overview of the pre-Augustan tradition on Carmentis. The only extant text mentioning Carmentis in some detail prior to Vergil is Pausanias (8.43.2; he reports that Carmentis' original Greek name was Nikostrate; cf. also Servius, ad *Aen.* 8.51), who briefly refers to her only as Evander's mother and a nymph of the mountains. Vergil's contemporary, Strabo, briefly informs us (at 5.230) that Carmentis/Nikostrate was also mentioned in the *ab urbe condita* work of the early Roman historiographer Coelius Antipater. Cicero (*Brut.* 56) reports that Carmentis was an earlier Italian deity whose cult had its own *flamen*.

12 Barchiesi 1997, 93; Pasco-Pranger 2002, 262–267.

13 Detailed comparative discussion of the two versions in Fantham 1982 and Barchiesi 1991.

Carmentis truly becomes an important figure of Roman history in Vergil; in the *Aeneid* she is introduced as the force and inspiration behind the foundation of Proto-Rome, being the divine mother of the actual founder of Pallanteum, Evander (*Aen.* 8.333–341):[14]

"me pulsum patria pelagique extrema sequentem
Fortuna omnipotens et ineluctabile fatum
his posuere locis, matrisque egere tremenda
Carmentis nymphae monita et deus auctor Apollo".
Vix ea dicta, dehinc progressus monstrat et aram
et Carmentalem Romani nomine portam
quam memorant, nymphae priscum Carmentis honorem,
vatis fatidicae, cecinit quae prima futuros
Aeneadas magnos et nobile Pallanteum.

"I came here, driven out of my own country by
almighty Fortune and inescapable Fate. As I was
sailing along the coast, my mother, the nymph Carmentis,
uttered frightening warnings inspired by the god Apollo".
Scarce had he uttered these words when they reached the alter and gate that
Romans now call the "Porta Carmentalis", an ancient
honor bestowed on the seer Carmentis, who was the first to
prophesy the greatness of Aeneas' descendants
and the future importance of Pallanteum.[15]

---

14 On the sources behind Carmentis' origin and various names see the reference bibliography cited earlier in n. 10; to those add Schmitzer 2007, 115–118; Marincič 2002, 151 with nn. 39–43. Carmentis' etymology from *carmen*, 'prophetic song', is recorded already in Daniel's Servius ad *Aen.* 836; in the same comment DServius associates Carmentis with two attendants, Porrima and Postvorta, whose names likewise are interpreted as significant of prophetic qualities, disclosing their ability to know, respectively, the past and the future; the full comment in DServius runs as follows: *ideo Carmentis appellata a suis quod divinatione fata caneret: nam antiqui vates carmentes dicebantur ... Alii huius comites Porrimam et Postvortam tradunt, quia vatibus et praeterita et futura sunt nota* ("Carmentis was so named by her own people because by divinatory power she would sing the fates: therefore, the ancient seers were called '*carmentes*' ... Others report of attendants to her, [named] Porrima and Postvorta, because the seers note both those things that happened before and those that are about to happen"); this information is clearly at odds with the association of Porrima and Postverta with fertility, added at the end of the second Carmentalia *aetion* in Ov. *F.* 1.633–636, and it may be explained only on account of Servius' confusion, and perhaps his desire to strength the thematic unity between the two Carmentalia causes as recorded in Ovid. Fantham 1982 and Labate 2003 also offer detailed overviews of the tradition on Carmentis and Evander.

15 Text Mynors 1972 (OCT); transl. Johnston 2012, 177–178.

In his re-writing of the Vergilian text Ovid elaborates on the brief, barely nine-line-long narrative of Evander's journey from Arcadia to Rome, and crafts a unit eight times longer: the *Fasti* account of Evander's migration and settlement at Pallanteum takes up seventy-two lines in total (*Fasti* 1.471–542). The great difference in size is a witty display of *referre idem aliter* – of a reproduction that follows very closely the Vergilian narrative and expands liberally on the minimal information recorded in the text of the *Aeneid*. The Vergilian passage records Carmentis' contribution to the foundation of Rome (a) at 8.335–336, where Evander attributes to divine advice (*monita*), coming jointly from his mother and Apollo, his decision to migrate from Arcadia to Italy; and (b) at 8.340–341; here Carmentis herself is called a "prophesying seer" (*fatidica vatis*). Further, Vergil / Evander drastically summarises the content of these prophecies, on the future greatness of Aeneas' descendants and the transformation of the humble Pallanteum into the glorious Rome (341).[16] Ovid uses respectively the *monita* and the *prophecies* of the Vergilian Carmentis, before and after she and Evander set out for Rome, as the starting point of two separate stretches of narrative, and elaborates, as if requested upon, on the content of the prophetess' two speeches. Ovid's audience is directed to observe how closely to the Vergilian Aeneas' model the Evander of the *Fasti* is described. In her first prophetic stretch (1.483–503) Carmentis does not just urge Evander to migrate to Italy but goes out of her way to parallel him to a refugee or an exile from the East (Cadmus, Tydeus, Jason) originally embarking on some great challenge against his will but under the auspices of the fates, and even becoming the founder of some celebrated city or the agent of some major deed. Evander's mission is shadowed by the dominant presence of Carmentis herself, who leads the Arcadian ship to Italy and upwards the Tiber all the way to the site of Proto-Rome. The experience of Vergil's Evander picks up the story from the point Ovid's Carmentis concluded her own: their arrival at Latium. Upon arriving at Latium, Carmentis, in anticipation (or imitation, in terms of meta-literariness) of the Vergilian sequence, prophesies the glories of Aeneas' people following their settlement at Latium, all the way to the time of Augustus and the impending apotheosis of Livia and Tiberius (1.515–537).[17] In

16 The association of Carmentis with Apollo (expressly in Vergil, at *Aen.* 8.336, *Carmentis nymphae monita et deus auctor Apollo*; "[brought this way] by my mother the Nymph Carmentis' / frightening warnings, and by the divine director Apollo", implicitly but clearly so in Ovid, at *F.* 1.473–474, *simul aetherios animo conceperat ignes, / ore dabat vero carmina plena dei*; "she, as soon as her spirit absorbed the heavenly fire, / spoke true prophecies, filled with the god"), is another point of proximity enforcing the interrelation of the two texts.

17 Discussion on the political recollection of Tiberius and Livia see in Schmitzer 2007; Green 2004, 235–237; Herbert-Brown 1994, 160–162.

Aeneid 8 Evander's brief recollection of Carmentis is prompted by the sight of an altar to the goddess at the *porta Carmentalis.* Ovid in response provides a detailed and dramatized account of Evander's relationship to his mother, thus transferring the leading role in the settlement of the Arcadians at Rome from Evander to Carmentis and her prophecy that reaches all the way to Augustus and his heir. As a result, Vergil's allusive recollection of the Augustan Rome of marble through the detailed description of the humble Palatine is replaced by an explicit celebration of the grand destiny of Aeneas and his descendants.[18]

In light of the drastic transformation of the Vergilian Evander narrative in the *Fasti,*[19] it is notable that the Hercules and Cacus episode has been repeated with no essential change – save the conclusion to the story / *aetion*, the foundation of the Ara Maxima. In the literary texts that came about after the publication of the *Aeneid,* including also Propertius 4.9 which comes before the *Fasti*, the founder of the altar was Hercules. In *Aeneid* 8, however, which precedes both Propertius and Ovid, the foundation of the Ara Maxima and the rites celebrating Hercules victory are attributed to Potitius. This chronological inaccuracy in Vergil is accompanied by yet another anachronism. In *Aeneid* 8 the fight between Hercules and Cacus is narrated as an aetiology for the prophetic Hercules' hymn, performed by the Salii at the close of the festivities. The anachronism is politically motivated and tied to Augustus: Vergil wished to insert the Salian priests into the foundational era of Rome and Hercules' momentous journey through Latium, possibly because Augustus was invoked in the Salian song.[20] In this way, the invocation of Augustus in the genuine *carmen Saliare* and the profile of Hercules the liberator as celebrated in the new *carmen Saliare* invented by Vergil would interfuse, and the former will be projected against the latter. Yet, in the tradition prior to the *Aeneid*, recorded at least in Livy 1.7, the arrival of Hercules

18 Cf. the discussion in Barchiesi 1997, 197–202, who reads Carmenta's speech and overall performance in the *Fasti* narrative as that of a "spokes-woman for the new dynastic requirements and claims of Augustus's family", for her wards "contain in concentrated form the theology of the principate" (p. 199).

19 First Fantham 1992, and subsequently Labate 2003 and Farrell 2013, 239–250, discuss the role of Carmenta and Evander in the *Fasti* in the context of the emphasis, throughout the poem, on the one hand, on Evander as true founder of Rome (vs. both Romulus and Vergil's Aeneas), and, on the other, on Evander's Arcadian origins which are so firmly attached to him that they never actually allow him to lose his Arcadian/Greek identity and become true Roman. The contradiction means to underscore Ovid's similar plight; in permanent exile in Tomis, his heart and mind have never left Rome.

20 On Vergil's attribution of the hymn to Hercules in *Aeneid* 8 to the Salian priests as a means to highlight the Augustan transformation of Evander's festival for Hercules see now Miller 2014; brief mention also in Marincič 2002, 151.

at Rome is known to Evander, for it has been foretold to him by Carmentis (whose prophetic credentials Livy further bolsters by linking her to the Sibyl[21]), as the hero later acknowledges upon welcoming Hercules to Pallanteum:

> *Evander tum ea, profugus ex Peloponneso, auctoritate magis quam imperio regebat loca, venerabilis vir miraculo litterarum, rei novae inter rudes artium homines, venerabilior divinitate credita Carmentae matris, quam fatiloquam ante Sibyllae in Italiam adventum miratae eae gentes fuerant. Is tum Evander concursu pastorum trepidantium circa advenam manifestae reum caedis excitus postquam facinus facinorisque causam audivit, habitum formamque viri aliquantum ampliorem augustioremque humana intuens rogitat qui vir esset. Ubi nomen patremque ac patriam accepit, "Iove nate, Hercules, salve", inquit; "te mihi mater, veridica interpres deum, aucturum caelestium numerum cecinit, tibique aram hic dicatum iri quam opulentissima olim in terris gens maximam vocet tuoque ritu colat".* (Livy 1.7.8–11)

> Evander, an exile from the Peloponnese, controlled that region in those days, more through personal influence than sovereign power. He was a man revered for his wonderful invention of letters, a new thing to men unacquainted with the arts, and even more revered because of the divinity which men attributed to his mother Carmenta, whom those tribes had admired as a prophetess before the Sibyl's coming into Italy. Now this Evander was then attracted by the concourse of shepherds, who, crowding excitedly about the stranger, were accusing him as a murderer caught red-handed. When he had been told about the deed and the reason for it, and had marked the bearing of the man and his figure, which was somewhat ampler and more august than a mortal's, he inquired who lie was. Upon learning his name, his father, and his birth-place, he exclaimed, "Hail, Hercules, son of Jupiter! You are he, of whom my mother, truthful interpreter of Heaven, foretold to me that you should be added to the number of the gods, and that an altar should be dedicated to you here which the nation one day to be the most powerful on earth should call the Greatest Altar, and should serve according to your rite".[22]

Against the authority of the Vergilian narrative on Evander, Ovid, in his picturing of Carmentis as she was delivering her prophecy in full, and reporting next to it the duel between Hercules and Cacus, restores the prophetic nymph to her prominent place of the spiritual ancestor in Roman prehistory, reasserts Livy's version that Evander already knew of Hercules' arrival, and questions the genuineness of the *carmen Saliare* in the version recorded in *Aeneid* 8. It may be suggested that in exposing Vergil's politically motivated playing with the Roman past, Ovid

---

21 Marincič 2002, 151 notes that in an obscure version going back to Eratosthenes, Evander is reported to have been the son of the Italian Sibyl (a tradition Vergil certainly knows given the way Livy introduces Carmentis as a proto-Sibyl at 1.7.8–9).

22 Text and transl. Foster 1919 (Loeb).

truly reaches beyond the objectives of the author of the *Aeneid*, onto Augustus' strategic manipulation of time.[23]

Carmentis from *carpenta* (?): Rationalising a Pseudo-etymological *Aetion*

The aetiology behind the second feast day of the *Carmentalia*, the fictitious story of the *carpenta*, the two-wheeled carriages, and the Roman matrons who took to the streets and aborted their fetuses in protest for being denied their ancestral privilege to ride on wagons, is commonly read today as politically motivated[24] – a witty allusion-in-protest to Augustus' social legislation of 18 BCE (the two Julian laws on marriage and adultery).[25] From a different perspective, strictly associated with the politics of controlling time and, as a result, manipulating historical and political memory, Ovid's initiative to embrace this clearly false etymology may expose a subtler intervention on the Roman calendar by Augustus, which concerns the association of Carmentis with childbirth. The text of the *Fasti* that records this second *Carmentalia* aetiology runs as follows (*Fasti* 1.619–628):

nam prius Ausonias matres carpenta vehebant
(haec quoque ab Euandri dicta parente reor);
mox honor eripitur, matronaque destinat omnis
ingratos nulla prole novare viros,
neve daret partus, ictu temeraria caeco
visceribus crescens excutiebat onus.
corripuisse patres ausas immitia nuptas,
ius tamen exemptum restituisse ferunt;
binaque nunc pariter Tegeaeae sacra parenti
pro pueris fieri virginibusque iubent.

23 On the particular *aetion* of the Ara Maxima foundation, critics see also the challenge raised against Augustus by the Fabii, the old aristocratic family that claimed ancestry from Hercules (*Maximus* was a title ascribed in Augustan Rome to the Fabii, who were Ovid's patrons no less); therefore Hercules' foundation of the Ara Maxima, and by association the control of the calendar by arranging the feasts accordingly, could be claimed by both Augustus and the Fabii; cf. Schmitzer 2007, 125–132; King 2006, 181–183.

24 On the second Carmentis aetiology according to Ovid see Green 2004, 282–290; Green aptly calls it "creative history".

25 Pasco-Pranger 2002, 265–267; on the *leges Iuliae* (including scholarship), see Treggiari 1991, 277–298.

> For formerly the Ausonian mothers drove in carriages (these I think were called after Evander's mother). Afterwards the honour was taken from them, and every woman vowed not to propagate the line of their ungrateful husband, by giving birth to offspring, and lest she should bear children, she unwisely by a secret thrust expelled the growing burden from her womb. They say the senate reprimanded the wives for their daring cruelty, but restored the right which had been taken from them, and they ordered that now two like festivals be held for the Tegean mother, to promote the birth of boys and girls.[26]

Thanks to the testimony of Varro (ap. Gell. 16.16.4 = *Antiquitates rerum divinarum* fr. 103–104 Cardauns) and Hyginus (Hyg. 277), both contemporaries to Augustus, and an earlier reference in Cicero *Brutus* 56, we know that Carmentis is an old goddess of childbirth, who protects married women, and in return the matrons in her honour have established a shrine at Rome.[27] The origins of this association Plutarch eagerly seeks to find out when, at his *Quaestiones Romanae* 56, rightly wonders: διὰ τί τὸ τῆς Καρμέντης ἱερὸν ἐξ ἀρχῆς δοκοῦσιν αἱ μητέρες ἱδρύσασθαι, καὶ νῦν καὶ μάλιστα σέβονται; ("why people believe that the temple of Carmentis had been originally founded by the mothers, who even today pay great respect to the goddess?"). He proceeds to explain the association from the identification of Carmentis as one of the fates: οἱ δὲ μοῖραν ἡγοῦνται τὴν Καρμένταν εἶναι, καὶ διὰ τοῦτο θύειν αὐτῇ τὰς μητέρας ("some believe that Carmentis is a fate, and this is why the mothers offer sacrifices to her"). This identification results from the etymological link of Carmentis and *carmen*, 'the prophetic song', domain also of the Fates/Parcae.[28] Plutarch's interpretation, however, does not specify the sources wherefrom he traced this belief. And, he omits altogether the story of the Roman matrons who denied their motherhood in such a gruesome way – a story which is the only extant tradition that associates Carmentis with motherhood. This omission means that Plutarch either was ignorant of the tradition reported in the *Fasti* or did not believe it – or, more likely, both: he could not identify any other source that would record the horrible story of the Roman matrons who collectively aborted their fetuses because they were denied their privilege to ride on carriages.[29] The lack of any other source on the *carpenta* story as recorded in the *Fasti*, and Plutarch's silence on it, suggest that the ad-

---

26 Text and transl. Frazer 1951.

27 Lipka 2009, 82. Hyginus and Isidore of Seville, further, report that Carmentis had invented the Roman alphabet (Isidore, *De grammatica* 1.4.1), or adapted the Greek alphabet to Italian use (Hyginus).

28 Pasco-Pranger 2002, 266; Fantham 2011, 436; August. *De Civ. D.* 4.11 refers to multiple Carmentae functionally identical with the Parcae (*fata nascentibus canunt*).

29 On abortion in the ancient world, a practice never illegal though controversial, see Kapparis 2002, 33–52, 167–194; pp. 148–151 discuss abortion and Augustus' legislation on the family.

dition of the massive abortions was a detail invented by Ovid. And the inspiration for Ovid must have been the identity of Carmentis as a goddess of childbirth, even though the poet in an ingenious, distorted use of the Alexandrian footnote begins his narrative by giving the impression that he would endorse the belief (*reor*), falsely,[30] that the name of Carmentis and the *carpenta* share common etymology.[31] Yet, this awkward, one-time mention of Carmentis in the *carpenta* narrative (and on a parenthetical line, 620), triggers, a few lines later, the memory of Carmentis the goddess of childbirth, when Ovid reports that January 15 commemorates one of the most horrible crimes against the idea of Roman family in the tradition of the *Respublica*, the collective decision of the Roman matrons in 195 BCE to abort their fetuses thus threatening the survival of their husbands' family line, in protest against the State's denial to give them back their ancestral privilege to ride on carriages. I would like to discuss in some detail this second form of manipulating the tradition on the name and origin of the *Carmentalia*, and in the course of my study illustrate how Ovid recreates Roman time by commenting on Augustan ideology.

Carmentis, Ovid, and the Augustan Social and Religious Policy

The imaginary etymological association of Carmenta with the *carpenta* has been devised by Ovid in order to justify the existence of two separate *Carmentalia* festivals rather than one festival interrupted and resuming a few days later, as it was the case for other Roman festivals, too, such as the *Lemuria* and *Lucaria*, both of which were 'split' and celebrated on non-consecutive days (the *Lucaria*

---

30 One of the most common forms of reflexive (including self-reflexive) annotation is the famously called 'Alexandrian footnote', where general appeals to some traditional source (most notably the verbs *fama est*, *dicitur* and *fertur*, but also other verbs of similar meaning, such as *reor*) seen to trigger an allusion to specific literary models of the past; the term was introduced by Ross 1975, 78; for the impact of Ross' formulation in the scholarship see Hinds 1998, 1–3. In Ovid, however, this trope is employed often in a witty fashion, to deceive the audience into believing that the poet presumes some literary predecessor, when he actually does not, but instead he advances an invention of his own. From a different perspective, Pasco-Pranger 2006, 193 argues that the use of '*reor*', meaning 'I believe', Ovid confesses that he expresses personal views and probably is lying; also Green 2004, 283.

31 Pasco-Pranger 2002, 266; Pasco-Pranger 2006, 193. Green 2004, 283: "Ovid is here applying one of the four standard techniques for etymologising, namely *immutatio* (the changing of letters, in this case 'm' for 'p')".

on July 19 and 21; the *Lemuria* on May 9, 11 and 13).[32] Ovid intervenes in the tradition probably because he found a vacuum in it: not even the Romans themselves could explain why instead of the typical one 'split' festival, two separate feasts of the same deity were celebrated so closely in time to each other. The *fasti Praenestini*, one of Ovid's major sources behind the composition of his *Fasti*, suggest that the second festival day to Carmentis was instituted already by Romulus. The Carmental gate received its name from its proximity to the sacred grove of Carmentis.[33]

The mention of Carmentis' two sisters or companions, Porrima and Postverta, at 633–36, reproduce an image of Carmenta, in Barchiesi's words, as "that of a goddess who looks both forward and backward (*porro* and *post*), an effective female counterpart to the two-faced Janus who had dominated this first part of the first book [of the *Fasti*]".[34] The association to Janus, however, is added at the end of the *carpenta* pseudo-aetiology, and after this imaginary story has been introduced as the leading *aetion* behind the festival. In doing so, Ovid subtly expressed his doubts about the credibility of the Janus-related tradition, and proposed an alternative one of his own, by taking advantage, on the one hand, of Carmentis' ties to fertility – recorded, as we saw earlier, in Plutarch, and also in Varro, and on the other, of the absence of an *aetion* for these ties.[35] The version about the two companions of Carmentis and their tenuous, if not enforced, association with childbirth inspired Ovid to come up with the para-etymological link between the similarly sounding '*Carmentis*' and '*carpenta*', and devise a story that would slyly comment on the appropriation of Rome's early history by the Augustan regime.

The narrative core of this new *aetion* involves an event that allegedly took place in 195 BCE, and is recorded in detail in Livy (34.1). Briefly, the story runs as follows. During the fourth century, the Roman Senate had granted patrician matrons the privilege of riding on two-wheeled carriages (*carpenta*) as reward for their donation of their golden jewellery in fulfilment of a vow to Apollo made by Camillus. The privilege was later to be temporarily revoked during the Second Punic War (215 BCE), by the *lex Oppia*, in order to save the horses

32 Lipka 2009, 39.

33 Green 2004, 288–289.

34 Barchiesi 1997, 93.

35 Varro (*Antiquitates rerum divinarum* fr. 103 Cardauns) etymologises the names of Porrima and Postverta from Carmentis' role as goddess of childbirth rather than prophecy. In Green's words (Green 2004, 290), "the two names ... represent the two ways in which a child can be born into the world, either feet first (considered backwards, increasing the mother's labour), or head first (considered forwards and natural)".

for the war effort.[36] But the Senate did not renew the privileges when the war ended. In 195 BCE tribunes Marcus Fundanius and Lucius Valerius called for the repeal of this *lex Oppia*, but they were opposed by the brothers Marcus and Publius Junius Brutus. Supporters for repealing the *lex Oppia*, and those who opposed its repealing, gathered daily on the Capitoline to argue over the matter, including women from the countryside. In Livy's story there is no mention of Carmentis whatsoever – the story however was selected because of the intriguing way in which the Roman matrons demanded what was rightfully theirs – and succeeded even against the famous Cato Maior: when Cato threatened to vote down the proposal to repeal the *lex Oppia*, the women "filled all the streets and blocked the approaches to the Forum" (Livy 34.1.5) until their privileges were restored:

> *matronae nulla nec auctoritate nec verecundia nec imperio virorum contineri limine poterant, omnes vias urbis aditusque in forum obsidebant, viros descendentes ad forum orantes ut florente re publica, crescente in dies privata omnium fortuna matronis quoque pristinum ornatum reddi paterentur. augebatur haec frequentia mulierum in dies; nam etiam ex oppidis conciliabulisque conveniebant.* (Livy 34.1.5–6)

> The matrons could not be kept at home by advice or modesty or their husbands' orders, but blocked all the streets and approaches to the Forum, begging the men as they came down to the Forum that, in the prosperous condition of the state, when the private fortunes of all men were daily increasing, they should allow the women too to have their former distinctions restored. The crowd of women grew larger day by day; for they were now coming in from the towns and rural districts.[37]

36 Liv. 5.25.7–10: (215 BCE)

Ita in aestimationem urbs agerque venit. Pecunia ex aerario prompta, et tribunis militum consularibus ut aurum ex ea coemerent negotium datum. Cuius cum copia non esset, matronae coetibus ad eam rem consultandam habitis communi decreto pollicitae tribunis militum aurum et omnia ornamenta sua, in aerarium detulerunt. Grata ea res ut quae maxime senatui unquam fuit; honoremque ob eam munificentiam ferunt matronis habitum ut pilento ad sacra ludosque, carpentis festo profestoque uterentur.

Thus the city and territory came into the estimate. The money was drawn from the treasury, and the consular tribunes were commissioned to purchase gold with it. As there was not a sufficient supply, the matrons, after meeting to talk the matter over, made themselves by common consent responsible to the tribunes for the gold, and sent all their trinkets to the treasury. The senate were in the highest degree grateful for this, and the tradition goes that in return for this munificence the matrons had conferred upon them the honour of driving to sacred festivals and games in a carriage, and on holy days and work days in a two-wheeled carriage. (text and transl. Foster 1925)

37 Text and transl. Sage 1935.

Ovid devises a horrible escalation to the story: the women agreed to abstain from sexual intercourse and prevent from getting pregnant, while those already pregnant resorted to abortion. Carmentis, in her identity as goddess of fertility and childbirth, criticises Augustus' social legislation on family and marriage, and more specifically, the legal pressure on the Roman elites to marry and have many children. The alleged 'commemoration' of a horrible protest that centres on massive abortions by the women of the elite, the violent termination of a whole future generation of Roman citizens on the anniversary of the festival of Carmentis, the goddess of fertility and childbirth, adds impiety or disregard for the divine laws, to civic disobedience or violation of the laws of the Princeps, which were promulgated for the restitution, morally but also numerically, of the aristocracy.[38] Augustus' moral legislation was realised in two parts. The main body of social laws was published in 18 BCE; the strong reaction among the members of the aristocracy to this legislation led to its revision 27 years later, in 9 CE, with the *lex Pappia-Poppaea*. Evidently, the protests to these social laws by those that stood to be affected were prolonged and notable (Suetonius, *Augustus* 34.1), and reasonably so, given that with the *leges Iuliae* the private life of all (upper-class) Romans was coming under total control of the State.[39] The composition of the *Fasti* during this era, itself a work that has been inspired by the Augustan revival of ancestral religious customs, rites and temples, reflects this Roman reaction to Augustus' effort not just to reawaken a long tradition of Roman customs that compelled citizens to marry, but more than that, to create an ideological link between this tradition and his legislation. The implementation of laws about morality was directly associated with the return of the Golden Age, or else, the complete and total, full-circle, restart of Roman historical time. This policy, however, run counter to the philosophy of the return of the Golden Age, which presupposed the people's embrace of moral discipline of their own accord, not because of legislative enforcement.[40]

---

38 Green 2004, 285 correctly argues that Ovid here alludes to his earlier poetry, specifically *Am.* 2.14.9–10, where he assumes the roles of *curator morum et legum*, in dialogue with the moral philosophy of the *princeps*, under development at the time, and rebukes Corinna for proceeding to an abortion, which the poet fiercely criticises as unnatural and murderous, and imagines "the demographic and historical consequence if women of antiquity had acted in a similar manner"; see also Gamel 1989, 189–197.

39 On Augustus' moral legislation see Galinsky 1981 and 1996, 128–140; Wallace-Hadrill 1981; Nicolet 1984; Badian 1985; Edwards 1993; Bauman 1992, 105–108 and notes.

40 Gatz 1967, 157–159; Galinsky 1996, 128–129; and more recently Feeney 2007, 136–137: the Golden Age was "perhaps never something that human beings could reach a hand out to and grasp".

And yet, protest against Augustus' canonisation of family life goes hand in hand with approval of another piece of Augustan moral legislation, the public declaration of one's social status by wearing certain insignia. Overall, dress functions as a semiotic system in a sense that within a given dress code each element of the system carries particular meaning when compared with and contrasted to others.[41] Its visual aspects are most important for the construction of one's identity, principally social rank, since they are readily perceived and understood.[42] In fact, the importance of one's dress, consisting both of clothing and other visible markers, such as the means by which someone ferried oneself around, as foremost an indicator of status and social rank, led to its becoming very early an object of regulation in the Roman world.[43] The *carpenta* in our story is to be seen precisely as one of those distinct markers of social identity. The Roman women of the elite in the *carpenta* story protest against a decision that deprives them from an important badge of privilege and class, and prevents them from advertising it in public, contrary to the relevant canonisation introduced with the *leges Iuliae!* The schizophrenic conduct of the Roman matrons, at once against and for the *leges Iuliae*, exposes the ethical flaws of Augustus' program of moral restoration, where essence is disregarded in order to maintain appearances.[44] The Roman matrons fervently demand the restoration of their status insignia, the *carpenta*, and thus desire to show off their identities as elite matrons, by upholding their obligation to bear offspring (by means of preventing pregnancies and introducing abortions). The absurdity emphasised in the reaction of the matrons is propounded also by the conduct of the womenfolk overall: in Livy's narrative, not only the matrons but women from various Italian towns flock to Rome to protest for the abolition of the *lex Oppia*. This spontaneously shaping organisation of a pan-Italian women's movement for the restoration of the privileges of the members of the upper class among them is obviously an exaggeration, yet it is precisely for its being perceived as a situation totally unrealistic, that it makes its way into the subtext of a story that represents reaction to the absurd on more than one level.

---

41 Lee 2015, 24; Sommer 2012, 257.

42 Lee 2015, 28.

43 Edmondson and Keith 2008, 4–5; Foxhall 2013, 112. On the messages conveyed by the dress see Sommer 2012, 258. On the regulations on dress in Roman city see Edmondson 2008, 32–37.

44 Note also that the *leges Juliae* for the first time provide for the distinction in appearance between the *matronae* and the *meretrices*, by ordering the latter to wear in public the toga, that is a masculine garment, which clearly distances them not just from the world of the Roman women but even from the feminine gender; see Dixon 2008.

Conclusion

The legendary association of Carmentis to fertility and the foundation of Rome (and the Roman nation) gives Ovid a brilliant opportunity to draw inspiration from the deity's cult, devise an association with Rome's legendary origins (the acceptable origins by the regime, established officially by Vergil in *Aeneid* 8) and use this approved version of the legendary past to criticise Augustus' moral legislation and the ideology of Roman rebirth tied with it. Ovid's conversation with Augustus' manipulation of Roman time is even more appropriate in light of acknowledging that the two-day celebration of Carmentis was part of a multi-day sequence of feasts associated with Augustus' policies. Ovid devises the story of the massive abortions in the context of a grand-scale criticism to Augustus' social program, which included restructuring of Roman religious practices through rewriting the calendar as to produce thematic patterns that would revolve around the Princeps' religious and social initiatives, and so give divine-clad approval to his legislation. The story of the *carpenta* is only one of the several inventive *aetia* throughout the *Fasti*, which are to be read as the poet's literary embrace and exposure at once of Augustus' ingenious control of the fluidity of time and the politics enmeshed with it.

Bibliography

Badian, E. 1985. A Phantom Marriage Law. *Philologus* 129: 82–98.
Barchiesi, A. 1991. Discordant Muses. *PCPhS* 37: 1–21.
Barchiesi, A. 1997. *The Poet and The Prince: Ovid and the Augustan Discourse.* Berkeley.
Bauman, R.A. 1992. *Women and Politics in Ancient Rome.* London.
Beard, M. 1987. A Complex of Times: No More Sheep on Romulus' Birthday. *PCPhS* 33: 1–15.
Dixon, J. 2008. Dressing the Adulteress. In *Greek and Roman Textiles and Dress: An Interdisciplinary Anthology*, eds. M. Harlow and M.-L. Nosch, 298–305. Oxford and Philadelphia.
Edmondson, J. 2008. Public Dress and Social Control in Late Republican and Early Imperial Rome. In *Roman Dress and the Fabrics of Roman Culture*, eds. J. Edmondson and A. Keith, 21–46. Toronto, Buffalo and London.
Edmondson, J. and A. Keith. 2008. Introduction: From Costume History to Dress Studies. In *Roman Dress and the Fabrics of Roman Culture*, eds. J. Edmondson and A. Keith, 1–17. Toronto, Buffalo and London.
Edwards, C. 1993. *The Politics of Immorality in Ancient Rome.* Cambridge.
Fantham, E. 1992. The Role of Evander in Ovid's *Fasti. Arethusa* 25: 155–171.
Fantham, E. 2011. The *Fasti* as a Source of Women's Participation in Roman Cult. In *Roman Readings: Roman Response to Greek Literature from Plautus to Statius and Quintilian*, ed. E. Fantham, 430–455. Berlin and New York.

Farrell, J. 2013. Complementarity and Contradiction in Ovidian Mythography. In *Writing Myth: Mythography in the Ancient World*, eds. S.M. Trzaskoma and R.S. Smith, 223–252. Leuven, Paris and Walpole, MA.

Feeney, D. 2007. *Caesar's Calendar: Ancient Time and the Beginnings of History.* Berkeley.

Foster, B.O. (transl.) 1919. *Livy, History of Rome*, vol. i: Books 1–2 (Loeb Classical Library). Cambridge, MA and London.

Foster, B.O. (transl.) 1925. *Livy, History of Rome*, vol. iii: Books 5–7 (Loeb Classical Library). Cambridge, MA and London.

Foxhall, L. 2013. *Studying Gender in Classical Antiquity.* New York.

Frazer, J.G. (transl.) 1951. *Ovid's Fasti* (Loeb Classical Library). Cambridge, MA and London.

Galinsky, K. 1981. Augustus' Legislation on Morals and Marriage. *Philologus* 125: 126–144.

Galinsky, K. 1996. *Augustan Culture: An Interpretative Introduction.* Princeton, NJ.

Galinsky, K. 2011. Continuity and Change: Religion in the Augustan Semi-Century. In *A Companion to Roman Religion*, ed. J. Rüpke, 71–82. Malden, MA.

Galinsky, K. 2012. *Augustus, Introduction to the Life of an Emperor.* Cambridge.

Gamel, M.-K. 1989. *Non sine caede:* Abortion Politics and Poetics in Ovid's *Amores. Helios* 16.2: 183–206.

Gatz, B. 1967. *Weltalter, Goldene Zeit und sinnverwandte Vorstellungen.* Hildesheim.

Green, S.J. 2004. *Ovid, Fasti I. A Commentary.* Leiden and Boston.

Herbert-Brown, G. 1994. *Ovid and The Fasti: A Historical Study.* Oxford.

Herbert-Brown, G. (ed.) 2002. *Ovid's Fasti. Historical Readings at its Bimillennium.* Oxford.

Hinds, S. 1988. *Allusion and Intertext: Dynamics of Appropriation in Roman Poetry.* Cambridge.

Johnston, P.A. (transl.) 2012. *The Aeneid of Vergil.* Norman, OK.

Kapparis, K. 2002. *Abortion in the Ancient World.* London.

King, R.J. 2006. *Desiring Rome: Male Subjectivity and Reading Ovid's Fasti.* Columbus, OH.

Kline, A.S. (transl.) 2004. *Ovid, Fasti (Roman Calendar).* Available online at http://www.poetryintranslation.com/PITBR/Latin/Fastihome.htm

Labate, M. 2003. Tra Grecia e Roma: l'identità culturale augustea nei Fasti di Ovidio. In *Fecunda licentia: tradizione e innovazione in Ovidio elegiac*, ed. R. Gazich, 71–118. Milan.

LaBua, G. (ed.) 2010. *Vates operose dierum: studi sui Fasti di Ovidio.* Pisa.

Lee, M.M. 2015. *Body, Dress and Identity in Ancient Greece.* New York.

Lipka, M. 2009. *Roman Gods: A Conceptual Approach.* Leiden and Boston.

Mackie, N. 1992. Reconsidering Ovid's Fasti. *Arethusa* (special issue) 25.1: 1–180. Baltimore.

Marinčič, M. 2002. Roman Archaeology in Vergil's Arcadia (Vergil *Eclogue* 4; *Aeneid* 8; Livy 1.7). In *Clio and The Poets: Augustan Poetry and the Traditions of Ancient Historiography*, eds. D.S. Levene and D.P. Nelis, 143–161. Leiden.

Miller, J.F. 1981. *Ovid's Elegiac Festivals. Studies in the Fasti.* Frankfurt.

Miller, J.F. 2014. Virgil's Salian Hymn to Hercules. *CJ* 109: 439–463.

Mynors, R.A.B. (ed.) 1972. *P. Vergili Maronis Opera* (Oxford Classical Texts; reprinted edition with corrections). Oxford.

Nagle, B.R. 2005. *Ovid's Fasti. Roman Holidays: Translated with Notes and Introduction.* Bloomington and Indianapolis.

Newlands, C.E. 1995. *Playing With Time. Ovid and The Fasti.* Ithaca, NY.

Nicolet, C. 1984. Augustus, Government and the Propertied Classes. In *Caesar Augustus: Seven Aspects*, eds. F. Millar and C. Segal, 89–128. Oxford.
Orlin, E.M. 2010. *Foreign Cults in Rome: Creating a Roman Empire.* New York.
Pasco-Pranger, M. 2002. Added Days: Calendrical Poetics and the Julio-Claudian Holidays. In *Ovid's Fasti. Historical Readings at its Bimillennium*, ed. G. Herbert-Brown, 251–274, Oxford.
Pasco-Pranger, M. 2006. *Founding the Year. Ovid's Fasti and the Poetics of the Roman Calendar.* Leiden.
Price, S.R.F. 1996. The Place of Religion: Rome in the Early Empire. In *The Cambridge Ancient History*, vol. 10, eds. A.K. Bowman, E. Champlin and A. Lintott, 812–847. Cambridge.
Radke, G. 1963. *Beobachtungen zum Romischen Kalender. RhM* 106: 313–335.
Ross, D.O. 1975. *Backgrounds to Augustan Poetry. Gallus, Elegy and Rome.* Cambridge.
Rüpke, J. 1995. *Kalender und und Öffentlichkeit: Die Geschichte der Repräsentation und religiösen Qualifikation von Zeit in Rom.* Berlin.
Sage, E.T. (transl.) 1935. *Livy, History of Rome*, vol. ix: Books 31–34 (Loeb Classical Library). Cambridge, MA and London.
Scheid, J. 2005. Augustus and Roman Religion: Continuity, Conservativism and Innovation. In *The Cambridge Companion to the Age of Augustus*, ed. K. Galinsky, 175–194. Cambridge.
Schmitzer, U. 2007. Ovids Carmentalia – oder: Kann man einem Dichter vertrauen? In *Ovid. Kultur – Werk – Wirkung*, eds. M. Janka, U. Schmitzer and H. Seng, 113–144. Darmstadt.
Sommer, C.M. 2012. Dress and Identity. A Social Psychologist's Perspective. In *Kleidung und Identität in religiösen Kontexten der römischen Kaiserzeit. Altertumwissenschaftliches Kolloquium in Verbindung mit der Arbeitsgruppe "Kleidung und Religion" Project Dress ID, Rheinische Friedrich-Wilhelms-Universität, Bonn, 30. und 31. Oktober 2009*, eds. S. Schrenk, K. Vössing, and M. Tellenbach, 257–263. Regensburg.
Stern, S. 2013. *Calendars in Antiquity: Empires, States and Societies.* Oxford.
Treggiari, S. 1991. *Roman Marriage: Iusti Coniuges from the Time of Cicero to the Time of Ulpian.* Oxford.
Wallace-Hadrill, A. 1981. Family and Inheritance in the Augustan Marriage Laws. *PCPhS* 27: 58–80.
Wallace-Hadrill, A. 1987. Time for Augustus: Ovid, Augustus and the *Fasti.* In *Homo Viator. Classical Essays for John Bramble*, eds. M. Whitby, P.R. Hardie and M. Whitby, 221–230. Bristol.
Wallace-Hadrill, A. 2005. *Mutatas Formas:* The Augustan Transformation of Roman Knowledge. In *The Cambridge Companion to the Age of Augustus*, ed. K. Galinsky, 55–84. Cambridge.
Wiseman, T.P. 2008. *Unwritten Rome.* Exeter.
Wiseman, T.P. 1998. *Roman Drama and Roman History.* Exeter.
Wissowa, G. 1912^2. *Religion und Kultus der Römer.* Munich (repr. 1971).
Zanker, P. 1990. *The Power of Images in the Age of Augustus* (transl. A. Shapiro). Ann Arbor.

Françoise Létoublon*

The Decisive Moment in Mythology: The Instant of Metamorphosis

Borrowing from French photographer Henri Cartier-Bresson the notion of the *decisive moment*,[1] we intend to show in this article that the ancient authors who narrate metamorphoses mainly seek to capture the fugitive moment when a person or animal is changed from one form to another, and sometimes from one realm of nature to another, most often from human form to animal, plant or watercourse. Artists also try to capture this instant through several devices, for instance Bernini in his well-known sculptural group of Apollo and Daphne. A girl who becomes a tree or a spring, or a young hunter who becomes a bird or another kind of animal are shown in visual arts with different features relevant to her/his previous and new form.[2]

The text corpus chosen for this study consists of the so-called Greek 'mythographers',[3] Antoninus Liberalis and Parthenius, collectors of myths living in the first or second century AD. We chose them because of the amount of myths they tell with a relative stylistic and linguistic unity. Although Antoninus Liberalis

* I want to express my warm gratitude to the organizers of the conference in Patras and the editors of this volume, particularly Athina Papachrysostomou and Menelaos Christopoulos. My thanks also go to Stephen Rojcewicz and to the anonymous reviewer who corrected my English. Every remaining error is my responsibility

1 In 1952, Cartier-Bresson published his book *Images à la sauvette*, whose English edition was entitled *The Decisive moment.* It included a portfolio of 126 of his photos from the East and the West. The book's cover was drawn by Henri Matisse. For his 'philosophical' (so called in the English wikipedia notice on Cartier-Bresson) preface, Cartier-Bresson took his keynote from the 17th century French author Cardinal de Retz: "Il n'y a rien dans ce monde qui n'ait un moment décisif" ("There is nothing in this world that does not have a decisive moment"). Cartier-Bresson applied this to his photographic style. He said: "Photographier: c'est dans un même instant et en une fraction de seconde reconnaître un fait et l'organisation rigoureuse de formes perçues visuellement qui expriment et signifient ce fait" ("To me, photography is the simultaneous recognition, in a fraction of a second, of the significance of an event as well as a precise organization of forms which give that event its proper expression").

2 Let us recall that Nicander's work, the model for both Ant. Lib. and Ov. *Met.*, was known under the title *Heteroioumena*, meaning – more or less – 'becoming other', 'made other'.

3 For mythography and mythographers, see mainly Henrichs 1987; Dowden 1992; Calame 2004; Higbie 2007; Dowden 2011; Dowden and Livingstone 2011; Bremmer 2011. Ant. Lib. and Parth. are available in good specific modern editions whereas, apart from [Apollodorus], other mythographers are fragmentary (see Fowler 2000; Higbie 2007). Boyle 2007 appears to be a very insightful study on Ovidian poetics of metamorphosis, with several echoes on our theme.

DOI 10.1515/9783110535150-021

seems a little later than Parthenius, we will take him as a point of departure because he is mostly interested in the theme of metamorphosis, and thus seems the most appropriate for a more general study of the poetics of metamorphosis, whereas the main interest of Parthenius is for different forms of passions (see the title *Erotica Pathemata*).[4]

As mythology consists mainly and fundamentally of narratives,[5] it is deeply linked to time, to chronology and succession of events. Therefore we think it is important to look very precisely at formal devices that Greek language may use for expressing this relation to time, starting from grammatical features, such as the contrast between imperfect and aorist, emphasizing aspect rather than tense. Later we'll deal with the importance of transitive use, and of active voice as revealing the role of the agent of metamorphosis.

Those aspects have not been studied in previous research on metamorphosis, as far as we know, apart from some insights in F. Frontisi's works,[6] and the studies of some mythic tales of pursuit and fall by C. Delattre.[7]

The Moment of Metamorphosis for Antoninus Liberalis

It may be remarked how often the Greek texts use the aspectual contrast between durative imperfect which shows the frame of the event, and the aorist which expresses the irruption of the metamorphosis into this generally quiet setting, since the metamorphosis usually occurs as the result of some violent attempt, or as a punishment. The imperfect appears then to be the form appropriate for description of the action frame, whereas the perfective aorist refers to the sudden events that occur in the narrative.[8] Let us look at two instances of this aspectual contrast met in the corpus as defined above.[9]

4 'Love passions', *Passions d'amour* in the French edition (Biraud 2008).

5 For the meaning of *muthos* vs *logos*, see Calame 2004 and Fowler 2011, the latter with a thorough study both of uses in ancient texts and of Nestle's influential article.

6 Frontisi-Ducroux 2000 and 2003.

7 Delattre 2010 and 2013.

8 On aspect (and tense) in Greek morphosyntax, see Ruipérez 1980; Comrie 1976; Hewson 1997; Hewson 2006; Hewson 2014.

9 The aorist and imperfect verbal forms to which we intend to draw attention are in bold letters. The opposition between imperfect and aorist sometimes appears rather subjective, and one may sometimes feel that the opposite form could have been used as well.

We choose two short narrative passages from Antoninus Liberalis' *Metamorphoses* for illustrating this aspectual contrasts:

Ἠμαθίδες 9.1–3[10]

ὑπὸ δὲ τὸν χρόνον τοῦτον ἐβασίλευε Πίερος αὐτόχθων Ἠμαθίας καὶ αὐτῷ θυγατέρες ἐγένοντο ἐννέα, καὶ χορὸν ἐναντίον ἔστησαν αὗται Μούσαις καὶ ἀγὼν ἐγένετο μουσικῆς ἐπὶ τῷ Ἑλικῶνι. ὅτε μὲν οὖν αἱ θυγατέρες ᾄδοιεν <αἱ> τοῦ Πιέρου, ἐπήχλυε πάντα καὶ οὐδὲν ὑπήκουε πρὸς τὴν χορείαν, ὑπὸ δὲ Μουσῶν ἵστατο μὲν οὐρανὸς καὶ ἄστρα καὶ θάλασσα καὶ ποταμοί, ὁ δ' Ἑλικὼν ηὔξετο κηλούμενος ὑφ' ἡδονῆς εἰς τὸν οὐρανόν, ἄχρις αὐτὸν βουλῇ Ποσειδῶνος ἔπαυσεν ὁ Πήγασος τῇ ὁπλῇ τὴν κορυφὴν πατάξας. ἐπεὶ δὲ νεῖκος ἤραντο θνηταὶ θεαῖς, μετέβαλον αὐτὰς αἱ Μοῦσαι καὶ ἐποίησαν ὄρνιθας <ἡμαθίδας> ἐννέα.

The imperfects ἐβασίλευε, ἐπήχλυε, ὑπήκουε, ἵστατο stand in strong contrast with the punctual aspect of the aorists ἔστησαν, ἔπαυσεν, μετέβαλον, ἐποίησαν.

The second story chosen for a comment on verbal aspect is called Ἀλώπηξ ("Fox") 41.8–10:[11]

ἐφάνη γὰρ ἐν χρόνῳ τούτῳ Καδμείοις ἀλώπηξ, χρῆμά τι ἐξηλλαγμένον· αὕτη συνεχῶς ἐκ τοῦ Τευμησσοῦ κατιοῦσα πολλάκις τοὺς Καδμείους ἡρπάζετο καὶ αὐτῇ προὐτίθεσαν παιδίον διὰ τριακοστῆς ἡμέρας, ἡ δὲ κατήσθιε λαμβάνουσα. καὶ ἐπειδὴ Ἀμφιτρύων εἰσιὼν τὰς <Θήβας> ἐδεήθη Καδμείων ἐπὶ Τηλεβόας αὐτῷ συστρατεῦσαι, καὶ οἵδ' ἔφασαν εἰ μὴ αὐτοῖς τὴν ἀλώπεκα συνεξέλοι, συντίθεται Ἀμφιτρύων ἐπὶ τούτοις πρὸς τοὺς Καδμείους. καὶ ἐλθὼν πρὸς τὸν Κέφαλον ἔλεγε τὴν συνθήκην καὶ ἔπειθε βῆναι εἰς <Θήβ>ας σὺν τῷ κυνί, ὁ δὲ Κέφαλος ἀποδέχεται <τὸν λόγον> καὶ ἐλθὼν κυνηγετεῖ τὴν ἀλώπεκα. ἦν δὲ θεμιτὸν οὔτε τὴν ἀλώπεκα καταληφθῆναι ὑπό τινος διώκοντος οὔτε τὸν κύνα ἐκφυγεῖν διωκόμενον οὐδέν. ἐπεὶ δὲ ἐγένοντο ἐν τῷ πεδίῳ τῶν Θηβαίων, Ζεὺς ἐποίησεν ἀμφοτέρους λίθους.

Imperfects ἡρπάζετο, προὐτίθεσαν, κατήσθιε, ἔλεγε, ἔπειθε vs. aorists ἐφάνη, ἐποίησεν: Through the use of aorist forms, the narrator puts the suddenly irrupting event of metamorphosis in evidence. Other formal devices going from gram-

10 This story concerns the rivalry between the Muses and Pierides or Emathides, the nine daughters of Pieros, who was then king in Emathia (imperfect ἐβασίλευε). On the Emathides, see Forbes Irving 1990, 238–239; Hard 2004, 207. The story apparently comes from Nicander, as Ant. Lib. himself says. Ov. tells this story in *Met.* 5.293, suggesting that it gives the aetiology for the use of Pierides as an epithet for the Muses. It could be interesting to study the theme of rivalry with the Muses as a whole, since we know at least two parallels: the Sirens who precipated themselves into the sea after having been defeated in the contest (Bettini and Spina 2010), and Thamyris, whom they punished with blindness (Pellizer 2004; Létoublon 2010).

11 This story – the last one in Antoninus Liberalis' collection – is recorded under the title Alopex, but is better known as Procris and Cephalus' story, particularly because of the version told by Ovid (*Met.* 7.763–93) which probably also derives from Nicander. See Pellizer 1981, 42–45; Forbes Irving 1990, 146 and 299; Delattre 2010 and 2013.

matical features to symbols may be noticed. We will come back later to the preceding moment in the tale.

From Devices of Metamorphosis to Its Symbols

The verbs used for describing the metamorphosis are most often transitive (active) verbs:[12]

> ἐποίησεν (3.4.2: Ποσειδῶν ... ἐποίησεν ὄρνιθα, ὅς ὀνομάζεται <ἔτι> ἱέραξ, 4.7.4: Ἀπόλλων δὲ κατ' ὀργὴν ἁψάμενος αὐτοῦ τῇ χειρῖ πέτρον ἐποίησεν ἵναπερ εἱστήκει, 5.5.8: Τιμάνδρην δὲ ἐποίησεν αἰγίθαλλον, 6.3.6: πιέσας ἀμφοτέραις ταῖς χερσὶν ἐποίησεν ὄρνιθα αἰετόν, 6.3.8: ἥν ἐποίησε φήνην, 11.9.3: Ζεὺς ... οἰκτείρας ἐποίησε πάντας ὄρνιθας, 15.4.3: ἡ δὲ αὐτὴν ἐποίησεν ὀρνίθιον γλαῦκα, 15.4.5: Ἑρμῆς δ' αὐτὸν ἐποίησε χαραδριόν, 15.4.7: ὁ δὲ κἀκεῖνον ἐποίησεν νυκτικόρακα κακάγγελον, 16.2.5: αὐτὴν ... γέρανον ἐποίησεν, 18.3.4: Ἀπόλλων ... ὄρνιθα ἐποίησεν τὸν παῖδα ἠέροπον, 19.3.3: καὶ ὁ Ζεὺς πάντας αὐτοὺς ἐποίησεν ὄρνιθας, 20.5.4: Ἅρπην μὲν καὶ Ἅρπασον ᾤκτειρε Ποσειδῶν καὶ ἐποίησεν αὐτοὺς ὄρνιθας τῷ αὐτῷ λεγόμενους ὀνόματι, 20.6.3: μεταβαλὼν ἐποίησεν πάντας ὄρνιθας, 22.4.7: Ποσειδῶν δὲ ... τὰς ἀδελφὰς ἐρρίζωσε καὶ ἐποίησεν αἰγείρους, 27.4.3: καὶ ἀλλάξασα ἐποίησεν αὐτὴν ἀγήρων καὶ ἀθάνατον δαίμονα καὶ ὠνόμασεν ἀντὶ Ἰφιγενείας Ὀρσιλοχίαν, 34.5.1: καὶ αὐτὴν ὁ Ζεὺς μεταβαλὼν ἐποίησε δένδρον καὶ ἐκάλεσεν ὁμώνυμον αὐτῇ σμύρναν, 35.4.4: Λητὼ δὲ μεταβαλοῦσα πάντας ἐποίησε βατράχους, 36.2.2: ἐποίησε τὴν μὲν αἶγα μεταβαλὼν ἀθάνατον, 36.3.4: Ζεὺς δὲ Πανδάρεον μὲν ἀντὶ τῆς κλοπῆς ἐποίησεν ὅθιπερ εἱστήκει πέτρον, 39.6.5: Ἀφροδίτη μετέβαλεν αὐτὴν καὶ ἐποίησεν ἐξ ἀνθρώπου λίθον καὶ τοὺς πόδας ἐρρίζωσεν ἐπὶ τὴν γῆν, 41.10.3: Ζεὺς ἰδὼν ἐποίησεν ἀμφοτέρους λίθους),[13] as well as several forms of μεταβάλλειν in the periphery of ἐποίησεν (see above 20.6.3, 34.5.1, 35.4.4, 36.2.2, 39.6.5) or independently (5.5.1: Ζεὺς δὲ μετέβαλεν εἰς ὄρνιθας, 9.3.2: μετέβαλον αὐτὰς αἱ Μοῦσαι καὶ ἐποίησαν ὄρνιθας ἐννέα, 10.4.4: αὐτὰς Ἑρμῆς ἁψάμενος ῥάβδῳ μετέβαλεν εἰς ὄρνιθας, 14.2.7: μετέβαλε δὲ πάντας εἰς ὄρνιθας, 17.6.3: καὶ μετέβαλε τὴν φύσιν τῆς παιδὸς εἰς κόρον, 22.5.5: νύμφαι δὲ μετέβαλον κατ' ὀργὴν Κέραμβον 23.6.4: Ἑρμῆς δὲ ... ἐρράπισεν αὐτὸν τῇ ῥάβδῳ καὶ μετέβαλεν εἰς πέτρον, 26.4.5: μετέβαλον τὸν Ὕλαν καὶ ἐποίησαν Ἠχώ), though we also find this verb as intransitive (more or less like *changer* in French, *to change* in English): see 21.5.5: Ἄγριος δὲ μετέβαλεν εἰς γῦπα, 37.5.5: αἱ ψυχαὶ δὲ μετέβαλον εἰς ὄρνιθας.

As less frequently transitive verb forms, we can also cite ἤλλαξαν, used once for Byblis 30.8: καὶ πολὺν ὕπνον ἐνέβαλον καὶ αὐτὴν ἤλλαξαν ἀπ' ἀνθρώπου εἰς δαίμονα καὶ ὠνόμασαν ἀμαδρυάδα νύμφην Βυβλίδα and μετεμόρφωσεν (2.6.5: αὐτὰς ἁψαμένη ῥάβδῳ μετεμόρφωσεν εἰς ὄρνιθας, 15.4.6: τὸν Ἑρμῆν ἐνείκεσε ὅτι μετε-

---

12 On the notions of transitive (verbs) and transitivity, see Conti 2014; on active verbs, see Rutger 2014: active transitive verbs are linked to the notion of agent (see below, n. 13).

13 We did not try to abbeviate this list because we will later meet several details given here, thus avoiding repeating the text.

μόρφωσεν αὐτοῦ τὸν υἱόν). We also meet the intransitive verbs ἐγένετο for Aegypios 5.5, for Askalabos 24.3: ὁ δὲ μεταβαλὼν ἐγένετο ποικίλος ἐκ τοῦ σώματος ἀσκάλαβος; for Dryope 32.4: Δρυόπη δὲ μετέβαλε καὶ ἀντὶ θνητῆς ἐγένετο νύμφη.

The transitive verbs clearly indicate that someone else is implied other than the metamorphosed person and his/her persecutor, generally a god who feels pity or anger: Zeus, Apollo, Hermes, Leto, Artemis, Athena are thus shown as the agents of the metamorphosis.[14]

The Touch and the Wand

In several cases, the active role of a god turns to a touch (expressed in the texts through several uses of the participle ἁψάμενος), which may be interpreted as a symbolic action. In some cases the god touches the human being with his/her hand(s) (χειρί, χερσίν), but in most cases, they use explicitly a stick or a magic wand (Gr. ῥάβδῳ), which well shows how close ancient Greek myths are to fairy tales:[15]

> Meleagrides 2.8: αἱ δὲ ἀδελφαὶ αὐτοῦ παρὰ τὸ σῆμα ἐθρήνουν ἀδιαλείπτως ἄχρις αὐτὰς Ἄρτεμις ἁψαμένη ῥάβδῳ μετεμόρφωσεν εἰς ὄρνιθας.
> Kragaleus 5.9: Ἀπόλλων δὲ κατ' ὀργὴν ἁψάμενος αὐτοῦ τῇ χειρὶ πέτρον ἐποίησεν.
> Periphas 6.3: πιέσας <αὐτὸν> ἀμφοτέραις ταῖς χερσὶν ἐποίησεν ὄρνιθα αἰετόν.
> Minyades 10.4: ἄχρις αὐτὰς Ἑρμῆς ἁψάμενος τῇ ῥάβδῳ μετέβαλεν εἰς ὄρνιθας.
> Battos 23.6: ἐρράπισεν αὐτὸν τῇ ῥάβδῳ.

In three cases, Zeus' lightning (κεραυνός) replaces the more usual wand, but the terrible touch by lightning is avoided by other gods who want to spare their protégés:

> Periphas 6.3.2: Ζεὺς δὲ νεμεσήσας ἐβούλετο μὲν σύμπασαν αὐτοῦ τὴν οἰκίαν κεραυνῷ συμφλέξαι, δεηθέντος δ' Ἀπόλλωνος μὴ αὐτὸν ἀπολέσθαι πανώλεθρον, ἐπεὶ περισσῶς αὐτὸν ἐτίμα, τοῦτο μὲν Ἀπόλλωνι δίδωσι Ζεύς, ἐλθὼν δ'εἰς τὰ οἰκία τοῦ Περίφαντος … πιέσας ἀμφοτέραις ταῖς χερσὶν ἐποίησεν ὄρνιθα αἰετόν.

> Φῶρες ('The Thives') 19.3.1: Ζεὺς δὲ βροντήσας ἀνέτεινε τὸν κεραυνόν, Μοῖραι δὲ καὶ Θέμις ἐκώλυσαν οὐ γὰρ ἦν ὅσιον ἀυτόθι θανεῖν καὶ ὁ Ζεὺς πάντας αὐτὸς ἐποίησεν ὄρνιθας.

14 On agency in Greek see Luraghi 2014.

15 On magic wand in fairy tales, see "Les objets magiques dans les contes" on the website of the Bibliothèque nationale de France (bnf.fr), and the discussion of Bettelheim 1975 by Jack Zipes 2002 and 2006. Particularly Zipes 2002, 179–204 (ch. 6 "On the Use and Abuse of Folk and Fairy Tales with Children. Bruno Bettelheim's Moralistic Magic Wand").

In one case only, for Typhon, Zeus' lightning becomes a lethal weapon; Typhon 28.3: ἔπει<τα> δὲ Τυφῶνα Ζεὺς βάλλει κεραυνῷ· καιόμενος δὲ ὁ Τυφὼν ἔκρυψεν ἑαυτὸν καὶ ἠφάνισε τὴν φλόγα τῇ θαλάσσῃ.[16]

We may also think of the importance of the spear given to Kephalos by Procris in the tale of the Teumessos fox (41), as some engravings for Ovid's *Metamorphoses* illustrate.[17]

Disappearing

The metamorphosis is sometimes shown as some incomprehensible change in the world through a succession of events without any visible link: this seems particularly striking in several cases of disappearance of a human person, usually surrounded by some metamorphosis in the landscape:[18] it might seem that the source replaces Lamia as a result of metamorphosis, but nothing is said explicitely.

> Ctesylla 1.5: ἐκ δὲ τῆς στρωμνῆς πελειὰς ἐξέπτη καὶ τὸ σῶμα τῆς Κτησύλλης ἀφανὲς ἐγένετο.
> Sybaris-Lamia 8.7: καὶ αὐτὴ μὲν τοῦ τραύματος ἀφανὴς ἐγένετο, ἐκ δὲ τῆς πέτρας ἐκείνης ἀνεφάνη πηγή, καὶ αὐτὴν οἱ ἐπιχώριοι καλοῦσι Σύβαριν.[19]
> Kyknos 12.8: ἀθυμήσας δὲ κατέβαλεν ἑαυτὸν εἰς τὴν Κωνώπην λεγομένην λίμνην καὶ ἠφανίσθη.[20]
> Metioche and Menippe 25.4: Φερσεφόνη δὲ καὶ Ἅιδης οἰκτείραντες τὰ μὲν σώματα τῶν παρθένων ἠφάνισαν, ἀντὶ δ' ἐκείνων ἀστέρας ἀνήνεγκαν ἐκ τῆς γῆς· οἱ δὲ φανέντες ἀνηνέχθη-

16 For Aarne-Thompson and Hansen (2002, 305–312) the tale of Typhon represents the Greek versions of an international tale, "Ogre Steals Thunder's Instruments" AT 1148B).

17 Whereas Ant. Lib.'s narrative is not very evocative. For the spear as a substitute for the magic wand, see for instance the illustration for the medieval manuscript of Othea, or later the Rusconi engraving for Lodovico Dolce's translation of Ovid's *Mets.* See below our analysis of the 'Instant before'.

18 We did not mention until now the importance of the landscape: as it is well shown by Cohen 2007 among others, Greeks had no words for referring to the landscape, but all the same it plays an important role in mythology (mountains, caves, trees and bodies of water, particularly).

19 The story comes from Nicander once more. On Lamia or Sybaris, Forbes Irving (1990, 303–304) refers to parallels in Pausanias, Strabo, and Aelian under the name of the "hero of Temesa" who disappears into the sea while Lamia is transformed into a spring. The contrast in the text between ἐγένετο ἀφανής and ἀνεφάνη πήγη is striking. The aim of the narration appears through the etymogical pun between ἀφανής and ὠνόμασαν αὐτὴν Ἀφαίαν.

20 On this Kyknos, son of Apollo, see Forbes Irving 1990, 257 (after the same Nicander; Ovid has a somehow different story from ours).

σαν εἰς οὐρανόν, καὶ αὐτοὺς ὠνόμασαν ἄνθρωποι κομήτας.[21]
Britomartis 40.4: ἡ δὲ Βριτόμαρτις ἀποβᾶσα ἐκ τοῦ πλοίου κατέφυγεν εἰς ἄλσος, ὅθιπέρ ἐστι νῦν αὐτῆς τὸ ἱερόν, κἀνταῦθα ἐγένετο ἀφανής, καὶ ὠνόμασαν αὐτὴν Ἀφαίαν.[22]

We cannot in this limited space refer to the Latin text in detail, but Antoninus Liberalis demonstrates how much Ovid's *Metamorphoses* owe to their Greek sources.[23] His text looks like a prosaic version, without the ornaments of poetry that are so meaningful in Ovid. The nature of the process of metamorphosis is thus as if denuded of its poetic milieu.

The Instant Before: Pursuit and Failing

If we take into account the time immediately preceding the metamorphosis, we see how frequently the character who will endure metamorphosis is pursued: another character, most often pushed by an erotic desire frequently through the metaphor of hunting,[24] tries to touch her/him, in the hope of possession.[25] The usual topos concerns a girl, more precisely a virgin (παρθένος) pursued by a would be male raper, but it corresponds to an obsessive theme of the pursuit that cannot touch the target, exemplified in a well-known nightmare told in a Homeric simile, considered by the specialist of the Irrational in Greece as an expression of frustration.[26] In our mythological corpus, among several examples, the best might be the story of Daphne in Parthenius 15: she was a very good hunter, active in the mountains of Peloponnesos with her dogs, and as she was dear to Artemis, she did not ever miss the target (15.1 καὶ αὐτὴν εὔστοχα βάλλειν

---

21 A tale found only in Ant. Lib. (referring to Nicander and Corinna) and Ov. *Met.* 13.685 – 699. The role of the shuttle (Gr. κερκίς) used by the Koronides Virgins for killing themselves is striking.

22 The story gives the *aition* for Artemis Aphaia's cult in Aigina. See Forbes Irving 1990, 287, who refers also to Dictynna in Crete.

23 Lafaye 1971 (1904); Cameron 2004.

24 On hunting as metaphor, see Schnapp 1997 and Delattre 2013.

25 On visual representations of this theme, see Sourvinou-Inwood 1987.

26 Hom. *Il.* 22. 199 – 201:

ὡς δ' ἐν ὀνείρῳ οὐ δύναται φεύγοντα διώκειν·
οὔτ' ἄρ' ὃ τὸν δύναται ὑποφεύγειν οὔθ' ὃ διώκειν·
ὣς ὃ τὸν οὐ δύνατο μάρψαι ποσίν, οὐδ' ὃς ἀλύξαι.

On the whole passage, see the excellent commentary by Richardson 1993, 127. In his famous *The Greeks and the Irrational*, Dodds quotes this passage as an example of anxiety-dreams: "The poet does not ascribe such nightmares to his heroes, but he knows well what they are like, and makes brilliant use of the experience to express frustration" (Dodds 2004 [1951], 106).

ἐποίει). A youth called Leukippos, despairing to seduce her as a man, disguised himself as a woman and took the habit of hunting with her. She appreciated this company and they were embracing each other (15.2 οὐ μεθίει τε αὐτὸν ἀμφιπεσοῦσά τε καὶ ἐξηρτημένη πᾶσαν ὁρᾶν). But Apollo was also in love with her, and he inspired Leukippos to go bathing to the spring with the girls, who snatched his clothes, discovered him and aimed their javelins at him. He then disappeared (15.4 ἀφανὴς γίνεται).[27] Seeing Apollo about to reach her, Daphne asked Zeus to remove her from the human world (ὡς δὲ συνεδιώκετο, παρὰ Διὸς αἰτεῖται ἐξ ἀνθρώπων ἀπαλλαγῆναι. Καὶ αὐτήν φασι γενέσθαι τὸ δένδρον τὸ ἐπικληθὲν ἀπ' ἐκείνης δάφνην).[28] The proper name thus becomes a common name.[29]

The metamorphosis appears then as a solution for the pursued person for escaping the pursuer, who feels disappointment and frustration, even if he may find some compensation in touching the girl who has been transformed into a tree, for instance: Parthenius leaves this aside but Ovid – whom we quote in translation, since the passage is extensive – narrates it in a very expressive manner.

Phoebus admired and loved the graceful tree (for still, though changed, her slender form remained), and with his right hand lingering on the trunk he felt her bosom throbbing in the bark. He clung to trunk and branch as though to entwine his form with hers, and fondly kissed the wood that shrank from every kiss. And thus the God; "Although thou canst not be my bride, thou shalt be called my chosen tree, and thy green leaves, O Laurel! shall forever crown my brows, be wreathed around my quiver and my lyre; the Roman heroes shall be crowned with thee, as long processions climb the Capitol and chanting throngs proclaim their victories; and as a faithful warden thou shalt guard the civic crown of oak leaves fixed between thy branches, and before Augustan gates. And as my youthful head is never shorn, so, also, shalt thou ever bear thy leaves unchanging to thy glory". Here the God, Phoebus Apollo, ended his lament, and unto him the Laurel bent her boughs, so lately fashioned; and it seemed to him her graceful nod gave answer to his love. (*Met.* 1.553–567, transl. Brookes More, e-text).

---

27 On disappearing, see the former paragraph.

28 Daphne's myth is known through several Greek and Latin sources (particularly Stat. *Theb.* 4.289–290, Paus. 10.7.8, Ov. *Met.* 1.452, 504, 525, 544–546, Hyg. 203). Parthenius' variant is very interesting because of his addition of the transvestite youth Leukippos who arouses Apollo's jealousy.

29 This is true only as far as mythology is concerned of course.

Procris and Kephalos' story in Antoninus Liberalis is interesting in the perspective opened with Parthenius' story of Leukippos below: as Kephalos and Procris successively challenge one another, Procris disguises herself as a youth so that she becomes his hunting companion; since he sees that she never misses the target, he desires to possess her javelin. She promises to give it with her dog moreover, in case he grants her his favors. After he accepted, she revealed herself and gave him the javelin and the dog. The pursuit that we see thereafter concerns the dog and the fox: the text says it was impossible for the dog to catch up to the fox, as well as for the fox to escape the dog (41.10: Ἦν δὲ θεμιτὸν οὔτε τῇ ἀλώπεκι καταληφθῆναι ὑπό τινος διώκοντος, οὐδὲ τὸν κύνα ἐκφυγεῖν διωκόμενον οὐδέν). Zeus transformed both dog and fox into stones: as if the obsession of pursuit and escaping the pursuer was transmitted from humans to animals, the sole solution for this infernal race being that each participant was fixed forever in his/ her position (petrification).[30]

Incestuous Loves

It would be interesting to look into psychoanalytical analysis for the probably very deep roots of this obsessive theme of metamorphoses in Greek myths. I will not develop this general aspect here,[31] but only mention how often incestuous love occurs in these simple and prosaic narratives, particularly if we add Parthenius of Nicaea to our corpus.[32]

Byblis and Caunus' story is told in Antoninus Liberalis 30, as well as by Parthenius 11 (who quotes two poems relating to them), Conon *Dieg.* 2, Ovid *Met.* 9. 441–665. The characters are sister and brother, and moreover they are twins, for Antoninus (30.2 δίδυμοι παῖδες).[33] For him, Byblis does not confess her love either to her brother or to her parents (30.3 ἄφατος ἔρως), whereas in Ovid she writes a letter to Caunus who flies away, as if Ovid seeks to show her guilt

---

30 Delattre 2010 and 2013. Once again, it would be interesting to compare with the Ovidian version: instead of a mere petrification, he suggests that both animals were transformed into statues rather than stones.

31 For a general insight on myth and psychoanalysis, see Armstrong 2011, without a special interest in metamorphosis. See also Armstrong 2005 and 2013 and the introduction to the volume by Zajko and O'Gorman 2013, 1–17 under the title "Myths and their Receptions: Narrative, Antiquity, and the Unconscious".

32 On Parthenius' style see the introduction in Biraud – Voisin – Zucker 2008. On Eros in Parthenius see Zucker 2008.

33 On twins in antiquity see Dasen 2005.

more clearly.[34] In Antoninus Liberalis' version, she does not dare to tell her passion to anyone, but she cannot bear it either, deciding to kill herself by precipitating her body from a cliff.[35] Sympathetic nymphs, however, save her, transforming her into a hamadryad nymph and giving rise to a spring called "Byblis' tears" (30.4 δάκρυον Βυβλίδος).

Smyrna (Ant. Lib. 34) is parallel to Ovid's Myrrha (*Met.* 10.298–518).[36] She also is in love with her own father (called Theias there, whereas in Ovid he is Kinyras, king of Cyprus). With her nurse's complicity, she manages to secretly gain entry into his father's bed in the dark, until she becomes pregnant, and he discovers her through the stratagem of a torch hidden in the room.[37] Eventually Theias kills himself, Smyrna gives birth to Adonis and becomes a tree, the liquid flowing from the trunk being assimilated to her tears.[38]

Though shorter than Antoninus Liberalis' corpus, Parthenius[39] shows eight cases of incestuous loves.[40] Apart from Byblis and Caunus (Parth. 11),[41] common

34 Following the text, some of the Ovidian engravings show Byblis writing her letter.
35 In other versions, she tries to hang herself to a tree (Parth., Conon).
36 Forbes Irving thinks this tale is among the oldest ones (1990, 128, 274–277). According to Athenaeus, the difference between both names depends on dialects. The main differences between the different sources are the name of the father, Theias vs Kinyras, the place they are living in, Cyprus vs Mount Liban, and the way Adonis is born from the tree or before his mother's metamorphosis.
37 34.4: καὶ ὡς ἐκύησε ἡ Σμύρνα, Θείαντα δὲ πόθος ἔλαβε ἐκμαθεῖν ἥτις ἦν ἡ κύουσα, καὶ ὁ μὲν κατέκρυπτε πῦρ εἰς τὸν οἶκον, Σμύρνα ὡς ἐξίκετο πρὸς αὐτὸν, ἐπάιστος ἐγένετο προενεχθέντος ἐξαπίνης τοῦ πυρός. The same stratagem is met in Ovid's version. Cf. also Procris and Kephalos' story in Ant. Lib. 41.3.5: ὁ δὲ Κέφαλος, ὅτε αὐτὴν ἔγνω παρελθοῦσαν ἐπὶ τὸν οἶκον καὶ κατακλινεῖσαν ὡς παρὰ τὸν ξένον, δᾷδα καιομένην παρήνεγκεν καὶ κατεφώρασεν αὐτήν.
38 34.5: τοῦτο λέγεται κατ᾽ ἔτος ἕκαστον δακρύειν τὸν ἀπὸ τοῦ ξύλου καρπόν.
39 36 stories in Parth., 41 in Ant. Lib.
40 We do not give the details for those stories. There are four adelphic love stories (Parth. 2 Polymele and her brother Diores, 5 Leukippos, 11 Byblis and Caunus; in 31, the incestuous love concerns Evopis and her brother Troizen, whereas Thymoïtes, Evopis' husband, experiments with another kind of sexual crime, necrophilia). Three tales deal with father-daughter incest (13 Harpalyke with her father Clymenos, 28 Cleite, 33 Assaon whose daughter is there Niobe, with an original version besides the most known versions). The story of Periander's mother (Parth. 17) tells the sole case of incest between a mother and her son.
41 Parthenius tells two versions of the tale and quotes some pieces of poetry. In the first version, Caunus is first in love with his sister, and he leaves the family home because of this love. In the second one, Byblis is in love with Caunus and tries to talk with him without convincing him (11.3: οἱ δὲ πλείους τὴν Βυβλίδα φασὶν ἐρασθεῖσαν τοῦ Καύνου λόγους αὐτῷ προσφέρειν καὶ δεῖσθαι μὴ περιιδεῖν αὐτὴν εἰς πᾶν κακὸν προελθοῦσαν); he flees to the Leleges where he founds a city called Caunus, and in her despair, she kills herself (τὴν δὲ ἄρα ὑπὸ τοῦ πάθους

to both authors as well as to Latin writers, we find in Parthenius another story of brother and sister, and the most original story, perhaps, that of a mother in love with her son.

Leukippos[42] (Parth. 5) is in love with his sister whose name is not given.[43] As he cannot any longer contain his desire, he confesses it to their mother, who complacently helps him to gain satisfaction.[44] The most surprising detail in this story is that brother and sister seem to have lived happy together for some time, until the girl's fiancé denounced her relation to her father, who strikes his daughter and is eventually killed by his son: Parth. 5.4 – 5 (Leukippos): κἀκ τούτου συνῆσαν οὐ μάλα τινὰ δεδοικότες, ἕως τις ἐξαγγέλλει τὴν κόρην τῷ μνηστῆρι.

The third main incestuous love story told by Parthenius concerns Periander's mother, once more a woman we do not know the name of, whereas Periander is well known from several sources as tyrant of Corinth. For Parthenius, this woman was in love with her son since his childhood, but she could then contain her desire by caressing him (17.1: καὶ τέως ἀνεπίμπλατο τῆς ἐπιθυμίας περιπλεκομένη τῷ παιδί). But later, she found a way for secretly having an affair with him, letting him believe she was a married woman. The pleasure of love was so strong for Periander that he wanted to know who she was, and recurred to the device of the hidden torch we have already met (17.6 ἐπεὶ δὲ ἡ μήτηρ ἀπεῖργεν αἰτιωμένη τὴν αἰσχύνην τῆς γυναικός, κελεύει τινὰ τῶν ἀμφ'αὐτὸν οἰκετῶν λύχνα κατακρύψαι, τῆς δὲ κατὰ τὸ συνηθὲς ἀφικομένης καὶ μελλούσης κατακλίνεσθαι ἀναδραμὼν ὁ Περίανδρος ἀναιρεῖ τὸ φῶς καὶ ἰδὼν τὴν μητέρα ὥρμησεν ἐπὶ τὸ διεργάσασθαι αὐτήν).[45] Periander did not succeed killing his mother, says Parthenius, but he became a fool[46] (17.7: κἀκ τούτου παραπλὴξ ἦν νοῦ τε καὶ φρενῶν κατέσκηψέ τε εἰς ὠμότητα καὶ πολλοὺς ἀπέσφαξε τῶν πολιτῶν), while his mother killed herself (ἀνεῖλε ἑαυτήν).

μὴ ἀνιεμένην, πρὸς δὲ καὶ δοκοῦσαν αἰτίαν γενομένην Καύνῳ τῆς ἀπαλλαγῆς, ἀναψαμένην ἀπό τινος δρυὸς τὴν μίτραν ἐνθεῖναι τὸν τράχηλον).

42 He is a different character from Leukippos in Ant. Lib. quoted above.

43 This is often the case for women in mythology. Let us recall that in Ovid Pygmalion does not give a name to the sculptor's creature. The name is deeply linked to the *person* (see Salvadore 1987).

44 5.2: ἐπεὶ μέντοι χρόνου διαγενομένου οὐδὲ ἐπ' ὀλίγον ἐλώφα τὸ πάθος, ἀνακοινοῦται τῇ μητρὶ καὶ πολλὰ καθικέτευε μὴ περιιδεῖν αὐτὸν ἀπολλύμενον.

45 The French commentary to Parthenius (Biraud, Voisin and Zucker 2008, 184) notes that another sexual crime is known about Periander, i.e. necrophilia (Hdt 3.50, 5.92).

46 A psychiatrist in Grenoble told me anonymously that he professionally met several cases of foolishness after an incestuous relation between son and mother.

These cases, with the troubling recurrence of the theme of a kind of complete happiness as long as the incestuous relation remains secret,[47] and of the hidden torch that reveals the truth, seem to correspond to metamorphosis as a solution for the transgression of the most important prohibitions for mankind.

It seems interesting to think about how Freudian theory would have been more powerful if Freud had used Parthenius' corpus rather than focusing on Sophocles' Oidipous:[48] as Vernant demonstrated, the latter is seeking for power, having married his mother as a consequence of solving the Sphinx's enigma, rather than as a realization of sexual desire,[49] whereas Parthenius seems to express the extreme pleasure taken in an incestuous love – apparently consciously in Leukippos' case, unconsciously in Periander's one.

Metamorphosis and Metaphor

A metamorphosis, summarily defined as a change from one realm of nature to another, is in some way a metaphor re-deployed from a spatial to a temporal frame of reference, from before to afterwards. Metamorphosis tells in a timeline what metaphor displays in a single moment of time. Metamorphosis is concerned with evolution whereas metaphor is a kind of equivalence in simultaneity. The tale of Aedon in Antoninus Liberalis is somehow clearer than the very poetic version of Procne and Philomele given by Ovid: as for the Odyssean passage where Penelope compares herself to Aedon or the nightingale (ἀηδών),[50] one may think

47 An episode even more explicit than Leukippos and his sister's happiness (Parth. 5.4–5, quoted above) occurs in 17.4 about Periander's erotic pleasure: τῇ δ' ὑστεραίᾳ ἀναπυνθανομένης αὐτῆς εἰ κατὰ νοῦν αὐτῷ γένοιτο καὶ εἰ αὖτις λέγοι αὐτὴν παρ' αὐτὸν ἀφικέσθαι, ὁ Περίανδρος σπουδάζειν τε ἔφη καὶ ἡσθῆναι οὐ μετρίως. ὡς δὲ ἐκ τούτου οὐκ ἀνίει φοιτῶσα πρὸς τὸν παῖδα καί τις ἔρως ἐπῄει τὸν Περίανδρον, ἤδη σπουδὴν ἐτίθετο γνωρίσαι τὴν ἄνθρωπον ἥτις ἦν.

48 On Freud and Sophocles' Oidipous, see Armstrong 2005, 2011, and 2013 quoted above.

49 Vernant's title is eloquent: "Œdipe sans complexe" ("Oidipous without complex"). Vernant 1967.

50 Hom. *Od.* 19.518–526:

ὡς δ' ὅτε Πανδαρέου κούρη, χλωρὶς ἀηδών,
καλὸν ἀείδῃσιν ἔαρος νέον ἱσταμένοιο,
δενδρέων ἐν πετάλοισι καθεζομένη πυκινοῖσιν,
ἥ τε θαμὰ τρωπῶσα χέει πολυηχέα φωνήν,
παῖδ' ὀλοφυρομένη Ἴτυλον φίλον, ὅν ποτε χαλκῷ
κτεῖνε δι' ἀφραδίας, κοῦρον Ζήθοιο ἄνακτος
ὣς καὶ ἐμοὶ δίχα θυμὸς ὀρώρεται ἔνθα καὶ ἔνθα.

Cf. *Od.* 20.66 alluding to sisters in the plural form: ὡς δ' ὅτε Πανδαρέου κούρας ἀνέλοντο θύελλαι.

of the complete literal inadequacy between her nocturnal laments and the well-known beauty of the bird's song.[51] She herself says that she weeps and laments *as* Aedon who lost her son Itylos (a name close to Itys in Ovid, but nevertheless different): if the Homeric simile refers to Aedon, which in this interpretation is a proper noun, according to our grammatical tradition, it points to an unfortunate woman who killed Itylos, be it voluntarily or not, and parallelism certainly exists. But the context implies that Aedon *sings* in spring: thus the bird is involved, which means the mother is transformed into a bird, and the word Aedon already is here a common noun. Therefore the Homeric comparison already conceals a metaphor. Antoninus Liberalis is not that subtle, but he also calls both sisters by birds' names, Aedon and Chelidonis ("swallow"), which indicates that metamorphosis has been intended from the beginning.

As fundamentally situated in space, painting and sculpture cannot express metamorphosis except through contiguity,[52] for instance by taking into account the moment that immediately precedes or follows it, the instance where the character undergoes the beginning of the change. The Daphne by Bernini has still a human face and body, but her arms and fingers are already being transformed into branches and palm leaves. Actaeon is often shown as a man with deer antlers. On an Etruscan black-figures hydria of around 530 BC we can see several characters, who are partly transformed: a human diver as well as dolphins who still have human legs. The Centaurs (half men and horses) and the god Pan (a human figure with goat feet) could embody an ongoing or stopped metamorphosis. Narrative literature, as an art of time, is able to tell how forms change, a coach becoming a pumpkin, a beast becoming a handsome young man, a man becoming a deer, or a girl becoming a tree or a spring. F. Frontisi-Ducroux explains this as a "goût vorace de la différence", "désir du différent, de l'étrange, de l'étranger, de l'ailleurs; désir de fusion avec l'autre".[53] In the same time, let us recall that girls who are pursued, be it by Pan, who does not look very appealing, or by Apollo who is often taken as a model of masculine beauty, generally refuse this fusion and prefer complete metamorphosis or disappearance rather than such a union with this lover.

In conclusion, we would like to open up the question of the meaning of metamorphosis, especially for young girls and youths, by linking it to the notion of

---

51 Létoublon 2004 on the myths of Aedon, Philomele and Procne in Homer, Antoninus Liberalis, and Ovid. In the passage quoted, note the etymological relation between χλωρὶς ἀηδών and καλὸν ἀείδησιν (see Chantraine 2009, 21 on ἀείδω, 25 on ἀηδών).

52 On metamorphosis in art see Sharrock 1996.

53 Frontisi-Ducroux 2003, 277.

"rites of passage" in anthropology and sociology.[54] Among those rites, several societies, ancient and modern, put in relief the difficulty of passing from childhood to adolescence, and then from adolescence to adulthood.[55] Its reality was demonstrated for Antiquity through the study of several literary genres and texts,[56] and we may even now hold firmly enough that the whole genre of Greek 'ideal' novel constitutes a kind of answer to this question: how young girls and youths may go through this period of their life, what tests or trials are they supposed to succeed in, encountering one another, falling in love, separations, and so on,[57] the existing rituals being interpreted as some kind of symbolic transposition of actual trials.[58] Some myths or mythological episodes that deal with transvestites (Achilles among Lycomedes' daughters in the Epic Cycle, Leukippos in Parthenius above) could receive an explanation as rites of passage.[59]

In his various works on the subject, Ken Dowden put aside myths, epic and novel genres.[60] In my "mythological paradigms" paper,[61] I show how often, in the idealistic novels written in Greek during the first centuries of Roman era, the characters themselves tell some myths for a *paradigmatic* use.[62] If in

54 The notion was discovered and analyzed by Arnold Van Gennep 1909 (see Van Gennep 1960 in English). It won success through such œuvres as Victor Turner 1997. We organised an international conference in Grenoble for the centenary of Van Gennep's book (edited by Philippe Hameau, Christian Abry, Françoise Létoublon 2010). The concern of rites of passage in antiquity has been studied by Padilla 1999.

55 Scars appear sometimes as marks of the initition rite that a character had endured as a youth: this might hold good for Odysseus' scar on the thigh (*Od.* 19.385–398) and for Orestes' scar near the eyebrow, if we follow Eur. *El.* (Létoublon 2010).

56 Probably first with tragedy and the character of Orestes, see for instance Vidal-Naquet 1986 (1981) and Padilla 1999.

57 On those *lieux communs* in Greek novels, see Létoublon 1993.

58 Those rituals are different for girls and youths, Dowden 1999 and 2011; Lalanne 2006 and 2010. The corespondences between myths and rituals are not easy to understand (Dowden 2011, 489–492 about Iphigeneia and *arkteia*).

59 On Achilles, see Dowden 2011, 493.

60 Particularly Dowden 1989, 1999.

61 Létoublon 2013.

62 The passages quoted are the following: [Longus], *Daphnis and Chloe* 1.23.3 (concerning Phatta, implied in a poetry and music contest, but including an allusion to a pursuit and flight tale concerning Pitys and Pan), 2.34–37 (concerning Syrinx and Pan), 3.23 (Echo and Pan): all those stories concern virgins (the word παρθένος is constantly recurring) who are pursued by Pan and escape him through metamorphosis: Pitys as a pine-tree, Syrinx as the eponymous music instrument, Echo as the eponymous natural phenomenon. Another paradigmatic myth is told in 2.6 about Eros in Philetas' garden. There are also examples of the device in Ach. Tat. *Leucippe and Clitophon* 1.16–18, 5.3.4–8, 5.5.2–9 (Tereus, the lark and the nightingale), 8.6.3–10 (the

myths metamorphosis appears as a a means for escaping the rape by the dangerous pursuer, narrating the myths might incarnate the symbolic transposition of the rite of passage: for instance in *Leucippe and Clitophon*, the youth hears a slave singing the story of Daphne and Apollo, and the text reads thus: τοῦτό μου μᾶλλον ᾆσθὲν τὴν ψυχὴν ἐξέκαυσεν ὑπέκκαυμα γὰρ ἐπιθυμίας λόγος ἐρωτικός ("this song inflamed my soul all the more, for erotic stories fuel the appetite"; 1.5.5). Thus, hearing (or reading?) love stories 'fuel' erotic desire and pulsions, which might be a powerful impulse for telling them.[63]

What about the pursuer in those myths of pursuit? Let us recall that he is often a god, and not the most awful of them; if Pan actually appears often in these myths, Zeus, Poseidon and the very handsome Apollo also appear very frequently. We quoted Apollo's feeling of frustration in Ovid's version of Daphne. In the novels, Pan's frustration is even more characteristic, especially as he realizes that Syrinx disappeared in the bulrushes.[64] But afterwards, he feels a kind of pleasure in making an instrument with the cut bulrushes, with a striking image of dismembering and reconstructing the body,[65] and the music he produces becomes the metaphor of her respiration.[66] The whole passage shows the in-

myth of Syrinx, parallel to *Daphnis and Chloe*, receives here the charge of proving Leucippe's virginity) and 8.12 (Styx's water, as a very sophistic test of virtue for the secondary character of Melite. The aetiology of the myth is the story of Rhodopis and Euthynicos, who both swore they would never "know Aphrodite").

63 Zajko and O'Gorman (2013, 13–14) well emphasize the importance of desire in both the author and the audience or readers.

64 Ach. Tat. 8.6.9: "He thought the maid had been changed into the reeds and wept that he had cut her, supposing his beloved had been slashed" (transl. B. Reardon).

65 Ach. Tat. 8.6.10:

> gathering up the severed bits of reed as if they were the limbs of her body and joining them together as a single body, he held in his hands the cut ends of the reeds and kissed them as if they were the maiden's wounds. He groaned as he put his lover's lips to them and so breathed into the flutes from above as he kissed them.

This passage seems very close to Ovid's description of the metamorphosis of Daphne into laurel, as Zajko and O'Gorman (2013, 8) analyze it:

> "The description lingers over each body part as it is translated into the parts of a tree which seems the most appropriate for it. Feet as roots, hair as leaves, et cetera. In one way the transformation is made to seem appropriate and 'natural'; this raises the question of whether a transformation takes place at all. For Apollo's sexual desire does not seem to be diminished; indeed, he persists in his fondling of the nymph/tree, who can no longer run away" (see above § 4).

66 "His breath flowed through the narrow reed passages and made flutelike sounds: the syrinx had a voice".

strument as a substitute for the girl.[67] We think it possible to establish a parallel with the psychoanalytic notion of sublimation, and to generalize this conclusion to other cases where a god feels frustrated in his erotic impulse when the person he was pursuing either disappears or undergoes metamorphosis.[68] Therefore the frustrated pursuer is himself transformed somehow in this adventure. In the mythographers' corpus, the cases of girls enduring metamorphosis after a sexually deviated adventure like that of Byblis and Smyrna might show such sublimation: their tears are flowing forever in the form of a spring or of myrrh.

These stories show that the relation of both would be raper and pursued girl to their own body is seen as a problem. Zajko and O'Gorman commenting Daphne's metamorphose in Ovid say it better than I could:[69] "The relationship between mind and body is centrally important here but it is also highly problematic: Ovid does not help the observer, who is left to surmise some inner state by means of physical phenomena. It is through the observer's interpretative choices that these phenomena may become signs of internal life, or, to use a psychoanalytic term, symptoms".

As Harris says, metamorphosis places the character "between two deaths": neither dying nor still living as a human person. Transformed into an unmoving tree, a still more motionless stone, a flowing watercourse or a flying bird, she will stay forever the symbol of the impossible flight. Daphne becoming as the laurel the symbol of Apollo who pursued her is the symbol of both the refuse of being touched by him and of accepting it.[70]

Bibliography

Aarne, A. and S. Thompson. 1981. *The Types of the Folktale: A Classification and a Bibliography.* Helsinki.

Armstrong, R.H. 2005. *A Compulsion for Antiquity: Freud and the Ancient World.* Ithaca.

Armstrong, R.H. 2011. Psychoanalysis: The Wellspring of Myth? In Dowden and Livingstone 2011, 471–485.

Armstrong, R.H. 2013. Myth, Religion, Illusion: How Freud Got His Fire Back. In *Classical Myth and Psychoanalysis. Ancient and Modern Stories of the Self*, eds. V. Zajko and E. O'Gorman, 59–74. Oxford.

67 See Zajko and O'Gorman 2013, 8: "The theme of metamorphosis itself de-naturalizes the process of human development and change, as the characters in the poem transform into flowers, animals, and trees".

68 Harris (2013, 258–260) recurs to Lacan for analyzing Daphne's impossible flight as a Lacanian *jouissance.*

69 Zajko and O'Gorman 2013, 9.

70 Harris 2013.

Bettelheim, B. 1975. *The Uses of Enchantment. The Meaning and Importance of Fairy Tales.* London.
Bettini, M. and L. Spina. 2007. *Il mito delle Sirene.* Turin.
Biraud, M., D. Voisin and A. Zucker. 2008. *Parthénios de Nicée, Passions d'amour.* Grenoble.
Bonnecase, D. and A.-M. Tatham. 2009. *La métamorphose. Définitions, formes, thèmes.* Saint-Pierre de Salerne.
Boyle, A.J. 2007. Ovid and Greek Myth. In Woodard 2007, 355–381.
Bremmer, J.N. 1987. *Interpretations of Greek Mythology.* London.
Bremmer, J.N. 2011. A Brief History of the Study of Greek Mythology. In Dowden and Livingstone 2011, 527–547.
Calame, C. 2000. *Poétique des mythes dans la Grèce antique.* Paris.
Calame, C. 2004. Du muthos des anciens Grecs au mythe des anthropologues. Entretien avec Claude Calame. In Mezzadri 2004, 9–37.
Cameron, A. 2004. *Greek Mythography in the Roman World.* Oxford.
Canfora, L. and D. Lanza. 1992. *Lo spazio letterario della Grecia antica.* Salerno.
Cartier-Bresson, H. 2007 (1973). *The Decisive Moment.* New York. (orig. ed. *Images à la sauvette.* Paris 1952)
Chantraine, P. 2009. *Dictionnaire étymologique de la langue grecque*, achevé par J. Taillardat, O. Masson et J.-L. Perpillou, avec en supplément les Chroniques d'étymologie grecque rassemblées par A. Blanc, Ch. de Lamberterie et J.-L. Perpillou. Paris.
Christopoulos, M. and M. Païzi-Apostolopoulou. 2014. *Crime and Punishment in Homeric and Archaic Epic* (Proceedings of the 12th International Symposium on the *Odyssey*). Ithaca.
Cohen, A. 2007. Mythic Landscapes of Greece. In Woodard 2007, 305–330.
Comrie, B. 1976. *Aspect: An Introduction to the Study of Verbal Aspect and Related Problems.* Cambridge.
Conti, L. 2014. Transitivity. In Giannakis, vol. iii, 423–427.
Darbo-Peschanski, C. 2000. *Constructions du temps dans le monde grec ancien.* Paris.
Dasen, V. 2005. *Jumeaux, jumelles dans l'Antiquité grecque et romaine.* Zürich.
Delattre, C. 2010. Le renard de Teumesse chez Antoninus Liberalis (Met. XLI): formes et structures d'une narration. *REG* 123: 91–111.
Delattre, C. 2013. Le temps suspendu dans les récits mythiques de poursuite et de chute. In Morel and Rouveret 2013, 43–57.
Dodds, E.R. 2004. *The Greeks and the Irrational* (Sather Classical Lectures 25, 1951). Berkeley.
Dowden, K. 1989. *Death and the Maiden. Girls' Initiation Rites in Greek Mythology.* London.
Dowden, K. 1992. *The Uses of Greek Mythology.* London.
Dowden, K. 1999. Fluctuating Meanings: 'Passage Rites' in Ritual, Myth, *Odyssey* and the Greek Romance. In Padilla 1999, 221–243.
Dowden, K. 2011. Telling the Mythology: From Hesiod to the Fifth Century. In Dowden and Livingstone 2011, 47–72.
Dowden, K. and N. Livingston. 2011. *A Companion to Greek Mythology.* Malden.
Elsner, J. 1996. *Art and Text in Roman Culture.* Cambridge.
Forbes Irving, P.M.C. 1990. *Metamorphosis in Greek Myths.* Oxford.
Fowler, R.L. 2000. *Early Greek Mythographers.* Oxford.
Fowler, R.L. 2011. Mythos and Logos. *JHS* 131: 45–66.

Frontisi-Ducroux, F. 2000. Figures du temps: la métamorphose. In Darbo-Pechanski 2000, 49–63.
Frontisi-Ducroux, F. 2003. *L'homme-cerf et la femme araignée. Figures grecques de la métamorphose.* Paris.
Gartziou-Tatti, A. 2014. Erotic Pursuit and Punishment in the *Odyssey.* The Eternally Punished of *Nekyia.* In Christopoulos and Païzi-Apostolopoulou 2014, 97–116.
Giannakis, G.K. 2014. *Encyclopedia of Ancient Greek Language and Linguistics.* 3 vols. Leiden and Boston.
Hameau, P., C. Abry and F. Létoublon. 2010. *Les rites de passage. De la Grèce d'Homère à notre XXIe siècle.* Grenoble.
Hansen, W. 2002. *Ariadne's Thread. A Guide to International Tales Found in Classical Literature.* Ithaca and London.
Hard, R. 2004. *The Routledge Handbook of Greek Mythology Based on H.J. Rose's Handbook of Greek Mythology.* London.
Harris, O. 2013. The Ethics of Metamorphosis or A Poet Between Two Deaths. In Zajko and O'Gorman 2013, 251–266.
Henrichs, A. 1987. Three Approaches to Greek Mythography. In Bremmer 1987, 242–277.
Hewson, J. 1997. The Greek Verbal System. In Hewson and Bubenik 1997, 24–45.
Hewson, J. 2006. Le système verbal du grec ancien: trois distinctions de temps ou deux? *Glotta* 82: 96–107.
Hewson, J. 2014. Aspect (and Tense). In Giannakis 2014, vol. I, 181–183.
Hewson, J. and V. Bubenik. 1997. *Tense and Aspect in Indo-European Languages: Theory, Typology, Diachrony.* Amsterdam and Philadelphia.
Higbie, C. 2007. Hellenistic Mythographers. In Woodard 2007, 237–254.
Karakantza, E. 2004. Literary Rapes Revisited. A Study in Literary Conventions and Political Ideologies. *Mètis* n.s. 2: 29–45.
Lafaye, G. 1971. *Les métamorphoses d'Ovide et leurs modèles grecs.* Mit einer Einführung von M. von Albrecht. Hildesheim and New York (orig. ed. Paris1904).
Lalanne, S. 2006. *Une éducation grecque. Rites de passage et construction des genres dans le roman grec.* Paris.
Lalanne, S. 2010. Le "Chasseur noir", le roman grec et le refus du passage. In Hameau, Abry and Létoublon 2010, 111–119.
Létoublon, F. 1993. *Les lieux communs du roman: stéréotypes grecs d'aventure et d'amour.* Leiden.
Létoublon, F. 2004. Le rossignol, l'hirondelle et l'araignée. In Mezzadri 2004, 73–102.
Létoublon, F. 2009. Poétique de la métamorphose. In Bonnecase and Tatham 2009, 21–32.
Létoublon, F. 2010. To See or not to See: Blind People and Blindness in Ancient Greek Myths. In *Light and Darkness in Ancient Greek Myth and Religion*, eds. M. Christopoulos, E. Karakantza and O. Levaniouk, 167–180. Lanham.
Létoublon, F. 2010. Des traces sur la peau. In Hameau, Abry, and Létoublon 2010, 259–268.
Létoublon, F. 2013. Mythological Paradigms in the Greek Novels. In Paschalis and Panayotakis 2013, 127–145.
Luraghi, S. 2014. Agency and Causation. In Giannakis 2014, vol. I, 65–72.
Mezzadri, B. 2004. *Mythe et Mythologie dans l'Antiquité gréco-romaine. Europe* 904–905.
Padilla, M.W. 1999. *Rites of Passage in Ancient Greece.* Lewisburg.

Paschalis, M. and S. Panayotakis. 2013. *The Construction of the Real and the Ideal in the Ancient Novel* (ANS 17). Groningen.
Pellizer, E. 1981. *Favole d'identità, favole di paura. Storie di caccia e altri racconti della Grecia antica.* Rome.
Pellizer, E. 1992. La mitografia. In Canfora and Lanza 1992, 283–303.
Pellizer, E. 2004. Epreuves d'amour, épreuves de mort. In Mezzadri 2004, 103–118.
Richardson, N.J. 1993. *The Iliad: A Commentary*, vol. vi: Books 21–24. Cambridge.
Ruipérez, M.S. 1980. *Structure du système des aspects et des temps du verbe en grec ancien.* Paris. (orig. ed. 1954)
Rutger, A. 2014. *Active.* In Giannakis 2014, vol. i, 24–25.
Salvadore, M. 1987. *Il nome, la persona. Saggio sull'etimologia antica.* Genova.
Schnapp, A. 1997. *Le chasseur et la cité. Chasse et érotique en Grèce ancienne.* Paris.
Sharrock, A.R. 1996. Representing Metamorphosis. In Elsner 1996, 103–130.
Sourvinou-Inwood, C. 1987. A Series of Erotic Pursuits: Images and Meanings. *JHS* 107: 131–153.
Turner, V. and R.D. Abrahams. 1997. *The Ritual Process: Structure and Anti-structure.* New York.
Van Gennep, A. 1960. *The Rites of Passage.* Chicago (orig. ed. Paris 1909, reedited 1981).
Vernant, J.-P. 1967. 'Œdipe' sans complexe. *Raison présente* 4: 3–20.
Vernant, J.-P. 1990. Oidipous Without the Complex. In Vernant and Vidal-Naquet 1990, 85–111.
Vernant, J.-P. and P. Vidal-Naquet. 1990. *Myth and Tragedy in Ancient Greece.* New York.
Vidal-Naquet, P. 1986. *The Black Hunter. Forms of Thought and Forms of Society in the Ancient World.* Baltimore. (orig. ed. Paris 1981)
Zajko, V. and E. O'Gorman. 2013. *Classical Myth and Psychoanalysis. Ancient and Modern Stories of the Self.* Oxford.
Zipes, J. 2002. *Breaking the Magic Spell: Radical Theories of Folk and Fairy Tales.* Lexington.
Zipes, J. 2006. *Fairy Tales and the Art of Subversion. The Classical Genre for Children and the Process of Civilization.* New York.
Zucker, A. 2008. *Littérature et érotisme dans les Passions d'amour de Parthénios de Nicée.* Grenoble.

Index Locorum

* Plays, fragments, passages, and line numbers are followed by page number

Index Nominum Notabiliorum

Notes on Contributors

James Andrews is Associate Professor of Classics at Ohio University in Athens, Ohio USA. His studies in the speeches of Thucydides have appeared in *Classical Quarterly*, *The American Journal of Philology*, *Classical Philology*, and other forums. Another paper originally presented to our dear colleagues at the University of Patras, "The *Protagoras* as a Comedy of Pleasure", will appear in print shortly. A book-length treatment of the same dialogue is in the works.

Constantin Antypas is a researcher specializing on Archaic Greek nautical history. He has published articles in collective volumes and journals, among (a few) others: "*Dike* in a Pre-polis Society: The evidence from Homeric Epic" (Proceedings of the 12th International Symposium on the *Odyssey*, Ithaca 2103) and "Η εξερεύνηση της Δυτικής Μεσογείου και η εφαρμογή νεωτερικών ναυτικών τεχνολογιών κατά την Αρχαϊκή Εποχή: Η περίπτωση της Φώκαιας" (43th Conference of Greek Philological Association, Athens 2016). A revised version of his PhD Dissertation "Ὑγρὰ Κέλευθα: Homeric Ships and Routes" (2014) is under publication process. C. Antypas is currently a High School teacher in Piraeus and a non-professional open sea sailboat skipper.

Anton Bierl is Professor for Greek Literature at the University of Basel. He served as Senior Fellow at Harvard's Center for Hellenic Studies (2005 – 2011) and is Member of the IAS, Princeton (2010/11). He is director and co-editor of *Homer's Iliad: The Basel Commentary* and editor of the series *MythosEikonPoiesis*. His research interests include Homeric epic, drama, song and performance culture, the ancient novel, Greek myth and religion. His books include *Dionysos und die griechische Tragödie* (1991); *Die Orestie des Aischylos auf der modernen Bühne* (1996); *Ritual and Performativity* (2009); and the co-edited volumes *Literatur und Religion I-II* (2007); *Gewalt und Opfer* (2010); *Intende Lector* (2013) and *The Newest Sappho* (2016).

David Bouvier is Professor of Ancient Greek at the University of Lausanne. He also teaches courses of mythology at the EPFL and is associate member of Anhima. He has been visiting Associate Professor at the University of Chicago and visiting Professor at the EHESS. His research and publications focus on forms of memory and knowledge in Ancient Greece. He is the author of the *Le sceptre et la lyre. L'Iliade ou la mémoire des héros* (2002).

Jonathan S. Burgess received his PhD from the University of Toronto, where he is now a Professor of Classics. He has published widely on Homer and the Epic Cycle, and travel literature. He is the author of *The Tradition of the Trojan War in Homer and the Epic Cycle* (Baltimore 2001), *The Death and Afterlife of Achilles* (Baltimore 2009), and *Homer* (London 2015). Currently he is working on a project involving the travels of Odysseus, before and after his return to Ithaca.

Menelaos Christopoulos is Professor of Ancient Greek Literature at the University of Patras and Director of the *Centre for Odyssean Studies* (http://cods.upatras.gr). He has published on Homeric Poetry, Greek Drama and Greek Myth, including (as coeditor with E. Karakantza and

O. Levaniouk) *Light and Darkness in Greek Myth and Religion* (Lanham 2010) and (as coeditor with M. Paizi-Apostolopoulou) *Crime and Punishment in Homeric and Archaic Poetry* (Ithaca 2014).

Myrto Garani is Assistant Professor in Latin Literature at the National and Kapodistrian University of Athens, Greece. She is the author of *Empedocles Redivivus: Poetry and Analogy in Lucretius* (London and New York 2007). She has published a series of articles on Ovid's reception of Empedocles in the *Fasti*. She has also co-edited with Professor David Konstan the volume entitled: *The Philosophizing Muse: The Influence of Greek Philosophy on Roman Poetry* (Newcastle 2014).

Elena Iakovou is a PhD student at the Georg-August-University Göttingen (Germany) under the supervision of Professor Heinz-Günther Nesselrath working on the Reception of the Oedipus myth from Classical to Imperial times (8th c. BC to 3rd c. AD) with a focus on the fragmentary tragedy *Oedipus* of Euripides.

Stratis Kyriakidis is Emeritus Professor of Latin Literature at the Aristotle University of Thessaloniki and Visiting Professor at the University of Leeds. He is the author of *Roman Sensitivity: A Contribution to the Study of the Artistic Receptiveness and Creativity of the Romans (146–31 BC)* (Thessaloniki 1986) [in Greek]; *Narrative Structure and Poetics in the* Aeneid: *The Frame of Book 6* (Bari 1998); and *Catalogues of Proper Names in Latin Epic Poetry: Lucretius – Virgil – Ovid*, Pierides I (Newcastle upon Tyne 2007). He is the editor (with Francesco De Martino) of *Middles in Latin Poetry* (Bari 2004) and of *Libera fama: An Endless Journey*, Pierides VI (Newcastle upon Tyne 2016). His publications mainly focus on Latin literature of the late Republican and Augustan periods, on Manilius' *Astronomica* and on the Latin centos. With Philip Hardie he is the co-editor of the *Pierides* series at Cambridge Scholars Publishing.

Françoise Létoublon is Professor of Greek Literature and Linguistics at the University of Grenoble (emerita). She is the author of *Il allait, pareil à la nuit. Les verbes de mouvement en grec: supplétisme et aspect verbal* (Paris 1985) and of *Les lieux communs du roman* (Leiden 1993). She has edited *La langue et les textes en grec ancien.* Colloque Pierre Chantraine (Amsterdam 1993), *Impressions d'îles* (Toulouse 1996), *Hommage à Milman Parry. Le style formulaire de l'épopée homérique et la théorie de l'oralité poétique* (Amsterdam 1997), *Homère en France après la Querelle* (Paris 1999). She is currently working on Homeric poetry, mythology, and their reception, up from antiquity (Greek novels). She recently published "Mythological Paradigms" in *The Construction of the Real and the Ideal in the Ancient Novel* (*ANS* 17, 2013), "The Magnetic Stone of Love. Greek Novel and Poetry", and "Respect these Breasts and Pity Me. Greek Novel and Theater" (in coll. with Marco Genre), in *A Companion to the Ancient Novel*, and "Le Palladion dans la Guerre de Troie: un talisman du Cycle épique, un tabou de l'*Iliade*", *Studies on the Greek Epic Cycle* I, *Philologia antiqua* 7, 2014.

Marion Meyer is a classical archaeologist. Ph.D. University of Bonn (1984); habilitation University of Hamburg (1997). Staff member at the universities of Munich (1985–1990) and Hamburg (1990–1996); visiting scholar at the University of Florida (1996); Professor of Classical Archaeology at the universities of Bonn (1997–2003) and Vienna (since 2003). Main interests: Greek culture, with a focus on Athens (cult and politics), images (creation, tradition, use, function and significance), iconography and narrative; phenomena of acculturation in the Eastern Mediterranean. Recent publications: (with Ralf von den Hoff, eds.), *Helden wie sie. Übermensch – Vorbild – Kultfigur in der griechischen Antike* (Freiburg 2010); (with Deborah Klimburg-Salter, eds.), *Visualisierungen von Kult* (Wien 2014); "Was ist ein Mädchen? Der Blick auf die weibliche Jugend im klassischen Athen", in: S. Moraw and A. Kieburg (eds.), *Mädchen im Altertum / Girls in Antiquity* (Münster 2014) 221–236; "Machtlose Aphrodite. Körper und Körpersprache in Bildern der jungfräulichen Göttinnen Athena und Artemis", in: O. Krüger and N. Weibel (Hrsg.), *Die Körper der Religion. Corps en religion.* CultuRel 7 (Zürich 2015) 65–94. Forthcoming: *Athena, goddess of Athens. Cult and myth on the Acropolis until Classical times.*

Andreas N. Michalopoulos is Associate Professor of Latin at the National and Kapodistrian University of Athens. He has published extensively on Latin literature including *Ancient Etymologies in Ovid's Metamorphoses: A Commented Lexicon* (Leeds 2001), *Ovid, Heroides 16 and 17: Introduction, Text and Commentary* (Cambridge 2006), and *Ovid, Heroides 20 and 21: Introduction, Text and Commentary* (Athens 2014).

Cecilia Nobili is Research Fellow at the Università degli Studi di Milano. She has published on archaic Greek poetry (elegy, epic, epinicians), including *L'Inno omerico a Ermes e le tradizioni locali* (Milano 2011) and *Corone di gloria. Epigrammi agonistici ed epinici dal VII al IV secolo a.C.* (Alessandria 2016).

Athina Papachrysostomou is Assistant Professor of Ancient Greek Literature at University of Patras (Department of Philology). Her research focuses on Greek drama (especially comedy and Euripidean tragedy) and ancient Greek political history / thought (especially Athenian Democracy). Apart from several articles and chapters in collective volumes, she has also published three monographs: (i) *Six Comic Poets. A Commentary on Selected Fragments of Middle Comedy* (Tübingen 2008); (ii) *Πορεία προς την Αθηναϊκή Δημοκρατία. Η μαρτυρία των αρχαίων πηγών* (Athens 2014); and (iii) *Amphis: Introduction, Translation, Commentary* (Heidelberg 2016; Verlag Antike, Fragmenta Comica 20). She has received scholarships from the Onassis Foundation, the Loeb Classical Library Foundation, and the Fulbright Foundation.

Sophia Papaioannou is Associate Professor of Latin Literature at the National and Kapodistrian University of Athens. She has published extensively on Latin epic (Ovid and Vergil) and Roman Comedy. Representative publications on Augustan epic include two volumes on Ovid (*Epic Succession and Dissension: Ovid, Metamorphoses 13.623–14.582, and the Reinvention of the Aeneid*, De Gruyter 2005; and *Redesigning Achilles: The 'Recycling' of the Epic Cycle in Ovid, Metamorphoses 12.1–13.620*, De Gruyter 2007). Her work in Roman Comedy includes the volume *Terence and Interpretation* (Cambridge Scholars Publishing 2014), the first translation of Plautus' *Miles Gloriosus* in Greek, and the first annotated edition of the play in any language since 1963 (Athens, Smili; second edition 2010); she has co-edited (with A.K. Petrides) *New Perspectives on Postclassical Comedy* (Cambridge Scholars Publishing 2009); and she is currently preparing a commentary on Plautus' *Curculio.*

Ezio Pellizer was full Professor of Greek Literature until 2010 at the University of Trieste. Since 2015 he is Professor of Anthropology of Ancient Greece at the University of Udine, and Director of the *Gruppo di Ricerca sul Mito e la Mitografia* (GRiMM, grmito.units.it) on line.

Eleni Peraki-Kyriakidou is a retired Assistant Professor of Latin Literature at the Aristotle University of Thessaloniki. Her main areas of interest are Virgil's *Eclogues* and *Georgics*, Roman epic and historiography. She has also published a number of articles on ancient etymology and etymologizing. Together with Stelios Phiorakis she has written a book on *The Law Code of Gortyn* (Herakleion, 1973).

Paolo Daniele Scirpo is Post-doc researcher in Classical Archaeology at National and Kapodistrian University of Athens (Greece). His research interests are Greek Sicily and the relations between colonies and the mother country. His contributions in scientific journals also include a collection of archaeological essays on Sicily (Triskeles, Athens, 2005) and the Greek translation of Ernesto De Miro's book (*L'arte greca in Sicilia*, Palermo, 2008).

Nereida Villagra holds an FCT postdoctoral scholarship at the Centro de Estudos Clássicos at the Universidade de Lisboa. Her research focuses on mythography and on mythographical texts transmitted by scholia, and she is currently working on a critical edition of the Mythographus Homericus of the *Odyssey*.

Christos A. Zafiropoulos is Assistant Professor of Classics at the University of Patras. He has published on the Aesopic corpus and on Plato's dialogues including *Ethics in Aesop's Fables* (Leiden 2001) and *Socrates and Aesop* (Sankt Augustin 2015). His research field is Greek thought, with emphasis on Plato, folktales and popular wisdom, and drama. Currently he prepares a book on the intellectual's portrait in Plato, especially as regards the *philosophos'* demand for leisure, as it is articulated in the *Apology*, but also throughout the dialogues.

Giuseppe Zanetto is Professor of Greek Language and Literature at the University of Milan. His main scientific interests are Greek epic, Attic theatre, Hellenistic epigram, Greek narrative. He published an edition, with Italian translation and commentary, of the *Homeric Hymns* (Milan 1996, 2nd edition 2000), critical editions of Aristophanes' *Birds* (Milan 1987; 4th edition 1997) and [Euripides]' *Rhesus* (Stuttgart and Leipzig 1993), translations with commentary of several Euripides' and Terence's plays. He studied the language and the style of the Greek novel, being co-author of a *Lexicon of Greek novelists*. He has also been studying the Greek epistolary collections, publishing a critical edition of Theophylactus Simocatta's *Epistles* (Leipzig 1985) and some contributions to Aristaenetus. He is now working at a new critical edition of Achilles Tatius' novel.